W9-AWB-549

FUNDAMENTALS OF
GAS DYNAMICS

Robert D. Zucker

Associate Professor
Department of Aeronautics
Naval Postgraduate School

10   9   8   7

Library of Congress catalog card number: 76-39743

Matrix Publishers Inc.
8437 Mayfield Road
Chesterland, OH 44026
(216) 729-2858

ISBN: 0-916460-12-6

CONTENTS

UNIT 9
FANNO FLOW   **235**

Introduction / Objectives / Analysis for a General
Fluid / Working Equations for a Perfect Gas / Reference
State and Fanno Tables / Applications / Correlation
with Shocks / Friction Choking / Summary / Problems /
Check Test

UNIT 10
RAYLEIGH FLOW   **273**

Introduction / Objectives / Analysis for a General
Fluid / Working Equations for a Perfect Gas / Reference
State and Rayleigh Tables / Applications / Correlation
with Shocks / Thermal Choking / Summary / Problems /
Check Test

UNIT 11
REACTION PROPULSION SYSTEMS   **311**

Introduction / Objectives / Brayton Cycle / Propulsion
Engines / Thrust, Power, and Efficiency / Supersonic
Diffusers / Summary / Problems / Check Test

APPENDICES   **365**

PREFACE

This book is written for the average student who wants to learn something about gas dynamics. It aims at the undergraduate level and thus requires a minimum of prerequisites. The writing style is informal and incorporates many new ideas in educational technology such as behavioral learning objectives, meaningful summaries, and check tests. Such features make this book well suited to self-study methods such as PSI (Personalized System of Instruction) as well as conventional lecture courses. Sufficient material is included for a typical one-quarter or semester course depending on the student's background.

The approach is to develop all basic relations on a rigorous basis with equations that are valid for the general case of unsteady, three-dimensional flow of an arbitrary fluid. These relations are then simplified to approach meaningful engineering problems in one- and two-dimensional steady flow. All basic internal and external flows are covered with practical applications interwoven throughout the text. Attention is focused on the assump-

tions made in every analysis; emphasis is placed on the usefulness of the T-s diagram and the significance of the loss term.

Examples and problems are provided in both the English Engineering and the SI systems of units. The homework problems range
from routine to complex with all charts and tables necessary for
their solution included.

The objectives are to master the fundamental concepts and
develop good problem solving ability. Once this is done it is
easy to return to the basic equations and include other factors
such as real gases, two-phase flow, electro-magnetic effects,
etc. After completing this book the student should be capable of
attacking the many references which are available on these more
advanced topics.

I owe a great deal to Newman Hall and Ascher Shapiro whose
books provided my first introduction to the area of compressible
flow. Many other texts have appeared in the last twenty years
but the influence of these two has been overwhelming. Another
significant contribution has been made by my students who have
provided the motivation for preparing this material.

In particular I would like to thank LCDR Ernest Lewis who
helped formulate the objectives, LT Allen Roessig who prepared
the original program for calculating the compressible flow tables,
and LT Joseph Strada for modifying the computer program and checking all the example problems. All diagrams are the work of
Scientific Illustrators and numerous grammatical errors were de-
tected by Mrs. Judy Kaitala. The entire manuscript was typed by
Miss Deborah Almquist who lavished tender loving care over each
equation.

Finally, I would like to thank my wife, Polly, who has had
to share me with this project for much too long a time.

Robert D. Zucker

Pebble Beach, CA
February, 1977

TO THE STUDENT

You don't need much background to enter the fascinating world of gas dynamics. However, it will be assumed that you have been exposed to basic courses in calculus and thermodynamics. Specifically, you are expected to know:

a. simple differentiation and integration.
b. the meaning of a partial derivative.
c. the significance of a dot product.
d. how to draw free body diagrams.
e. how to resolve a force into its components.
f. Newton's Second Law of Motion.
g. about properties of fluids, particularly perfect gases.
h. the Zeroth, First, and Second Laws of Thermodynamics.

The first six prerequisites are very specific; the last two cover quite a bit of territory. In fact, a background in thermodynamics is so important to the study of gas dynamics that a review of the necessary concepts for control mass analysis is contained in Unit One. If you have recently completed a course in thermodynamics you can skip most of this unit but <u>read</u> <u>the</u> <u>questions</u> in Section 1.5. If you can answer these, press on! If any difficulties arise, refer back to the material in the unit. Many of the equations will be used throughout the rest of the book.

Units Two and Three convert the fundamental laws into a form needed for control volume analysis. If you have had a course in fluid mechanics much of this material should also be familiar to you. A section on constant density fluids is included to show the general applicability in that area and to tie in with any previous work you have done in this area. If you haven't studied fluid mechanics, don't panic! All material that you need to know in this area is included. Also, several special concepts are developed which are not treated in many thermo and fluid courses. Thus, even if you have an excellent background, <u>read</u> <u>these</u> <u>units</u>! They form the backbone of gas dynamics and are frequently referred to in later units.

In Unit Four you are introduced to the characteristics of compressible fluids. Then in the following units, various basic flow phenomena are analyzed one by one — varying area, normal and oblique shocks, expansions, duct friction, and heat transfer. A wide variety of practical engineering problems can be solved with these concepts and many of these problems are covered throughout the text. Examples of these are the off-design operation of supersonic nozzles, supersonic wind tunnels, blast waves, supersonic airfoils, some methods of flow measurement, and choking from friction or thermal effects. You will find that supersonic flow has its special problems in that it does not seem to follow your intuition. Propulsion systems (with their air inlets, afterburners, and exit nozzles) represent an interesting application of nearly all the basic gas dynamic flow situations. Thus, Unit Eleven describes and analyzes common propulsion systems including turbojets, turbofans, ramjets, pulsejets and rockets.

This book has been written especially for you, the student. I hope that its informal style will put you at ease and motivate you to read on. Once you have passed the review unit the remaining units follow a similar format. The following suggestions may help you to optimize your study time. When you start each unit read the introduction as this will give you a general idea of what the unit is all about. The next section contains a set of learning objectives. These tell exactly what you should be able to do after successfully completing the unit. Some objectives are marked optional as they are only for the most serious students. Merely scan the objectives as they won't mean much at this time. However, they will indicate important things to look for. Move right on to the next section. As you read the material you may occasionally be asked to do something — complete a derivation, fill in a chart, draw a diagram, etc. Make an honest attempt to follow these instructions before proceeding further. You will not be asked to do something that you haven't the background to do and your active participation will help solidify important concepts.

As you complete each section look back to see if any of the objectives have been covered. If so, make sure you can do them. Write out the answers; these will help you in later studies. You may wish to make your own review of each unit as you go along; then see if it agrees with the summary provided. After having worked a representative group of problems you are ready to check your knowledge by taking the test at the end of the unit. This should always be treated as a closed book affair with the exception of tables and charts which give conversion factors, properties of gases, and compressible flow functions. If you have any difficulties with this test you should go back and restudy appropriate sections. Do not proceed to the next unit without satisfactorily completing the previous one.

Not all units are the same length and in fact most of them are a little long to tackle all at once. You might find it easier to break them into "bite size" pieces according to the table shown below. Work some problems on the first group of objectives and sections <u>before</u> proceeding to the next group. "Crisis management" is <u>not</u> recommended. You should spend some time each day working through the material.

Learning can be fun and it should be! However, knowledge doesn't come free. You must expend time and effort to accomplish the job. I hope that this book will make the task of exploring gas dynamics more enjoyable. Any suggestions you might have to improve this material will be most welcome.

Unit	Sections	Objectives	Problems
2	1 - 5 6 - 7	1 - 5 6 - 9	1 - 6 7 - 14
3	1 - 7 8 - 9	1 - 8 9 - 12	1 - 12 13 - 20
4	1 - 7	1 - 10	1 - 16
5	1 - 6 7 - 10	1 - 7 8 - 12	1 - 8 9 - 22
6	1 - 5 6 - 8	1 - 7 8 - 10	1 - 6 7 - 17
7	1 - 3 4 - 9	1 - 2 3 - 10	1 - 5 6 - 18
8	1 - 5 6 - 8	1 - 6 7 - 10	1 - 6 7 - 16
9	1 - 6 7 - 9	1 - 7 8 - 11	1 - 12 13 - 22
10	1 - 6 7 - 9	1 - 7 8 - 11	1 - 8 9 - 21
11	1 - 3 4 - 5 6 - 7	1 - 4 5 - 11 12	1 - 5 6 - 18 19 - 22

UNIT 1

REVIEW OF ELEMENTARY PRINCIPLES

1.1 INTRODUCTION

It is assumed that before entering the world of gas dynamics you have had a reasonable background in mathematics (through calculus) together with a course in elementary thermodynamics. An exposure to basic fluid mechanics would be helpful but is not absolutely essential. The concepts used in fluid mechanics are relatively straightforward and can be developed as we need them. On the other hand, some of the concepts of thermodynamics are more abstract and we must assume that you already understand the fundamental laws of thermo as they apply to stationary systems. The extension of these laws to flow systems is so vital that they will be covered in depth in Units 2 and 3.

This unit is not intended to be a formal review of the above mentioned courses but rather it should be viewed as a collection of the bare minimum of concepts and facts which will be used later. It should be understood that a great deal of background will be omitted in this review and no attempt will be made to prove

each statement. Thus, if you have been away from this material
for any length of time, you may find it necessary occasionally
to refer to your notes or other textbooks to supplement this review. At the very least, the remainder of this unit may be considered an assumed common ground of knowledge from which we shall
venture forth.

At the end of the unit a number of questions are presented
for you to answer. No attempt should be made to continue further
until you feel that you can satisfactorily answer all of these
questions.

1.2 UNITS AND NOTATION

Dimension: a qualitative definition of a physical entity
             (such as time, length, force, etc.).

Unit: an exact magnitude of a dimension
             (such as feet, inches, meters, etc.).

In the United States most work in the area of thermo-gas
dynamics is currently done in the English Engineering System of
Units. However, the rest of the world is operating on the metric
or International System of Units (SI). Thus, we shall review
both of these systems.

Dimension	English Engineering	International System
Time	second (sec)	second (s)
Length	foot (ft)	meter (m)
Force	pound force (lbf)	newton (N)
Mass	pound mass (lbm)	kilogram (kg)
Temperature	Fahrenheit ($^{\circ}$F)	Celsius ($^{\circ}$C)

Caution: Never say pound as this is ambiguous. It is either
          a pound force or a pound mass.

FORCE AND MASS

In either system the force and mass units are related through
Newton's Second Law of Motion which states that

$$\Sigma \vec{F} \propto \frac{d(\overrightarrow{momentum})}{dt} \tag{1.1}$$

The proportionality factor is expressed as $K = 1/g_c$ and thus

$$\Sigma \vec{F} = \frac{1}{g_c} \frac{d(\overrightarrow{mom})}{dt} \tag{1.2}$$

For a given mass this becomes

$$\Sigma \vec{F} = \frac{m\vec{a}}{g_c} \qquad (1.3)$$

where $\Sigma \vec{F}$ is the vector force summation acting on mass m, and $\vec{a}$ is the vector acceleration of the mass.

In the English system we use the definition:

> A one-pound force will give a one-pound mass an acceleration of 32.174 ft/sec^2.

With the above definition we have

$$1(lbf) = \frac{1(lbm) \ 32.174(ft/sec^2)}{g_c}$$

and thus

$$g_c = 32.174 \ \frac{lbm-ft}{lbf-sec^2} \qquad (1.4a)$$

Note that g_c is <u>not</u> standard gravity (check the units). It is a proportionality factor the value of which depends on the units being used. In further discussions we shall take the numerical value of g_c to be 32.2 when using the English Engineering system.

In the SI system we use the definition:

> A one-Newton force will give a one-kilogram mass an acceleration of one meter/second^2.

Now equation 1.3 becomes

$$1(N) = \frac{1(kg) \ 1(m/s^2)}{g_c}$$

and

$$g_c = 1 \ \frac{kg-m}{N-s^2} \qquad (1.4b)$$

Since g_c has the numerical value of unity many authors using the SI system drop this factor from all equations. However, we shall leave g_c in the equations so that you may use any system of units with less likelihood of making errors.

DENSITY AND SPECIFIC VOLUME

Density is the mass per unit volume and is given the symbol ρ .
It has the units of (lbm/ft^3) or (kg/m^3) .

Specific Volume is the volume per unit mass and is given the symbol v . It has the units of (ft^3/lbm) or (m^3/kg).

Thus
$$\rho = \frac{1}{v} \tag{1.5}$$

Specific Weight is the weight (force of gravity) per unit volume
and is given the symbol γ . If we take a unit volume under the
influence of gravity, its weight will be γ . Thus, from equation 1.3 we have:

$$\gamma = \rho \frac{g}{g_c} \quad (lbf/ft^3) \text{ or } (N/m^3) \tag{1.6}$$

Note that mass, density, and specific volume do not depend on
the value of local gravity. Weight and specific weight do depend
on gravity. We shall not refer to specific weight in this book;
it is mentioned here only to show the distinction between it and
density.

PRESSURE

Pressure is the normal force per unit area and is given the symbol p . It has the units of (lbf/ft^2) or (N/m^2) . In the
SI system, 1 pascal (Pa) = 1 N/m^2 , 1 bar = $1x10^5$ N/m^2 .

Absolute Pressure is measured with respect to a perfect vacuum.

Gage Pressure is measured with reference to the surrounding
(ambient) pressure.

$$p_{abs} = p_{amb} + p_{gage} \tag{1.7}$$

If the gage pressure is negative (i.e., the absolute pressure is
below ambient) it is usually called a (positive) vacuum reading.

$$p_{abs} = p_{amb} - p_{vac} \tag{1.8}$$

Two cases of pressure readings are shown in Figure 1.1.
Case 1 shows the use of equation 1.7 and case 2 illustrates equation 1.8. It should be noted that the surrounding (ambient) pressure is not necessarily standard atmospheric. However, if no
other information is available it may be assumed that the sur-

roundings are at 14.69 psia (1.013 bar). Most equations require
the use of absolute pressure and in future conversions we shall
use the numerical value of 14.7 when using the English Engineering
system.

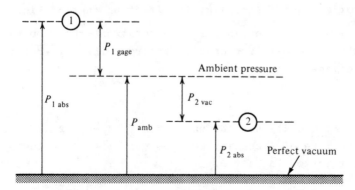

Figure 1.1 Absolute and gage pressures

TEMPERATURE

Degrees Fahrenheit (or Celsius) can safely be used when <u>dif-
ferences</u> in temperature are involved. However, most equations
require the use of absolute temperatures in Rankine (or Kelvin).

$$^{O}R \text{ (Rankine)} = {}^{O}F + 459.67^{O} \qquad (1.9a)$$

$$^{O}K \text{ (Kelvin)} = {}^{O}C + 273.15^{O} \qquad (1.9b)$$

The values of 460 and 273 will be used in further discussions.

VISCOSITY

We shall be dealing with fluids which are defined as:

> Any substance that will continuously deform
> when subjected to a shear stress.

Thus, the amount of deformation is of no significance (as it is
with a solid); but rather the <u>rate of deformation</u> is characteristic of each individual fluid and is indicated by the viscosity.

$$\text{viscosity} \equiv \frac{\text{shear stress}}{\text{rate of angular deformation}} \qquad (1.10)$$

Viscosity, sometimes called absolute viscosity, is given the symbol μ and has the units $lbf\text{-}sec/ft^{2}$ or $N\text{-}s/m^{2}$.

For most common fluids the viscosity is a function of the
fluid and varies with its state. Temperature has by far the
greatest effect on viscosity and therefore most charts and tables
use only this variable. Pressure has a slight effect on the viscosity of gases but a negligible effect on that of liquids.

In a number of engineering computations the combination of
(absolute) viscosity and density occurs. The "kinematic viscosity" is defined as:

$$\nu \equiv \frac{\mu g_c}{\rho} \tag{1.11}$$

Kinematic viscosity has the units of ft^2/sec or m^2/s . We
shall see more of viscosity when we deal with flow losses caused
by duct friction in Unit 9.

EQUATIONS OF STATE

In this book we shall consider all liquids as having constant density and all gases as following the perfect gas equation
of state. Thus, for liquids we have the relation

$$\rho = \text{constant} \tag{1.12}$$

The perfect gas equation of state is derived from kinetic
theory and neglects molecular volume and inter-molecular forces.
Thus, it is accurate only under conditions of relatively low density which would be at low pressure and/or high temperatures.
The form of the equation normally used in gas dynamics is

$$p = \rho R T \tag{1.13}$$

where			
$p \equiv$ absolute pressure	lbf/ft^2	N/m^2
$\rho \equiv$ density	lbm/ft^3	kg/m^3
$T \equiv$ absolute temperature	^oR	^oK
$R \equiv$ <u>individual</u> gas constant	$\dfrac{ft\text{-}lbf}{lbm\text{-}^oR}$	$\dfrac{N\text{-}m}{kg\text{-}^oK}$

The individual gas constant can be approximated in the English
Engineering system of units by dividing 1545 by the molecular
weight of the gas. In the SI system, R is approximated by dividing 8314 by the molecular weight.

Example 1.1 The (equivalent) molecular weight of air is 28.97.

$$R \approx 1545/28.97 = 53.3 \text{ ft-lbf/lbm-}^{\circ}R$$

or $$R \approx 8314/28.97 = 287 \text{ N-m/kg-}^{\circ}K$$

Example 1.2 Compute the density of air at 50 psia and $100^{\circ}F$.

$$\rho = \frac{p}{RT} = \frac{50(144)}{53.3(460+100)} = \underline{0.241} \text{ lbm/ft}^3$$

Properties of various gases may be found in a table in the Appendix. The major portion of this text will use the English Engineering system of units. However, a few examples and home-work problems will utilize the SI system. Some helpful conversion factors may be found in the Appendix.

1.3 SOME MATHEMATICAL CONCEPTS

VARIABLES: The equation

$$y = f(x) \tag{1.14}$$

indicates that a functional relation exists between the variables x and y. Further, it denotes that

   x is an <u>independent</u> variable the value of which can be chosen anyplace within some given range;

   y is a <u>dependent</u> variable the value of which is fixed once x has been selected.

In most cases it is possible to interchange the roles of the dependent and independent variables and write

$$x = f(y) \tag{1.15}$$

Frequently, a variable will depend on more than one other variable. One might write

$$P = f(x, y, z) \tag{1.16}$$

indicating that the value of the dependent variable P is fixed once values have been selected for the independent variables x, y, and z.

INFINITESIMAL: A quantity which eventually is allowed to approach zero in the limit is called an infinitesimal. It should be noted that the quantity, say Δx, can initially be chosen at a rather large finite value. If at some later stage in the analysis we let Δx approach zero, which is indicated by

$$\Delta x \to o$$

then Δx is called an infinitesimal.

DERIVATIVES: If $y = f(x)$, we define the derivative dy/dx as the limit of $\Delta y/\Delta x$ as Δx is allowed to approach zero. This is indicated by

$$\frac{dy}{dx} \equiv \lim_{\Delta x \to o} \frac{\Delta y}{\Delta x} \qquad (1.17)$$

For a unique derivative to exist, it is immaterial how Δx approaches zero.

If more than one independent variable is involved, then partial derivatives must be used. Say that $P = f(x, y, z)$. We can determine the partial derivative $\partial P/\partial x$ by taking the limit of $\Delta P/\Delta x$ as Δx approaches zero but in so doing we <u>must</u> hold the values of all other independent variables constant. This is indicated by

$$\frac{\partial P}{\partial x} \equiv \lim_{\Delta x \to o} \frac{\Delta P}{\Delta x}\bigg)_{y,z} \qquad (1.18)$$

where the subscripts y , z denote that these variables remain fixed during the limiting process. We could formulate other partial derivatives such as

$$\frac{\partial P}{\partial y} \equiv \lim_{\Delta y \to o} \frac{\Delta P}{\Delta y}\bigg)_{x,z} \quad \text{etc.} \qquad (1.19)$$

DIFFERENTIALS: For functions of single variables such as $y = f(x)$, the differential of the dependent variable is defined as

$$dy \equiv \frac{dy}{dx} \Delta x \qquad (1.20)$$

The differential of an independent variable is defined as its increment; thus, $dx \equiv \Delta x$ (1.21)

and one can write $dy = \frac{dy}{dx} dx$ (1.22)

For functions of more than one variable, such as $P = f(x, y, z)$, the differential of the dependent variable is defined as

$$dP \equiv \frac{\partial P}{\partial x}\bigg)_{y,z} \Delta x \;+\; \frac{\partial P}{\partial y}\bigg)_{x,z} \Delta y \;+\; \frac{\partial P}{\partial z}\bigg)_{x,y} \Delta z \qquad (1.23a)$$

or $$dP = \frac{\partial P}{\partial x}\bigg)_{y,z} dx \;+\; \frac{\partial P}{\partial y}\bigg)_{x,z} dy \;+\; \frac{\partial P}{\partial z}\bigg)_{x,y} dz \qquad (1.23b)$$

It is important to note that quantities such as ∂P, ∂x, ∂y, ∂z are <u>never</u> defined and <u>do</u> <u>not</u> <u>exist</u>. Under no circumstances can one "separate" a partial derivative. This is an error frequently made by students when integrating partial differential equations.

MAXIMUM AND MINIMUM: If a plot is made of the functional relationship $y = f(x)$, maximum or minimum points may be exhibited. At these points $dy/dx = 0$. If it is a maximum point d^2y/dx^2 will be negative, whereas d^2y/dx^2 will be positive at a minimum point.

BINOMIAL THEOREM (or Series):

$$(a + x)^n = a^n + na^{n-1} x + \frac{n(n-1)}{2!} a^{n-2} x^2 +$$

$$\frac{n(n-1)(n-2)}{3!} a^{n-3} x^3 + \ldots$$

(1.24)

The above expansion holds for $x^2 < a^2$. If n is a positive integer there are $(n + 1)$ terms in the series, otherwise the series is infinite.

TAYLORS SERIES: If the functional relationship $y = f(x)$ is not known, but the values of y together with those of its derivatives are known at a particular x (say x_1), then the value of y may be found at any other point (say x_2) through the use of the Taylor series expansion.

$$f(x_2) = f(x_1) + \frac{df}{dx} (x_2-x_1) + \frac{d^2f}{dx^2} \frac{(x_2-x_1)^2}{2!}$$

(1.25)

$$+ \frac{d^3f}{dx^3} \frac{(x_2-x_1)^3}{3!} + \ldots$$

In order to use this expansion the function must be continuous and possess continuous derivatives throughout the interval x_1 to x_2. It should be noted that all derivatives in the above expression must be evaluated at $x = x_1$, since the expansion is about this point.

If the increment $\Delta x = x_2-x_1$ is small, then only a few terms need be evaluated to obtain an accurate answer for $f(x_2)$.

If Δx is allowed to approach zero, then all higher order terms
may be dropped and

$$f(x_2) \approx f(x_1) \; + \; \frac{df}{dx}\bigg)_{x=x_1} dx \qquad \text{for dx} \to \text{o} \qquad (1.26)$$

1.4 THERMODYNAMIC CONCEPTS FOR CONTROL MASS ANALYSIS

I must apologize for the length of this section but a good understanding of thermodynamic principles is essential to the study of
Gas Dynamics.

GENERAL DEFINITIONS

Microscopic approach: Deals with molecules, their motion, and
   behavior on a statistical basis. It depends on our knowledge
   of the structure and behavior of matter. Thus, this view is
   continually being modified.

Macroscopic approach: Deals directly with the average behavior
   of molecules through observable and measurable properties
   (temperature, pressure, etc.). This classical approach involves no assumptions regarding the molecular structure of
   matter; thus, no modifications of the basic laws are necessary.
   The macroscopic approach will be used in this book.

Control Mass: A fixed quantity of mass which is being analyzed.
   It is separated from its surroundings by a boundary. A control mass is sometimes referred to as a "closed system." Although no matter crosses the boundary, energy may enter or
   leave the system.

Control Volume: A region in space which is being analyzed. The
   boundary separating it from its surroundings is called the
   "control surface." Matter, as well as energy may cross the
   control surface; and thus a control volume is sometimes referred to as an "open system." The analysis of control volumes
   is discussed in Units 2 and 3.

Properties: Characteristics which describe the state of a system.
   Any quantity which has a definite value for each definite state
   of the system (pressure, temperature, color, entropy, etc.).

Intensive Property: Depends only on the state of the system and
   is independent of its mass (e.g., temperature, pressure, etc.).

Extensive Property: Depends on the mass of the system (e.g., internal energy, volume, etc.).

Types of Properties:
1. Observable: readily measured (pressure, temperature, velocity, mass, etc.).
2. Mathematical: defined from combinations of other properties (density, specific heats, enthalpy, etc.).
3. Derived: arrived at as the result of an analysis.
   Internal energy (from 1st Law)
   Entropy (from 2nd Law)

State Change: Comes about as the result of a change in any property.

Path or Process: Represents a series of adjacent states which define a unique path from one state to another.

Types of Processes: Adiabatic   →   no heat transfer
                   Isothermal   → T = const
                   Isobaric   → p = const
                   Isentropic   → s = const

Cycle: A sequence of processes in which the system is returned to its original state.

Point Functions: Another word for properties, since they depend only on the state of the system and are independent of the history or process by which the state was obtained.

Path Functions: Quantities which are not functions of the state of the system but rather depend on the path taken to move from one state to another. Heat and work are path functions. They can be observed crossing the boundaries of a system during a process.

LAWS OF CLASSICAL THERMODYNAMICS

0^2 Relation among properties
0   Thermal equilibrium
1   Conservation of energy
2   Degradation of energy (irreversibilities)

The 0^2 Law (sometimes called the 00 Law) is seldom listed as a formal law of thermodynamics; however, one should realize that without such a statement our entire thermodynamic structure

would collapse. This law assumes that a relationship exists
among the properties of the system, i.e., an "Equation of State."
Such an equation may be extremely complicated, or even completely
unknown, but as long as we know it exists we can continue our
studies.

For a single component substance only three <u>independent</u> properties are required to fix the state of the system. Care must
be taken in the selection of these properties; e.g., temperature
and pressure are not independent if the substance exists in more
than one phase. When dealing with a unit mass only two independent properties are required to fix the state. Thus, one can
express any property in terms of two other known properties with
a relation such as

$$P = f(x, y)$$

If two systems are separated by a non-adiabatic wall (one
which permits heat transfer) the state of each system will change
until a new equilibrium state for the combined system is reached.
The two systems are then said to be in "thermal equilibrium" with
each other and will then have one property in common which we
call temperature.

The Zeroth Law states that two systems in thermal equilibrium with a third system are in thermal equilibrium with each
other (and thus have the same temperature).

FIRST LAW OF THERMODYNAMICS

The First Law deals with conservation of energy and it can
be expressed in many ways. Heat and work are two types of energy
in transit.

Heat is transferred from one system to another when an effect occurs solely as a result of a temperature difference between two systems. Heat is always transferred from the system
at the higher temperature to the one at the lower temperature.

Work is transferred from a system if the total external effect of the given action can be reduced to raising a weight.

For a closed system that executes a complete cycle

$$\Sigma Q = \Sigma W \tag{1.27}$$

where   Q   is positive for heat transferred <u>into</u> the system
and   W   is positive for work transferred <u>from</u> the system.

Other sign conventions are sometimes used but we shall adopt the above for this book.

For a closed system that executes a process :

$$Q = W + \Delta E \tag{1.28}$$

where E represents the total energy of the system. On a unit mass basis equation 1.28 is written as

$$q = w + \Delta e \tag{1.29}$$

The total energy may be broken down into (at least) three types:

$$e \equiv u + \frac{v^2}{2g_c} + \frac{g}{g_c} z \tag{1.30}$$

where u is the intrinsic internal energy manifested by the motion of the molecules within the system,

$\frac{v^2}{2g_c}$ is the kinetic energy represented by the gross movement of the system as a whole,

$\frac{g}{g_c} z$ is the potential energy caused by the position of the system in a field of gravity.

It is sometimes necessary to include other types of energy but these are the only ones that we shall be concerned with in this book.

For an infinitesimal process one could write equation 1.29 as

$$\delta q = \delta w + de \tag{1.31}$$

Note that since heat and work are path functions, infinitesimal amounts of these quantities are not exact differentials and thus they are written as δq and δw . The infinitesimal change in internal energy is an exact differential since internal energy is a point function or property. For a stationary system equation 1.31 becomes

$$\delta q = \delta w + du \tag{1.32}$$

Work done by pressure forces during a change in volume for a stationary system is:

$$\delta w = pdv \tag{1.33}$$

The combination of the terms u and pv enters into equations and thus it is convenient to define enthalpy

$$h \equiv u + pv \tag{1.34}$$

Enthalpy is a property since it is defined in terms of other properties. It is frequently used in its differential form

$$dh = du + d(pv) = du + pdv + vdp \qquad (1.35)$$

Other examples of defined properties are the specific heats at constant pressure (c_p) and constant volume (c_v) .

$$c_p \equiv \left(\frac{\partial h}{\partial T}\right)_p \qquad (1.36)$$

$$c_v \equiv \left(\frac{\partial u}{\partial T}\right)_v \qquad (1.37)$$

SECOND LAW OF THERMODYNAMICS

The Second Law has been expressed in many forms. Perhaps the most classic is the statement by Kelvin and Planck stating that it is impossible for an engine operating in a cycle to produce net work output if exchanging heat with only one temperature source. While in itself this may not appear to be a profound statement, it leads the way to several corallaries and eventually to the establishment of a most important property (entropy).

The Second Law also recognizes the degradation of energy by irreversible effects such as internal fluid friction, heat transfer through a finite temperature difference, lack of pressure equilibrium between a system and its surroundings, etc. All real processes have some degree of irreversibilities present. In some cases these effects are very small and we can envision an ideal limiting condition which has none of these effects and thus is "reversible." A reversible process is one in which both the system and its surroundings can be restored to their original states.

By prudent application of the second law it can be shown that the integral of $\delta Q/T$ for a reversible process is independent of the path. Thus, this integral must represent the change of a property, which is called entropy.

$$\Delta S \equiv \int \frac{\delta Q_R}{T} \qquad (1.38)$$

where the subscript R indicates that it must be applied to a reversible process. An alternate expression on a unit mass basis for a differential process is:

$$ds \equiv \frac{\delta q_R}{T} \qquad (1.39)$$

Although you have no doubt used entropy for many calculations,
plots, etc., you probably do not have a good feel for this property. In Unit 3 we shall divide entropy into two parts and by
using it in this fashion for the remainder of this book we hope
to gain a better understanding of this elusive "animal."

PROPERTY RELATIONS: Some extremely important relations come from
combinations of the First and Second Laws. Two that we shall use
throughout this book are:

$$Tds = du + pdv \qquad (1.40)$$

$$Tds = dh - vdp \qquad (1.41)$$

Although some special assumptions were made to derive these, the
results are equations which contain only properties and thus are
valid relations to use between any equilibrium states.

PERFECT GASES

Recall that for a unit mass of a single component substance
one property can be expressed as a function of any <u>two</u> other independent properties. However, for substances that follow the perfect gas equation of state

$$p = \rho RT \qquad (1.13)$$

it can be shown (see pg 173 of ref. 4) that <u>the</u> <u>internal</u> <u>energy</u>
<u>and</u> <u>also</u> <u>the</u> <u>enthalpy</u> <u>are</u> <u>functions</u> <u>of</u> <u>temperature</u> <u>only</u>. These
are extremely important facts as they permit us to make many in-
teresting simplifications for perfect gases.

Consider the specific heat at constant volume.

$$c_v \equiv \left(\frac{\partial u}{\partial T}\right)_v \qquad (1.37)$$

If $u = f(T)$ only, then it does not matter whether the volume
is held constant when computing c_v ; also the partial derivative
is now an ordinary derivative.

Thus $\qquad\qquad c_v = \dfrac{du}{dT} \qquad\qquad (1.42)$

or $\qquad\qquad du = c_v dT \qquad\qquad (1.43)$

Similarly, from the specific heat at constant pressure, we have
for a perfect gas:

$$dh = c_p dT \qquad (1.44)$$

It is important to realize that equations 1.43 and 1.44 are applicable to <u>any</u> and <u>all</u> processes (as long as we are dealing with a
perfect gas). If the specific heats remain constant (normally
assumed for perfect gases) then one can easily integrate equations 1.43 and 1.44.

$$\Delta u = c_v \Delta T \tag{1.45}$$

$$\Delta h = c_p \Delta T \tag{1.46}$$

In gas dynamics one simplifies calculations by introducing an
arbitrary base for internal energy. We let $u = 0$ when $T = 0$
absolute. Then from the definition of enthalpy, h also equals
zero when $T = 0$. Equations 1.45 and 1.46 can now be written
as:

$$u = c_v T \tag{1.47}$$

$$h = c_p T \tag{1.48}$$

Typical values of the specific heats for air at normal temperatures and pressures are: $c_p = 0.240$ and $c_v = 0.171$ Btu/lbm-^{O}R .
Learn these numbers (or their SI equivalents)! You will use them
often.

Other frequently used relations in connection with perfect
gases are

$$\gamma = c_p / c_v \tag{1.49}$$

$$c_p - c_v = R/J \tag{1.50}$$

Notice that the conversion factor

$$J = 778 \text{ ft-lbf/Btu} \tag{1.51}$$

has been introduced in the last equation since specific heats are
normally given in units of Btu/lbm-^{O}R. This factor will be omitted in all future equations and it will be left for you to consider when it is required. It is hoped that by this procedure
you will develop careful habits of checking units in all of your
work. What units are used for specific heat and R in the SI
system (see table on gas properties in the Appendix)? Would this
require a J factor in equation 1.50?

ENTROPY CHANGES: The change in entropy between any two points
can be obtained by integrating equation 1.39 along any reversible
path connecting the points, with the following results for perfect gases:

$$\Delta s_{1-2} = c_p \ln \frac{v_2}{v_1} + c_v \ln \frac{p_2}{p_1} \qquad (1.52)$$

$$\Delta s_{1-2} = c_p \ln \frac{T_2}{T_1} - R \ln \frac{p_2}{p_1} \qquad (1.53)$$

$$\Delta s_{1-2} = c_v \ln \frac{T_2}{T_1} + R \ln \frac{v_2}{v_1} \qquad (1.54)$$

Remember, absolute values of pressures and temperatures must be used in these equations; volumes may be either total or specific but both volumes must be the same. Watch the units on c_p , c_v , and R .

PROCESS DIAGRAMS

Many processes in the gaseous region can be represented as a polytropic process, that is, one that follows the relation

$$pv^n = \text{const} = C_1 \qquad (1.55)$$

where n is the polytropic exponent and can be any positive number. If the fluid is a perfect gas the equation of state can be introduced into 1.55 to show that

$$Tv^{n-1} = \text{const} = C_2 \qquad (1.56)$$

$$Tp^{\frac{1-n}{n}} = \text{const} = C_3 \qquad (1.57)$$

Keep in mind that C_1 , C_2 , and C_3 in the above equations are different constants. It is interesting to note that certain values of n represent particular processes.

$$n = 0 \quad \rightarrow \quad p = \text{const}$$
$$n = 1 \quad \rightarrow \quad T = \text{const}$$
$$n = \gamma \quad \rightarrow \quad s = \text{const}$$
$$n = \infty \quad \rightarrow \quad v = \text{const}$$

These plot in the p-v and T-s diagrams as shown in Figure 1.2. Learn these diagrams! You should also be able to figure out how temperature and entropy vary in the p-v diagram, and how pressure and volume vary in the T-s diagram. (e.g., draw several T = const lines in the p-v plane. Which one represents the highest temperature?)

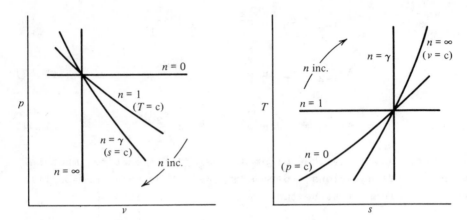

Figure 1.2 General polytropic process plots for perfect gases

1.5 QUESTIONS FOR REVIEW

A number of questions follow which are based on concepts
you have covered in previous calculus and thermodynamics courses.
Write down your answers as clearly and concisely as possible
using any source you wish (although all of the material has been
covered in this review). Do not proceed to Unit 2 until you completely understand the answers to all questions and can answer
them without reference to your notes.

1. How is an ordinary derivative such as dy/dx defined? How
   does this differ from a partial derivative?

2. Give the binomial theorem for expanding $(a + x)^n$.

3. What is the Taylor series expansion, it's applications and
   limitations?

4. State Newton's Second Law as you would apply it to a control
   mass.

5. Define a one-pound force in terms of the acceleration it will
   give a one-pound mass. Give a similar definition for a newton in the SI system.

6. Explain the significance of g_c in Newton's Second Law. What
   are the magnitude and units of g_c in the English Engineering
   system? In the SI system?

7. How is the Zeroth Law of Thermodynamics related to temperature?

8. How are degrees Fahrenheit and Rankine related? Degrees Celsius and Kelvin?

9. Explain the difference between absolute and gage pressures.

10. What is the distinguishing characteristic of a fluid (as compared to a solid)? How is this related to viscosity?

11. Explain the difference between a microscopic and a macroscopic approach to the analysis of fluid behavior.

12. Describe the control volume approach to problem analysis and compare it to the control mass approach. What kinds of systems are these also called?

13. Describe a property and give three or more examples.

14. Properties may be categorized as either intensive or extensive. Define what is meant by each and list examples of each type of property.

15. When dealing with a unit mass of a single component substance, how many independent properties are required to fix the state?

16. What does the term "specific" imply? (e.g., specific enthalpy).

17. What is the relationship between density and specific volume?

18. Of what value is an equation of state? Recall one with which you are familiar.

19. Define a point function and path function. Give examples of each.

20. What is a process? What is a cycle?

21. State the First Law of Thermodynamics for a closed system when executing a process.

22. What are the sign conventions used for heat and work?

23. State any form of the Second Law of Thermodynamics.

24. Define a reversible process for a thermodynamic system. Is any real process ever reversible? Of what practical value is the concept of reversible processes?

25. What are some effects which cause processes to be irreversible?

26. What is an adiabatic process? An isothermal process? An isentropic process?

27. Give equations that define enthalpy and entropy.

28. Give differential expressions that relate entropy to
    a. internal energy
    b. enthalpy

29. Define (in the form of partial derivatives) the specific heats c_v and c_p . Are these expressions valid for a material in any state?

30. State the perfect gas equation of state. When is it valid? (i.e., what are its limitations?) Give a consistent set of units for each term in the equation.

31. For a perfect gas, specific internal energy is a function of which state variables? How about specific enthalpy?

32. Give expressions for Δu and Δh which are valid for perfect gases. Do these hold for any process?

33. For perfect gases, at what temperature do we arbitrarily assign $u = 0$ and $h = 0$?

34. State any expression for the entropy change between two arbitrary points which is valid for a perfect gas.

35. If a perfect gas undergoes an isentropic process what equation relates the pressure and volume? Temperature and volume? Temperature and pressure?

36. Consider the general polytropic process (pv^n = const) for a perfect gas. In the p-v and T-s diagrams shown, label each process line with the correct value for n and identify which fluid property is held constant.

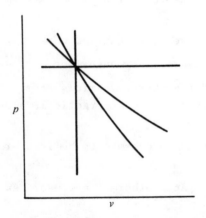
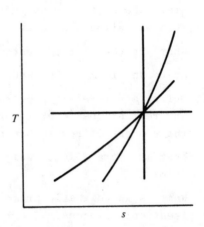

UNIT 2

CONTROL VOLUME ANALYSIS—PART I

2.1 INTRODUCTION

In the study of gas dynamics we are interested in fluids that are <u>flowing</u>. The analysis of flow problems is based on the same fundamental principles that you have used in previous courses in thermodynamics or fluid dynamics, namely:

  a. Conservation of Mass
  b. Conservation of Energy
  c. Newton's Second Law of Motion

When applying these principles to the solution of specific problems you must also know something about the properties of the fluid.

In the previous unit the above concepts were reviewed in a form applicable to a control mass. However, it is extremely difficult to approach flow problems from the control mass point of view. Thus, it will first be necessary to develop some fundamental expressions which can be used to analyze control volumes.

A technique will be developed to transform our basic laws
for a control mass into integral equations which are applicable
to finite control volumes. Simplifications will be made for
special cases such as steady one-dimensional flow, etc. We will
also analyze differential control volumes which will produce some
valuable differential relations.

In this unit we tackle mass and energy and in Unit 3 we shall
discuss momentum concepts.

2.2 OBJECTIVES

After successfully completing this unit you should be able to:

1. State the basic concepts from which the study of gas dynamics
    proceeds.

2. Explain what is meant by one-, two-, and three-dimensional
    flow.

3. Define steady flow.

4. Compute the flow rate and average velocity from a multi-
    dimensional velocity profile.

5. State the equation used to relate the material derivative of
    any extensive property to the properties inside of and cross-
    ing the boundaries of a control volume. Interpret in words
    the meaning of each term in the equation.

6. Starting with the basic concepts or equations which are valid
    for a control mass, obtain the integral forms of the continu-
    ity and energy equations for a control volume.

7. Simplify the integral forms of the continuity and energy
    equations for a control volume for the conditions of steady,
    one-dimensional flow.

8. Apply the simplified forms of the continuity and energy equa-
    tions to differential control volumes. (Optional.)

9. Demonstrate the ability to apply continuity and energy con-
    cepts in the analysis of control volumes.

2.3 FLOW DIMENSIONALITY AND AVERAGE VELOCITY

As we observe fluid moving around, the various properties
can be expressed as functions of location and time. Thus, in an
ordinary rectangular cartesian coordinate system we could say, in

general, that

$$V = f(x,y,z,t) \tag{2.1}$$

or

$$p = g(x,y,z,t) \tag{2.2}$$

Since it is necessary to specify three spatial coordinates and time, this is called three-dimensional unsteady flow.

Two-dimensional unsteady flow would be represented by

$$V = f(x,y,t) \tag{2.3}$$

and one-dimensional unsteady flow by

$$V = f(x,t) \tag{2.4}$$

The assumption of one-dimensional flow is a simplification normally applied to flow systems and the single coordinate is usually taken in the direction of flow. This is not necessarily "unidirectional flow" as the direction of the flow duct might change. Another way of looking at one-dimensional flow is to say that at any given section (x-coordinate) all fluid properties are constant across the cross-section. Keep in mind that the proper-ties can still change from section to section (as x changes).

The fundamental concepts reviewed in the previous unit were expressed in terms of a given mass of material; i.e., the control mass approach. When using the control mass approach we observe some property of the mass such as temperature, pressure, velocity, internal energy, etc. The (time) rate at which this property changes is called a "material derivative" (sometimes called a total or substantial derivative). It is written by various au-thors as $D(\ )/Dt$ or $d(\ )/dt$. Note that it is computed <u>as</u> <u>we follow the material around</u> and thus it involves two contri-butions.

First, the property may change because the mass has moved to a new position (e.g., at the same instant of time the temperature in Tucson is different from that in Anchorage). This contribu-tion to the material derivative is sometimes called the "convec-tive derivative."

Second, the property may change with time at any given posi-tion (e.g., even in Tucson the temperature varies from morning to night). This latter contribution is called the local or "partial derivative" with respect to time and is written as $\partial(\ )/\partial t$.

As an example, for a typical three-dimensional unsteady flow the material derivative of the pressure would be represented as:

$$\frac{dp}{dt} = \underbrace{\frac{\partial p}{\partial x}\frac{dx}{dt} + \frac{\partial p}{\partial y}\frac{dy}{dt} + \frac{\partial p}{\partial z}\frac{dz}{dt}} + \frac{\partial p}{\partial t} \qquad (2.5)$$

convective derivative

local time derivative

If the fluid properties at every point are <u>independent</u> of time we call this steady flow. Thus, in steady flow the <u>partial</u> derivative of any property with respect to time is zero.

$$\frac{\partial(\ \ )}{\partial t} = 0 \qquad \text{for steady flow.} \qquad (2.6)$$

Notice that this does not prevent properties from being different in different locations. Thus, the material derivative may be non-zero for the case of steady flow due to the contribution of the convective portion.

We shall now examine the problem of computing mass flow rates when the flow is not one-dimensional. Consider the flow of a real fluid in a circular duct. At low Reynolds numbers — where the viscous forces predominate — the fluid tends to flow in layers without any energy exchange between adjacent layers. This is termed "laminar flow" and we could easily establish (see page 185 of ref. 9) that the velocity profile for this case would be a paraboloid of revolution, a cross-section of which is shown in Figure 2.1.

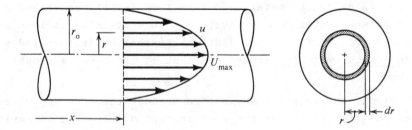

Figure 2.1 Velocity profile for laminar flow

At any given cross-section the velocity can be expressed as

$$u = U_{max}\left[1 - \left(\frac{r}{r_o}\right)^2\right] \qquad (2.7)$$

To compute the mass flow rate we integrate

$$\dot{m} = \text{mass flow rate} = \int_A \rho u dA \qquad (2.8)$$

where $\qquad\qquad\qquad dA = 2\pi r dr \qquad\qquad\qquad (2.9)$

Carry out the indicated integration and <u>show</u> that

$$\dot{m} = \rho(\pi r_o^2) \frac{U_m}{2} = \rho A \frac{U_m}{2} \qquad (2.10)$$

Note that for a multi-dimensional flow problem, when the flow rate is expressed as

$$\dot{m} = \rho A V \qquad (2.11)$$

the so called velocity V is an <u>average</u> velocity which for this case is $U_m/2$.

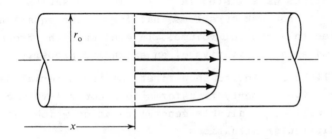

Figure 2.2 Velocity profile for turbulent flow

As we move to higher Reynolds numbers the large inertia forces cause irregular velocity fluctuations in all directions which in turn cause mixing between adjacent layers. The resulting energy transfer causes the fluid particles near the center to slow down while those particles next to the wall speed up. This produces the relatively "flat" velocity profile shown in Figure 2.2 which is typical of "turbulent flow." Notice that for this type of flow all particles at a given section have very nearly the same velocity which closely approximates a one-dimensional flow picture. Since most flows of engineering interest are well into this turbulent regime we can see why the assumption of one-dimensional flow is reasonably accurate.

STREAMLINES AND STREAMTUBES

As we progress through this book we will occasionally mention the following:

Streamline – a line which is everywhere tangent to the
           velocity vectors of those fluid particles
           which are on the line.

Streamtube – a flow passage which is formed by adjacent
            streamlines.

By virtue of these definitions no fluid particles ever cross a
streamline. Hence, fluid flows through a streamtube much like a
physical pipe.

2.4 TRANSFORMATION OF A MATERIAL DERIVATIVE
TO A CONTROL VOLUME APPROACH

In most gas dynamics problems it will be more convenient to
examine a fixed region in space, or a control volume. The fundamental equations were listed in Unit 1 for the analysis of a control mass. We now ask ourselves what form these equations take
when applied to a control volume. In each case the troublesome
term is a material derivative of an extensive property.

It will be simpler to show first how the material derivative
of any extensive property transforms to a control volume approach.
The result will be a valuable general relation which can be used
for many particular situations.

Let $N \equiv$ the total amount of any extensive property
            in a given mass.
        $\eta \equiv$ the amount of N per unit mass.

Thus: $$N = \int \eta \, dm = \iiint \rho \eta \, dv = \int_V \rho \eta \, dv \qquad (2.12)$$

where $dm \equiv$ incremental element of mass.
                   $dv \equiv$ incremental volume element.

Note that for simplicity we are indicating the triple volume
integral as $\int_V$.

Now let us consider what happens to the material derivative
dN/dt. Recall that a material derivative is the (time) rate of
change of a property computed as the mass moves around. Figure
2.3 shows an arbitrary mass at time t and the same mass at time
$t + \Delta t$. Remember that this system is at all times composed of
the same mass particles.

If Δt is small, then there will be an overlap of the two
regions as shown in Figure 2.4 with the common region identified

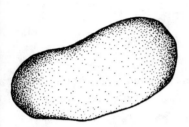
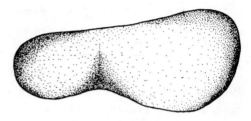

At time t the given mass occupies this region.

At time $t + \Delta t$ the same mass particles occupy this region.

Figure 2.3 Identification of control mass

as No. 2. At time t the given mass particles occupy regions 1 and 2. At time t + Δt the same mass particles occupy regions 2 and 3.

We construct our material derivative from the mathematical definition:

$$\frac{dN}{dt} \equiv \lim_{\Delta t \to o} \left[\frac{(\text{Final value of } N)_{t+\Delta t} - (\text{Initial value of } N)_t}{\Delta t} \right] \quad (2.13)$$

where the final value of N is the N of regions 2 and 3
        computed at time t + Δt
and the initial value of N is the N of regions 1 and 2
        computed at time t .

A more specific expression would be:

$$\frac{dN}{dt} = \lim_{\Delta t \to o} \left[\frac{(N_2 + N_3)_{t+\Delta t} - (N_1 + N_2)_t}{\Delta t} \right] \quad (2.14)$$

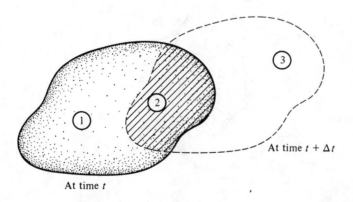

At time $t + \Delta t$

At time t

Figure 2.4 Control mass for small Δt

Look at the first and last terms:

$$\lim_{\Delta t \to o} \left[\frac{N_2(t+\Delta t) - N_2(t)}{\Delta t} \right] \quad \text{which by definition is} \quad \frac{\partial N_2}{\partial t} \, .$$

Note that the partial derivative notation is used since the region of integration is fixed and time is the only independent
parameter allowed to vary. Also note that as $\Delta t \to o$, region 2
approaches the original confines of the mass. We shall call this
region in space the "control volume."

Thus,
$$\lim_{\Delta t \to o} \left[\frac{N_2(t+\Delta t) - N_2(t)}{\Delta t} \right] = \frac{\partial N_{cv}}{\partial t} = \frac{\partial}{\partial t} \int_{cv} \rho \eta \, dv \qquad (2.15)$$

where cv stands for the control volume.

Next, consider the term $\lim_{\Delta t \to o} \dfrac{N_3(t+\Delta t)}{\Delta t}$.

The numerator represents the amount of N in region 3 at time
$t + \Delta t$ and by definition region 3 is formed by the fluid moving
out of the control volume. Let $\hat{n}$ be a unit normal, positive
when pointing outward from the control volume. Also let dA be
an increment of the surface area which separates regions 2 and 3
as shown in Figure 2.5.

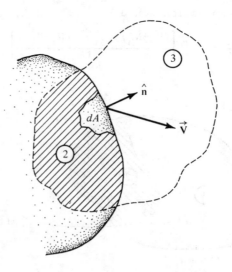

Figure 2.5 Flow out of control volume

$$\vec{V} \cdot \hat{n} = \text{component of } \vec{V} \perp \text{ to } dA$$
$$(\vec{V} \cdot \hat{n})dA = \text{incremental volumetric flow rate}$$
$$\rho(\vec{V} \cdot \hat{n})dA = \text{incremental mass flow rate}$$
$$\rho(\vec{V} \cdot \hat{n})dA\Delta t = \text{amount of mass that crossed } dA \text{ in time } \Delta t$$
$$\eta\rho(\vec{V} \cdot \hat{n})dA\Delta t = \text{amount of } N \text{ that crossed } dA \text{ in time } \Delta t$$

Thus, $\displaystyle\int_{S_{out}}\eta\rho(\vec{V}\cdot\hat{n})dA\,\Delta t \approx$ total amount of N in region 3   (2.16)

where $\displaystyle\int_{S_{out}}$ is a double integral over the surface where fluid

leaves the control volume. The term in question becomes:

$$\lim_{\Delta t \to o} \frac{N_3(t+\Delta t)}{\Delta t} = \int_{S_{out}}\eta\rho(\vec{V}\cdot\hat{n})dA \qquad (2.17)$$

This integral is called a flux or rate of N flow <u>out</u> of the control volume.

Since the Δt cancels, one might question the limit process. Actually, the integral expression in equation 2.16 is only approximately correct. This is because all of the properties in this integral are going to be evaluated at the surface S and at the time t. Thus, equation 2.16 is only approximate as written but <u>becomes</u> <u>exact</u> <u>in</u> <u>the</u> <u>limit</u> as $\Delta t \to o$.

Now let us consider the remaining term $\displaystyle\lim_{\Delta t\to o}\frac{N_1(t)}{\Delta t}$.

What does the numerator represent? We evaluate N_1 by the following procedure. Let $\hat{n}'$ be a unit normal, positive when pointing <u>inward</u> to the control volume as shown in Figure 2.6.

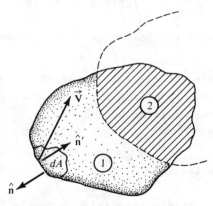

Figure 2.6 Flow into control volume

Complete the following in words:

$$\vec{V} \cdot \hat{n}' =$$
$$(\vec{V} \cdot \hat{n}') dA =$$
$$\rho(\vec{V} \cdot \hat{n}') dA =$$
$$\rho(\vec{V} \cdot \hat{n}') dA \Delta t =$$
$$\eta\rho(\vec{V} \cdot \hat{n}') dA \Delta t =$$

It should be clear that

$$\int_{S_{in}} \eta\rho(\vec{V} \cdot \hat{n}') dA \Delta t \approx \text{total amount of } N \text{ in region 1} \qquad (2.18)$$

and

$$\lim_{\Delta t \to o} \frac{N_1(t)}{\Delta t} = \int_{S_{in}} \eta\rho(\vec{V} \cdot \hat{n}') dA \qquad (2.19)$$

where $\int_{S_{in}}$ is a double integral over the surface where fluid enters the control volume. This term represents the "N flux" into the control volume.

We now substitute into equation 2.14 all the terms that we have developed in equations 2.15, 2.17, and 2.19.

$$\frac{dN}{dt} = \frac{\partial}{\partial t} \int_{cv} \rho\eta dv + \int_{S_{out}} \eta\rho(\vec{V} \cdot \hat{n}) dA - \int_{S_{in}} \eta\rho(\vec{V} \cdot \hat{n}') dA \qquad (2.20)$$

Noting that $\hat{n} = -\hat{n}'$ we can combine the last two terms into:

$$\int_{S_{out}} \eta\rho(\vec{V} \cdot \hat{n}) dA - \int_{S_{in}} \eta\rho(\vec{V} \cdot \hat{n}') dA =$$

$$\int_{S_{out}} \eta\rho(\vec{V} \cdot \hat{n}) dA + \int_{S_{in}} \eta\rho(\vec{V} \cdot \hat{n}) dA = \int_{cs} \eta\rho(\vec{V} \cdot \hat{n}) dA \qquad (2.21)$$

where cs represents the entire control surface surrounding the control volume.

This term represents the net rate at which N passes out of the control volume (i.e., flow rate out minus flow rate in). The final transformation equation becomes:

$$\left.\frac{dN}{dt}\right]_{\substack{\text{material} \\ \text{derivative}}} = \frac{\partial}{\partial t} \int_{cv} \eta\rho dv + \int_{cs} \eta\rho(\vec{V} \cdot \hat{n}) dA \qquad (2.22)$$

Triple integral   Double integral

This relation is known as Reynolds Transport Theorem and it can be interpreted in words as:

> The rate of change of N for a given mass as it is moving around is equal to the rate of change of N inside the control volume plus the net efflux (flow out minus flow in) of N from the control volume.

It is essential to note that we have not placed any restriction on N other than it must be a mass dependent (extensive) property. Thus N may be a scalar or a vector quantity. Examples of the application of this powerful transformation equation are found in the next two sections and in Unit 3.

2.5 CONSERVATION OF MASS

If we exclude the possibility of nuclear reactions from consideration, then we can separately account for the conservation of mass and energy. Thus, if we observe a given quantity of mass as it moves around we can say by definition that the mass will remain fixed. Another way of stating this is that the material derivative of the mass is zero.

$$\frac{d(Mass)}{dt} = 0 \qquad (2.23)$$

This is the "continuity equation" for a control mass. What corresponding expression can we write for a control volume? To find out we must transform the material derivative according to the relation developed in Section 2.4.

If N represents the total mass, then η is the mass per unit mass, or 1 . Substitution into equation 2.22 yields:

$$\frac{d(mass)}{dt} = \frac{\partial}{\partial t}\int_{cv}\rho\,dv + \int_{cs}\rho(\vec{V}\cdot\hat{n})dA \qquad (2.24)$$

But we know by equation 2.23 that this must be zero; thus, the transformed equation is:

$$0 = \frac{\partial}{\partial t}\int_{cv}\rho\,dv + \int_{cs}\rho(\vec{V}\cdot\hat{n})dA \qquad (2.25)$$

This is the "continuity equation" for a control volume. State in words what each term represents.

For steady flow any partial derivative with respect to time is zero and the equation becomes:

$$0 = \int_{cs} \rho(\vec{V} \cdot \hat{n})dA \tag{2.26}$$

Let us now consider the evaluation of the remaining integral for the case of one-dimensional flow. Figure 2.7 shows fluid crossing a portion of the control surface. Recall that for one-dimensional flow any fluid property will be constant over an entire cross section.

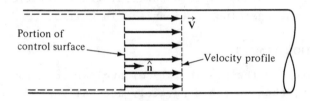

Figure 2.7 One-dimensional velocity profile

Thus, both the density and the velocity can be brought out from under the integral sign. If the surface is always chosen perpendicular to V, then the integral is very simple to evaluate.

$$\int \rho(\vec{V} \cdot \hat{n})dA = \rho\vec{V} \cdot \hat{n} \int dA = \rho VA$$

This integral must be evaluated over the entire control surface which yields:

$$\int_{cs} \rho(\vec{V} \cdot \hat{n})dA = \Sigma\rho VA \tag{2.27}$$

This summation is taken over all sections where fluid crosses the control surface and is positive where fluid leaves the control volume (since $\vec{V} \cdot \hat{n}$ is positive here) and negative where fluid enters the control volume. For steady, one-dimensional flow the continuity equation for a control volume becomes:

$$\Sigma\rho AV = 0 \tag{2.28}$$

If there is only one section where fluid enters and one section where fluid leaves the control volume, then this becomes

$$\left(\rho AV\right)_{out} - \left(\rho AV\right)_{in} = 0$$

or

$$\left(\rho AV\right)_{out} = \left(\rho AV\right)_{in} \tag{2.29}$$

We usually write this as:

$$\dot{m} = \rho AV = \text{constant} \qquad (2.30)$$

Implicit in this expression is that V is the component of velocity perpendicular to the area A. If the density ρ is in lbm per cubic foot, the area A is in square feet, and the velocity V is in feet per second, what are the units of the mass flow rate $\dot{m}$? What will each of these be in SI units?

Note that as a result of steady flow the mass flow rate into a control volume is equal to the mass flow rate out of the control volume. The converse of this is not true; i.e., just because it is known that the flow rates into and out of a control volume are the same this does not insure that the flow is steady.

Example 2.1 Air flows steadily through a 1" diameter section with a velocity of 1096 ft/sec. The temperature is $40°F$ and the pressure is 50 psia. The flow passage expands to 2" diameter and at this section the pressure and temperature have dropped to 2.82 psia and $-240°F$, respectively. What is the average velocity at this section?

Knowing: $\qquad p = \rho RT \quad \text{and} \quad A = \pi D^2/4$

For steady, one-dimensional flow:

$$\rho_1 A_1 V_1 = \rho_2 A_2 V_2$$

$$\left[\frac{p_1}{RT_1}\right]\left[\frac{\pi D_1^2}{4}\right]V_1 = \left[\frac{p_2}{RT_2}\right]\left[\frac{\pi D_2^2}{4}\right]V_2$$

$$V_2 = V_1 \frac{D_1^2 p_1 T_2}{D_2^2 p_2 T_1} = 1096\left(\frac{1}{2}\right)^2\left(\frac{50}{2.82}\right)\left(\frac{220}{500}\right)$$

$$V_2 = \underline{2138} \text{ ft/sec}$$

An alternate form of the continuity equation can be obtained by differentiating equation 2.30. For steady, one-dimensional flow this means that

$$d(\rho AV) = AVd\rho + \rho VdA + \rho AdV = 0 \qquad (2.31)$$

Dividing by $\rho A V$ yields

$$\frac{d\rho}{\rho} + \frac{dA}{A} + \frac{dV}{V} = 0$$ (2.32)

This expression can also be obtained by first taking the natural logarithm of equation 2.30 and then differentiating the result. This is called "logarithmic differentiation." Try it.

This differential form of the continuity equation is useful in interpreting the changes that must occur as fluid flows through a duct, channel or streamtube. It indicates that if mass is to be conserved, the changes in density, velocity and cross-sectional area must compensate for one another. For example, if the area is constant (dA=0), then any increase in velocity must be accompanied by a corresponding decrease in density. We shall also use this form of the continuity equation in several future derivations.

2.6 CONSERVATION OF ENERGY

The First Law of Thermodynamics is a statement of conservation of energy. For a system composed of a given quantity of mass that undergoes a process we can say that

$$Q = W + \Delta E$$ (1.28)

where Q is the net heat transferred into the system,
         W is the net work done by the system,
         ΔE is the change in total energy of the system.

This can also be written on a rate basis to yield an expression that is valid at any instant of time.

$$\frac{\delta Q}{dt} = \frac{\delta W}{dt} + \frac{dE}{dt}$$ (2.33)

We must carefully examine each term in this equation to clearly understand its significance. $\delta Q/dt$ and $\delta W/dt$ represent instantaneous rates of heat and work transfer between the system and its surroundings. They are rates of energy transfer across the boundaries of the system. These terms are not material derivatives. (Recall that heat and work are not properties of a system.) On the other hand, energy is a property of the system and dE/dt is a material derivative.

We now ask what form the energy equation takes when applied
to a control volume. To answer this we must first transform the
material derivative in equation 2.33 according to the relation
developed in Section 2.4.

If we let N be E , the total energy of the system, then
η represents e , the energy per unit mass

$$e = u + \frac{V^2}{2g_c} + \frac{g}{g_c} z \tag{1.30}$$

Substitution into equation 2.22 yields:

$$\frac{dE}{dt} = \frac{\partial}{\partial t} \int_{cv} e\rho dv + \int_{cs} e\rho(\vec{V}\cdot\hat{n})dA \tag{2.34}$$

and the transformed equation which is applicable to a control
volume is:

$$\boxed{\frac{\delta Q}{dt} = \frac{\delta W}{dt} + \frac{\partial}{\partial t} \int_{cv} e\rho dv + \int_{cs} e\rho(\vec{V}\cdot\hat{n})dA} \tag{2.35}$$

In this case $\delta Q/dt$ and $\delta W/dt$ represent instantaneous rates of
heat and work transfer across the surface that surrounds the control volume. <u>State</u> in words what the other terms represent.
(See discussion after equation 2.22.)

For one-dimensional flow the last integral in equation 2.35
is simple to evaluate as e, ρ, and V are constant over any
given cross section. Assuming that the velocity V is perpendicular to the surface A we have:

$$\int_{cs} e\rho(\vec{V}\cdot\hat{n})dA = \Sigma e\rho V \int dA = \Sigma e\rho VA = \Sigma\dot{m}e \tag{2.36}$$

The summation is taken over all sections where fluid crosses the
control surface and is positive where fluid leaves the control
volume and negative where fluid enters the control volume.

In using equation 2.35 we must be careful to include all
forms of work whether done by pressure forces (from normal
stresses) or shear forces (from tangential stresses). Figure 2.8
shows a simple control volume. Note that the control surface is
carefully chosen so that there is no fluid motion at the boundary
except: (a) where fluid enters and leaves the system, or
          (b) where a mechanical device such as a shaft crosses
              the boundaries of the system.

This prudent choice of the system boundary simplifies calculation
of the work quantities. For example, the pressure and shear
forces along the side walls do no work since they do not move
through any distance.

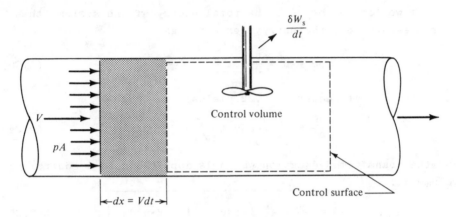

Figure 2.8 Identification of work quantities

The rate at which work is transmitted out of the system by
the mechanical device is called $\delta W_s/dt$ and is accomplished by
shear stresses between the device and the fluid. (Think of the
subscript s for shear stresses or shaft work.)

The other work quantities to be considered are where fluid
enters and leaves the system. Here the pressure forces do work
to push fluid into or out of the control volume. The shaded area
at the inlet represents the fluid which enters the control volume
during time dt . The work done here is

$$\delta W' = \vec{F} \cdot \vec{dx} = pAdx = pAVdt \qquad (2.37)$$

The rate of doing work is $\dfrac{\delta W'}{dt} = pAV$ (2.38)

This is called "flow work" or "displacement work." It can be
expressed in a more meaningful form by introducing

$$\dot{m} = \rho AV \qquad (2.11)$$

Thus, Rate of Doing Flow Work $= pAV = p\dfrac{\dot{m}}{\rho} = \dot{m}pv$ (2.39)

This represents work done <u>by</u> the system (positive) to force fluid
<u>out</u> of the control volume and represents work done <u>on</u> the system
(negative) to force fluid <u>into</u> the control volume.

Thus, the total work $\quad \dfrac{\delta W}{dt} = \dfrac{\delta W_s}{dt} + \Sigma \dot{m} pv$

We may now rewrite our energy equation in a more useful form which is applicable to one-dimensional flow. Notice how the flow work has been included in the last term.

$$\frac{\delta Q}{dt} = \frac{\delta W_s}{dt} + \frac{\partial}{\partial t} \int_{cv} e\rho \, dv + \Sigma \dot{m}(e + pv) \qquad (2.40)$$

If we consider steady flow, the term involving the partial derivative with respect to time is zero. Thus, for steady one-dimensional flow the energy equation for a control volume becomes:

$$\boxed{\frac{\delta Q}{dt} = \frac{\delta W_s}{dt} + \Sigma \dot{m}(e + pv)} \qquad (2.41)$$

If there is only one section where fluid leaves and one section where fluid enters the control volume we have (from continuity)

$$\dot{m}_{in} = \dot{m}_{out} = \dot{m} \qquad (2.42)$$

We may now divide equation 2.41 by $\dot{m}$

$$\frac{1}{\dot{m}} \frac{\delta Q}{dt} = \frac{1}{\dot{m}} \frac{\delta W_s}{dt} + (e + pv)_{out} - (e + pv)_{in} \qquad (2.43)$$

We now define $\qquad\qquad q \equiv \dfrac{1}{\dot{m}} \dfrac{\delta Q}{dt} \qquad (2.44)$

$$w_s \equiv \frac{1}{\dot{m}} \frac{\delta W_s}{dt} \qquad (2.45)$$

where q and w_s represent quantities of heat and shaft work crossing the control surface per unit mass of fluid flowing. What are the units of q and w_s?

Our equation has now become

$$q = w_s + (e + pv)_{out} - (e + pv)_{in} \qquad (2.46)$$

This can be directly applied to the finite control volume shown in Figure 2.9 with the result

$$q = w_s + (e_2 + p_2 v_2) - (e_1 + p_1 v_1) \qquad (2.47)$$

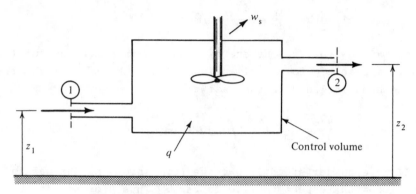

Figure 2.9 Finite control volume for energy analysis

Detailed substitution for e (from equation 1.30) yields

$$u_1 + p_1 v_1 + \frac{V_1^2}{2g_c} + \frac{g}{g_c} z_1 + q = u_2 + p_2 v_2 + \frac{V_2^2}{2g_c} + \frac{g}{g_c} z_2 + w_s \quad (2.48)$$

If we introduce the definition of enthalpy

$$h \equiv u + pv \quad (1.34)$$

the equation can be shortened to

$$h_1 + \frac{V_1^2}{2g_c} + \frac{g}{g_c} z_1 + q = h_2 + \frac{V_2^2}{2g_c} + \frac{g}{g_c} z_2 + w_s \quad (2.49)$$

This is the form of the energy equation that may be used to solve
many problems. Can you list the assumptions that have been made
to develop equation 2.49?

Example 2.2   Steam enters an ejector at the rate of 0.1 lbm/sec
with an enthalpy of 1300 Btu/lbm and negligible velocity.
Water enters at the rate of 1.0 lbm/sec with an enthalpy of
40 Btu/lbm and negligible velocity. The mixture leaves the ejector with an enthalpy of 150 Btu/lbm and a velocity of 90
ft/sec . All potentials may be neglected. Determine the magnitude and direction of the heat transfer.

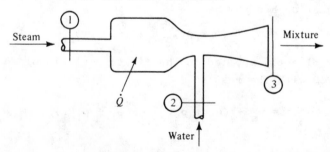

$$\dot{m}_1 = 0.1 \text{ lbm/sec} \qquad V_1 \approx 0 \qquad h_1 = 1300 \text{ Btu/lbm}$$
$$\dot{m}_2 = 1.0 \text{ lbm/sec} \qquad V_2 \approx 0 \qquad h_2 = 40 \text{ Btu/lbm}$$
$$V_3 = 90 \text{ ft/sec} \qquad h_3 = 150 \text{ Btu/lbm}$$

Continuity:
$$\dot{m}_3 = \dot{m}_1 + \dot{m}_2 = 0.1 + 1.0 = 1.1 \text{ lbm/sec}$$

Energy:
$$\dot{m}_1\left[h_1 + \frac{V_1^2}{2g_c} + \frac{g}{g_c}z_1\right] + \dot{m}_2\left[h_2 + \frac{V_2^2}{2g_c} + \frac{g}{g_c}z_2\right] + \dot{Q} =$$

$$\dot{m}_3\left[h_3 + \frac{V_3^2}{2g_c} + \frac{g}{g_c}z_3\right] + \dot{W}_s$$

$$\dot{m}_1 h_1 + \dot{m}_2 h_2 + \dot{Q} = \dot{m}_3\left[h_3 + \frac{V_3^2}{2g_c}\right]$$

$$0.1(1300) + 1.0(40) + \dot{Q} = 1.1\left[150 + \frac{90^2}{2(32.2)778}\right]$$

$$130 + 40 + \dot{Q} = 1.1(150 + .162) = 165.2$$

$$\dot{Q} = 165.2 - 130 - 40 = -\underline{4.8} \text{ Btu/sec}$$

The minus sign indicates that heat is lost from the ejector.

Example 2.3 A horizontal duct of constant area contains CO_2 flowing isothermally. At a section where the pressure is 14 bars absolute the average velocity is known to be 50 m/s . Farther downstream the pressure has dropped to 7 bars abs. Find the heat transfer.

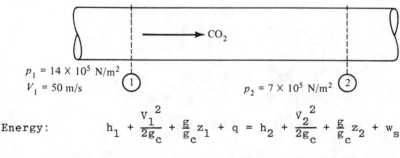

$$p_1 = 14 \times 10^5 \text{ N/m}^2$$
$$V_1 = 50 \text{ m/s} \quad \textcircled{1} \qquad\qquad p_2 = 7 \times 10^5 \text{ N/m}^2 \quad \textcircled{2}$$

Energy:
$$h_1 + \frac{V_1^2}{2g_c} + \frac{g}{g_c}z_1 + q = h_2 + \frac{V_2^2}{2g_c} + \frac{g}{g_c}z_2 + w_s$$

Since perfect gas and isothermal: $\Delta h = c_p \Delta T = 0$

and thus
$$q_{1-2} = (V_2^2 - V_1^2)/2g_c$$

State:
$$\frac{p_1}{\rho_1 T_1} = \frac{p_2}{\rho_2 T_2} \quad \rightarrow \quad \frac{p_1}{p_2} = \frac{\rho_1}{\rho_2}$$

Continuity: $$\rho_1 A_1 V_1 = \rho_2 A_2 V_2$$

$$\frac{V_2}{V_1} = \frac{\rho_1}{\rho_2} = \frac{p_1}{p_2}$$

and thus $$V_2 = \frac{p_1}{p_2} V_1 = \frac{14 \times 10^5}{7 \times 10^5} (50) = 100 \text{ m/s}$$

Returning to the energy equation, we have

$$q = (V_2^2 - V_1^2)/2g_c = (100^2 - 50^2)/2(1) = \underline{3750} \text{ joules/kg}$$

Example 2.4 Air at $2200^\circ R$ enters a turbine at the rate of 1.5 lbm/sec. The air expands through a pressure ratio of 15 and leaves at $1090^\circ R$. Velocities entering and leaving are negligible and there is no heat transfer. Calculate the horsepower output of the turbine.

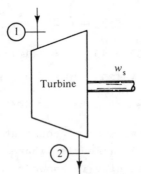

$T_1 = 2200^\circ R$ $T_2 = 1090^\circ R$ $\dot{m} = 1.5$ lbm/sec

$V_1 \approx 0$ $V_2 \approx 0$ $q = 0$

Energy: $$h_1 + \frac{V_1^2}{2g_c} + \frac{g}{g_c} z_1 + q = h_2 + \frac{V_2^2}{2g_c} + \frac{g}{g_c} z_2 + w_s$$

$$w_s = h_1 - h_2 = c_p(T_1 - T_2)$$

$$w_s = 0.24(2200-1090) = \underline{266} \text{ Btu/lbm}$$

$$HP = \dot{m} \, w_s \frac{778}{550} = 1.5(266) \frac{778}{550} = \underline{564} \text{ HP}$$

DIFFERENTIAL FORM OF ENERGY EQUATION

One can also apply the energy equation to a differential control volume as shown in Figure 2.10. We assume steady one-dimensional flow. The properties of the fluid entering the control volume are designated as ρ, u, p, V, etc. Fluid leaves

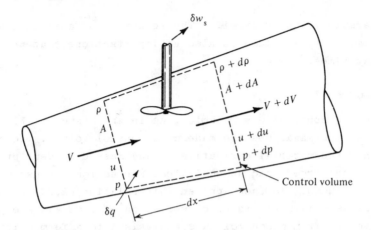

Figure 2.10 Energy analysis on infitesimal control volume

the control volume with properties that have changed slightly as indicated by $\rho + d\rho$, $u + du$, etc.

Application of equation 2.46 to this differential control volume will produce:

$$\delta q = \delta w_s + \left[(p+dp)(v+dv) + (u+du) + \frac{(V+dV)^2}{2g_c} + \frac{g}{g_c}(z+dz) \right]$$

$$- \left[pv + u + \frac{V^2}{2g_c} + \frac{g}{g_c}z \right] \qquad (2.50)$$

Expand equation 2.50, cancel like terms and <u>show</u> that:

$$\delta q = \delta w_s + pdv + vdp + \overset{\text{H.O.T.}}{\cancel{dpdv}} + du + \frac{2VdV + \overset{\text{H.O.T.}}{\cancel{(dV)^2}}}{2g_c} + \frac{g}{g_c}dz \qquad (2.51)$$

As dx is allowed to approach zero we can neglect the higher order terms (indicated by H.O.T.).

Noting that

$$2VdV = dV^2$$

and

$$pdv + vdp = d(pv)$$

we obtain

$$\delta q = \delta w_s + d(pv) + du + \frac{dV^2}{2g_c} + \frac{g}{g_c}dz \qquad (2.52)$$

and since

$$dh = du + d(pv)$$

we have

$$\boxed{\delta q = \delta w_s + dh + \frac{dV^2}{2g_c} + \frac{g}{g_c}dz} \qquad (2.53)$$

This can be directly integrated to produce equation 2.49 for a finite control volume but the differential form is frequently of

considerable value by itself. The technique of analyzing a differential control volume is also an important one that we shall
use many times.

2.7 SUMMARY

In the study of Gas Dynamics, as in any branch of Fluid
Dynamics, most analyses are made on a control volume. We have
shown how the material derivative of any mass dependent property
can be transformed into an equivalent expression for use with
control volumes. We then applied this relation (2.22) to show
how the basic laws regarding conservation of mass and energy can
be converted from a control mass analysis into a form suitable
for control volume analysis. Most of the work in this course
will be done assuming steady, one-dimensional flow; thus, each
general equation was simplified for these conditions.

Care should be taken to approach each problem in a consistent and organized fashion. For a typical problem the following
steps should be taken:

1. Sketch the flow system and identify the control volume.
2. Label sections where fluid enters and leaves the control
   volume.
3. Note where energy (Q and W_s) crosses the control surface.
4. Record all known quantities with their units.
5. Solve for the unknowns by a systematic application of
   the basic equations.

The basic concepts that we have used so far are few in number:

STATE   - A simple density relation such as $p = \rho RT$ or ρ = constant

CONTINUITY - Derived from conservation of mass

ENERGY   - Derived from conservation of energy.

Some of the most frequently used equations that were developed in this unit are summarized below. Some are restricted to
steady one-dimensional flow; others involve additional assumptions. You should know under what conditions each may be used.

a. Mass Flow Rate past a section

$$\dot{m} = \int_A \rho u \, dA \qquad (2.8)$$

u = velocity perpendicular to dA

b. Transformation of Material Derivative to Control Volume
   Analysis

$$\frac{dN}{dt} = \frac{\partial}{\partial t}\int_{cv}\eta\rho\,dv + \int_{cs}\eta\rho(\vec{V}\cdot\hat{n})dA \qquad (2.22)$$

If one-dimensional $\qquad \int_{cs}\eta\rho(\vec{V}\cdot\hat{n})dA = \Sigma\dot{m}\eta \qquad (2.54)$

If steady $\qquad \frac{\partial(\ )}{\partial t} = 0 \qquad (2.6)$

c. Mass Conservation - Continuity Equation $\begin{cases} N = Mass \\ \eta = 1 \end{cases}$

$$\frac{\partial}{\partial t}\int_{cv}\rho\,dv + \int_{cs}\rho(\vec{V}\cdot\hat{n})dA = 0 \qquad (2.25)$$

For steady, one-dimensional flow

$$\dot{m} = \rho AV = const \qquad (2.30)$$

$$\frac{d\rho}{\rho} + \frac{dA}{A} + \frac{dV}{V} = 0 \qquad (2.32)$$

d. Energy Conservation - Energy Equation $\begin{cases} N = E \\ \eta = e = u + \dfrac{V^2}{2g_c} + \dfrac{g}{g_c}z \end{cases}$

$$\frac{\delta Q}{dt} = \frac{\delta W}{dt} + \frac{\partial}{\partial t}\int_{cv}e\rho\,dv + \int_{cs}e\rho(\vec{V}\cdot\hat{n})dA \qquad (2.35)$$

$$w = shaft\ work\ (w_s) + flow\ work\ (pv)$$

For steady, one-dimensional flow

$$h_1 + \frac{V_1^2}{2g_c} + \frac{g}{g_c}z_1 + q = h_2 + \frac{V_2^2}{2g_c} + \frac{g}{g_c}z_2 + w_s \qquad (2.49)$$

$$\delta q = \delta w_s + dh + \frac{dV^2}{2g_c} + \frac{g}{g_c}dz \qquad (2.53)$$

2.8 PROBLEMS

Problem statements may occasionally give some irrelevant
information or, on the other hand, sometimes logical assumptions
have to be made before a solution can be carried out. For instance, unless specific information is given on potential differences it is logical to assume these are negligible; if no machine
is present it is reasonable to assume $w_s = 0$, etc. However,

think carefully before arbitrarily eliminating terms from any
equation — you may be eliminating a vital element from the problem. Check to see if there isn't some way to compute the desired
quantity (such as calculating the enthalpy of a gas from its temperature). Properties of selected gases may be found in the Appendix.

1. There is three-dimensional flow of an incompressible fluid in
    a duct of radius R . The velocity distribution at any sec-
    tion is hemispherical, with the maximum velocity U_m at the
    center and zero velocity at the wall. Show that the average
    velocity is 2/3 U_m .

2. A constant density fluid flows between two flat parallel
    plates which are separated by a distance δ . Sketch the ve-
    locity distribution and compute the average velocity if the
    velocity u is given by the following relation:

    (a) $u = k_1 y$,   (b) $u = k_2 y^2$,   (c) $u = k_3 \left[\delta y - y^2 \right]$.

    In each case express your answer in terms of the maximum ve-
    locity U_m .

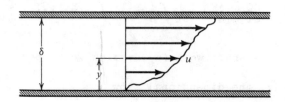

3. An incompressible fluid is flowing in a rectangular duct whose
    dimensions are 2 units in the Y-direction and 1 unit in the
    Z-direction. The velocity in the X-direction is given by the
    equation $u = 3y^2 + 5z$. Compute the average velocity.

4. Evaluate the integral $\int \rho e(\vec{V} \cdot \hat{n})dA$ over the surface shown for
    the velocity and energy distributions indicated. You may
    assume that the density is constant.

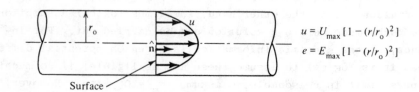

$$u = U_{max}[1 - (r/r_o)^2]$$
$$e = E_{max}[1 - (r/r_o)^2]$$

5. In a ten-inch-diameter duct the average velocity of water is 14 ft/sec.
   (a) What is the average velocity if the diameter changes to 6 inches?
   (b) Express the average velocity in terms of an arbitrary diameter.

6. Nitrogen flows in a constant area duct. Conditions at section one are as follows: p_1 = 200 psia , T_1 = 90°F , and V_1 = 10 ft/sec. At section two we find: p_2 = 45 psia and T_2 = 90°F . Determine the velocity at section 2.

7. Steam enters a turbine with an enthalpy of 1600 Btu/lbm and a velocity of 100 ft/sec, at a flow rate of 80,000 lbm/hr. The steam leaves the turbine with an enthalpy of 995 Btu/lbm and a velocity of 150 ft/sec. Compute the power output of the turbine assuming it to be 100% efficient. Neglect any heat transfer and potential energy changes.

8. A flow of 2.0 lbm/sec of air is compressed from 14.7 psia and 60°F to 200 psia and 150°F. Cooling water circulates around the cylinders at the rate of 25 lbm/min. The water enters at 45°F and leaves at 130°F. (Specific heat of water is 1.0 Btu/lbm-°F.) Calculate the power required to compress the air assuming negligible velocities at inlet and outlet.

9. Hydrogen expands isentropically from 15 bars absolute and 340°K to 3 bars absolute in a steady flow process without heat transfer.
   (a) Compute the final velocity if the initial velocity is negligible.
   (b) Compute the flow rate if the final duct size is 10 cm in diameter.

10. At a section where the diameter is 4 inches, methane flows with a velocity of 50 ft/sec and a pressure of 85 psia. At a downstream section, where the diameter has increased to 6 inches, the pressure is 45 psia. Assuming the flow to be iso-thermal, compute the heat transfer between the two locations.

11. Carbon dioxide flows in a horizontal duct at 7 bars abs and 300°K with a velocity of 10 m/s. At a downstream location the pressure is 3.5 bars abs and the temperature is 280°K. If 1.4×10^4 J/kg of heat is lost by the fluid between these loca-tions,

(a) Determine the velocity at the second location and

(b) Compute the ratio of initial to final areas.

12. Hydrogen flows through a horizontal insulated duct. At section 1 the enthalpy is 2400 Btu/lbm, the density is 0.5 lbm/ft^3, and the velocity is 500 ft/sec. At a downstream section h_2 = 2240 Btu/lbm and ρ_2 = 0.1 lbm/ft^3. No shaft work is done. Determine the velocity at section 2 and the ratio of areas.

13. Nitrogen traveling at 12 m/s with a pressure of 14 bars abs, temperature of 800°K, and an area of 0.05 m^2 enters a device where no work or heat transfer takes place. The temperature at the exit, where the area is 0.15 m^2, has dropped to 590°K. What are the velocity and the pressure at the outlet section?

14. Cold water with an enthalpy of 8 Btu/lbm enters a heater at the rate of 5 lbm/sec with a velocity of 10 ft/sec, and at a potential of 10 feet with respect to the other connections shown. Steam enters at the rate of 1 lbm/sec with a velocity of 50 ft/sec and an enthalpy of 1350 Btu/lbm. These two streams mix in the heater and hot water emerges with an enthalpy of 168 Btu/lbm and a velocity of 12 ft/sec. Determine the heat lost from the apparatus. What percentage error is involved if both kinetic and potential energy changes are neglected?

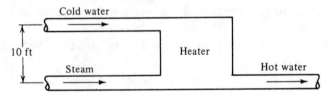

2.9 CHECK TEST

You should be able to complete this test without reference to material in the unit.

1. Name the basic concepts (or equations) from which the study of gas dynamics proceeds.

2. Define steady flow. Explain what is meant by one-dimensional flow.

3. An incompressible fluid flows in a duct of radius r_o. At a particular location the velocity distribution is

$u = U_m\left[1 - (r/r_o)^2\right]$ and the distribution of an extensive

property is $\beta = B_m\left[1 - (r/r_o)\right]$. Evaluate the integral

$\int \rho\beta(\vec{V}\cdot\hat{n})dA$ at this location.

4. Write the equation used to relate the material derivative of any mass dependent property to the properties inside of and crossing the boundaries of a control volume. State in words what the integrals actually represent.

5. Simplify the integral $\int_{cs}\rho\beta(\vec{V}\cdot\hat{n})dA$ for the control volume shown below if the flow is steady and one-dimensional. (Careful, β and ρ may vary from section to section.)

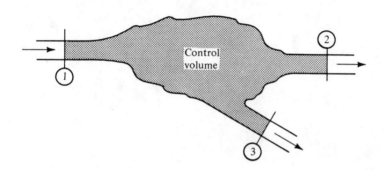

6. Write the simplest form of the energy equation that you would use to analyze the control volume below. You may assume steady, one-dimensional flow.

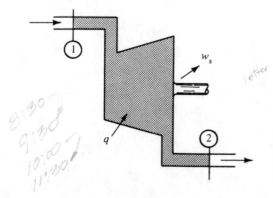

7. Work problem #13 in Section 2.8.

UNIT 3

CONTROL VOLUME ANALYSIS—PART II

3.1 INTRODUCTION

We begin this unit with a discussion of entropy which is one of the most useful thermodynamic properties in the study of Gas Dynamics. Entropy changes will be divided into two categories and it is hoped that treatment in this fashion will give you a better understanding of this important property. Next, we shall introduce the concept of a "stagnation" process. This leads to the stagnation state as a reference condition which will be used throughout our remaining discussions.

The above ideas permit rewriting our energy equation in alternate forms from which interesting observations can be made.

We then investigate some of the consequences of a constant-density fluid. This leads to special relations which not only can be used for liquids but under certain conditions are excellent approximations for gases.

At the close of the unit we complete our basic set of equations by transforming Newton's Second Law for use in the analysis of control volumes. This is done for both finite and differential volume elements.

3.2 OBJECTIVES

After successfully completing this unit you should be able to:

1. Explain how entropy changes can be divided into two categories. Define and interpret each part.

2. Define an isentropic process and explain the relationship among reversible, adiabatic and isentropic processes.

3. Show that by introducing the concept of entropy and the definition of enthalpy, the path function heat (δQ) may be removed from the energy equation to yield an expression called the "pressure-energy equation."

$$\frac{dp}{\rho} + \frac{dV^2}{2g_c} + \frac{g}{g_c}dz + Tds_i + \delta w_s = 0$$

4. Simplify the pressure-energy equation to obtain Bernoulli's equation. Note all assumptions or restrictions that apply to Bernoulli's equation.

5. Explain the stagnation state concept and the difference between static and stagnation properties.

6. Define stagnation enthalpy by an equation which is valid for any fluid.

7. Draw an h-s diagram representing a flow system and indicate static and stagnation points for an arbitrary section.

8. Introduce the stagnation concept into the energy equation and derive the "stagnation pressure-energy" equation.

$$\frac{dp_t}{\rho_t} + ds_e(T_t - T) + T_tds_i + \delta w_s = 0$$

9. Demonstrate the ability to apply continuity and energy concepts to the solution of typical flow problems with constant-density fluids.

10. Starting with the basic concept or equation which is valid for a control mass, obtain the integral form of the momentum equation for a control volume.

11. Simplify the integral form of the momentum equation for a control volume for the conditions of steady, one-dimensional flow.

12. Apply the simplified form of the momentum equation to a differential control volume. (Optional.)

13. Demonstrate the ability to apply momentum concepts in the analysis of control volumes.

3.3 COMMENTS ON ENTROPY

In Section 1.4 entropy changes were defined in the usual manner in terms of reversible processes:

$$\Delta S \equiv \int \frac{\delta Q_R}{T} \qquad (1.38)$$

The term δQ_R is related to a fictitious reversible process (a rare animal indeed) and consequently is not the heat transfer involved in the process under consideration. It would seem more appropriate to work with the actual heat transfer for the process. To accomplish this it is necessary to divide the entropy changes of any system into two categories. We shall follow the notation of Hall (reference 15).

Let
$$dS \equiv dS_e + dS_i \qquad (3.1)$$

The term dS_e represents that portion of entropy change caused by the actual heat transfer between the system and its (external) surroundings. It can be evaluated easily from

$$dS_e = \frac{\delta Q}{T} \qquad (3.2)$$

One should note that dS_e can be positive or negative depending on the direction of heat transfer. With the sign convention that we have adopted, if heat is added to a system, δQ is positive and thus dS_e is positive. If heat is removed from a system, δQ is negative and thus dS_e will be negative. Obviously, $dS_e = 0$ for an adiabatic process.

The term dS_i represents that portion of entropy change caused by irreversible effects. These effects may be internal in nature such as temperature and pressure gradients within the system or external in nature such as friction along the boundaries

of the system. Furthermore, from our thermodynamic studies we
know that all irreversibilities generate entropy; i.e., cause
the entropy of the system to increase. Thus, dS_i is always
positive. Obviously, $dS_i = 0$ for a reversible process.

Recall that an isentropic process is one of constant entropy.
This is also represented by $dS = 0$. The equation

$$dS = dS_e + dS_i \qquad\qquad (3.1)$$

confirms the well known fact that a reversible-adiabatic process
is also isentropic. It also clearly shows that the converse is
not necessarily true; an isentropic process does not have to be
reversible and adiabatic. If isentropic, we merely know that

$$dS = 0 = dS_e + dS_i \qquad\qquad (3.3)$$

If an isentropic process is known to contain irreversibilities,
what can be said about the direction of heat transfer?

Another familiar relation can be developed by taking the
cyclic integral of equation 3.1

$$\oint dS = \oint dS_e + \oint dS_i \qquad\qquad (3.4)$$

Since a cyclic integral must be taken around a closed path and
entropy (S) is a property, then

$$\oint dS = 0 \qquad\qquad (3.5)$$

We know that irreversible effects always generate entropy so

$$\oint dS_i \gneq 0 \qquad\qquad (3.6)$$

with the equal sign holding only for a reversible cycle.

Thus $$0 = \oint dS_e + (\gneq 0) \qquad\qquad (3.7)$$

and since $$dS_e = \delta Q/T \qquad\qquad (3.2)$$

then $$\oint \frac{\delta Q}{T} \leq 0 \qquad\qquad (3.8)$$

which is the "Inequality of Clausius."

The above expressions can be written for a unit mass in
which case we have:

$$ds = ds_e + ds_i \qquad (3.9)$$

$$ds_e = \delta q/T \qquad (3.10)$$

3.4 PRESSURE-ENERGY EQUATION

We are now ready to develop a very useful equation. Starting with the thermodynamic property relation

$$Tds = dh - vdp \qquad (1.41)$$

We introduce $ds = ds_e + ds_i$ and $v = 1/\rho$ to obtain

$$Tds_e + Tds_i = dh - \frac{dp}{\rho}$$

or

$$dh = Tds_e + Tds_i + \frac{dp}{\rho} \qquad (3.11)$$

Recalling the energy equation from Section 2.6

$$\delta q = \delta w_s + dh + \frac{dV^2}{2g_c} + \frac{g}{g_c} dz \qquad (2.53)$$

We now substitute for dh from (3.11) and obtain

$$\cancel{\delta q} = \delta w_s + \left[T\cancel{ds}_e + Tds_i + \frac{dp}{\rho} \right] + \frac{dV^2}{2g_c} + \frac{g}{g_c} dz \qquad (3.12)$$

Recognize (from Eq. 3.10) that $\delta q = Tds_e$ and we obtain a form of the energy equation which is often called the "pressure-energy equation."

$$\frac{dp}{\rho} + \frac{dV^2}{2g_c} + \frac{g}{g_c} dz + \delta w_s + Tds_i = 0 \qquad (3.13)$$

Notice that even though the heat term (δq) does not appear in this equation it is still applicable to cases which involve heat transfer.

Equation 3.13 can be readily simplified for special cases. For instance if no shaft work crosses the boundary $(\delta w_s = 0)$ and if there are no losses $(ds_i = 0)$, then

$$\frac{dp}{\rho} + \frac{dV^2}{2g_c} + \frac{g}{g_c} dz = 0 \qquad (3.14)$$

This is called "Euler's equation" and it can be integrated only
if we know the functional relationship that exists between the
pressure and density.

Example 3.1   Integrate Euler's equation for the case of isothermal
flow of a perfect gas.

$$\int_1^2 \frac{dp}{\rho} + \int_1^2 \frac{dV^2}{2g_c} + \int_1^2 \frac{g}{g_c}\,dz = 0$$

For isothermal flow: $pv = \text{const}$ or $p/\rho = c$

Thus $$\int_1^2 \frac{dp}{\rho} = c\int_1^2 \frac{dp}{p} = c\ln\frac{p_2}{p_1} = \frac{p}{\rho}\ln\frac{p_2}{p_1} = RT\ln\frac{p_2}{p_1}$$

and $$RT\ln\frac{p_2}{p_1} + \frac{V_2^2 - V_1^2}{2g_c} + \frac{g}{g_c}(z_2 - z_1) = 0$$

The special case of incompressible fluids will be considered
in Section 3.7.

3.5   THE STAGNATION CONCEPT

When we speak of the thermodynamic state of a flowing fluid
and mention its properties (e.g., temperature, pressure) there
may be some question as to what these properties actually represent or how they can be measured.   Imagine that you have been
miniaturized and put aboard a small submarine which is drifting
along with the fluid.   (An alternative might be to "saddle-up" a
small fluid particle and take a ride.)   If you had a thermometer
and pressure gage with you, they would indicate the temperature
and pressure corresponding to the static state of the fluid although the word "static" is usually omitted.   Thus, the static
properties are those which would be measured if you moved with
the fluid.

It is convenient to introduce the concept of a "stagnation"
state.   This is a reference state defined as that thermodynamic
state which would exist if the fluid were brought to zero velocity and zero potential.   To yield a consistent reference state we

must qualify how this "stagnation process" should be accomplished. The stagnation state must be reached:

(1) without any energy exchange   (Q = W = 0)

(2) without losses.

By virtue of (1), $ds_e = 0$; and from (2), $ds_i = 0$

Thus, the stagnation process is isentropic!

We can imagine the following example of actually carrying out the stagnation process. Consider fluid which is flowing and has the static properties shown as (a) in Figure 3.1. At location (b) the fluid has been brought to zero velocity and zero potential under the above restrictions.

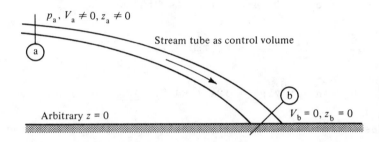

$p_a, V_a \neq 0, z_a \neq 0$

Stream tube as control volume

Arbitrary $z = 0$

$V_b = 0, z_b = 0$

Figure 3.1 Stagnation process

If we apply the energy equation to the control volume indicated for steady, one-dimensional flow, we have

$$h_a + \frac{V_a^2}{2g_c} + \frac{g}{g_c}z_a + \cancel{q} = h_b + \frac{\cancel{V_b^2}}{2g_c} + \frac{g}{g_c}\cancel{z_b} + \cancel{w}_s \qquad (2.49)$$

which simplifies to

$$h_a + \frac{V_a^2}{2g_c} + \frac{g}{g_c}z_a = h_b \qquad (3.15)$$

But condition (b) represents the "stagnation state" corresponding to the "static state" (a). Thus, we call h_b the stagnation or total enthalpy corresponding to state a and designate it as h_{ta}.

Thus,
$$h_{ta} = h_a + \frac{V_a^2}{2g_c} + \frac{g}{g_c}z_a \qquad (3.16)$$

Or, for any state we have in general:

$$h_t = h + \frac{V^2}{2g_c} + \frac{g}{g_c} z \qquad\qquad (3.17)$$

This is an important relation that is always valid. Learn it!
When dealing with gases, potential changes are usually neglected
and we write:

$$h_t = h + \frac{V^2}{2g_c} \qquad\qquad (3.18)$$

Example 3.2 Nitrogen at $500^\circ R$ is flowing at 1800 ft/sec. What
are the static and stagnation enthalpies?

$$h = c_p T = 0.248(500) = 124 \text{ Btu/lbm}$$

$$\frac{V^2}{2g_c} = \frac{(1800)^2}{2(32.2)778} = 64.7 \text{ Btu/lbm}$$

$$h_t = h + V^2/2g_c = 124 + 64.7 = \underline{188.7} \text{ Btu/lbm}$$

The introduction of the stagnation (or total) enthalpy makes
it possible to write equations in a simplified form. For example,
the one-dimensional steady-flow energy equation

$$h_1 + \frac{V_1^2}{2g_c} + \frac{g}{g_c} z_1 + q = h_2 + \frac{V_2^2}{2g_c} + \frac{g}{g_c} z_2 + w_s \qquad (2.49)$$

becomes

$$h_{t1} + q = h_{t2} + w_s \qquad\qquad (3.19)$$

and

$$\delta q = \delta w_s + dh + \frac{dV^2}{2g_c} + \frac{g}{g_c} dz \qquad\qquad (2.53)$$

becomes

$$\delta q = \delta w_s + dh_t \qquad\qquad (3.20)$$

Equation 3.19 (or 3.20) shows that in any adiabatic, no-work,
steady, one-dimensional flow system the stagnation enthalpy remains constant, irrespective of the losses. What else can be
said if the fluid is a perfect gas?

You should note that the stagnation state is a <u>reference</u> state which may or may not actually exist in the flow system. Also, in general, each point in a flow system has a different stagnation state as shown in Figure 3.2. Remember that, although the hypothetical process from 1 to 1_t must be reversible and adiabatic (as well as the process from 2 to 2_t), this in <u>no</u> way restricts the actual process that exists in the flow system between 1 and 2.

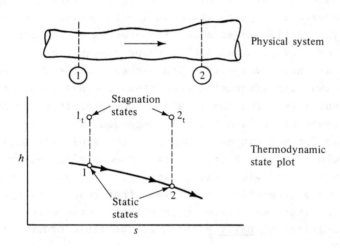

Figure 3.2 h–s diagram showing static and stagnation states

Also one must realize that when the frame of reference is changed, then stagnation conditions change, although the static conditions remain the same. (Recall that static properties are defined as those that would be measured if the measuring devices move with the fluid.)

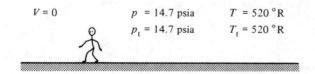

$$V = 0 \qquad \begin{array}{ll} p = 14.7\ \text{psia} & T = 520\,°\text{R} \\ p_t = 14.7\ \text{psia} & T_t = 520\,°\text{R} \end{array}$$

Figure 3.3 Earth as frame of reference

Consider still air with the earth as reference frame (See Figure 3.3). In this case, since the velocity is zero (with respect to the frame of reference), the static and stagnation conditions are the same.

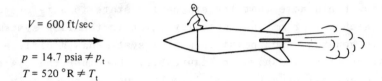

$V = 600$ ft/sec

$p = 14.7$ psia $\neq p_t$

$T = 520\,°R \neq T_t$

Figure 3.4 Missile as frame of reference

Now let's change the frame of reference by flying through this same air on a missile at 600 ft/sec (See Figure 3.4). As we look forward it appears that the air is coming at us at 600 ft/sec. The static pressure and temperature of the air remain <u>unchanged</u> at 14.7 psia and $520°R$, respectively. However, in this case, the air has a velocity (with respect to the frame of reference) and thus the stagnation conditions are different from the static conditions. You should always remember that the stagnation reference state is completely dependent on the frame of reference used for velocities. (Changing the arbitrary $z = 0$ reference would also affect the stagnation conditions but we shall not become involved with this situation.) You will soon learn how to compute stagnation properties other than enthalpy. Incidentally, is there any place in this last system where the stagnation conditions <u>actually</u> exist? Is the fluid brought to rest any place?

3.6 STAGNATION PRESSURE-ENERGY EQUATION

Consider the two section locations on the physical system shown in Figure 3.2. If we let the distance between these locations approach zero, then we are dealing with an infinitesimal control volume with the thermodynamic states differentially separated as shown in Figure 3.5. Also shown are the corresponding stagnation states for these two locations.

Figure 3.5 Infinitesimally—separated static states
with associated stagnation states

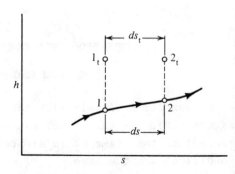

We may write the following property relation between points 1 and 2.

$$Tds = dh - vdp \qquad (1.41)$$

Note that even though the stagnation states do not actually exist, they represent legitimate thermodynamic states and thus any valid property relation or equation may be applied to these points. Thus, we may also apply equation 1.41 between states 1_t and 2_t .

$$T_t ds_t = dh_t - v_t dp_t \qquad (3.21)$$

However, $$ds_t = ds \qquad (3.22)$$

and $$ds = ds_e + ds_i \qquad (3.9)$$

Thus we may write:

$$T_t(ds_e + ds_i) = dh_t - v_t dp_t \qquad (3.23)$$

Recall the energy equation written in the form:

$$\delta q = \delta w_s + dh_t \qquad (3.20)$$

By substituting dh_t from equation 3.23 into equation 3.20 we obtain:

$$\delta q = \delta w_s + T_t(ds_e + ds_i) + v_t dp_t \qquad (3.24)$$

Now also recall that $$\delta q = Tds_e \qquad (3.10)$$

<u>Substitute</u> equation 3.10 into 3.24 and note that $v_t = 1/\rho_t$ (from 1.5) and you should obtain the following equation which is called the "stagnation pressure-energy equation:"

$$\boxed{\frac{dp_t}{\rho_t} + ds_e(T_t - T) + T_t ds_i + \delta w_s = 0} \qquad (3.25)$$

Consider what happens under the following assumptions:

(a) There is no shaft work $\rightarrow$ $\delta w_s = 0$
(b) There is no heat transfer $\rightarrow$ $ds_e = 0$
(c) There are no losses $\rightarrow$ $ds_i = 0$

Under these conditions (3.25) becomes

$$\frac{dp_t}{\rho_t} = 0 \qquad (3.26)$$

and since ρ_t cannot be infinite,

$$dp_t = 0$$

or $$p_t = \text{constant} \qquad (3.27)$$

Note that in general the total pressure will <u>not</u> remain constant; only under a special set of circumstances will equation 3.27 hold true. What are these circumstances?

Many flow systems are adiabatic and contain no shaft work. For these systems:

$$\frac{dp_t}{\rho_t} + T_t ds_i = 0 \qquad (3.28)$$

and the losses are clearly reflected by a change in stagnation pressure. This point will be discussed many times as we examine various flow systems in the remainder of the book.

3.7 CONSEQUENCES OF CONSTANT DENSITY

The density of a liquid is nearly constant and we shall soon see (in Unit 4) that under certain circumstances gases change their density very little. Thus, it will be interesting to see the form some of our equations take for the limiting case of constant density.

ENERGY RELATIONS

We start with the pressure-energy equation

$$\frac{dp}{\rho} + \frac{dV^2}{2g_c} + \frac{g}{g_c}dz + \delta w_s + Tds_i = 0 \qquad (3.13)$$

If $\rho = \text{const}$ we can easily integrate (3.13) between points 1 and 2 of a flow system.

$$\frac{p_2-p_1}{\rho} + \frac{V_2^2-V_1^2}{2g_c} + \frac{g}{g_c}(z_2-z_1) + w_s + \int_1^2 Tds_i = 0$$

or $$\frac{p_1}{\rho} + \frac{V_1^2}{2g_c} + \frac{g}{g_c}z_1 = \frac{p_2}{\rho} + \frac{V_2^2}{2g_c} + \frac{g}{g_c}z_2 + \int_1^2 Tds_i + w_s \qquad (3.29)$$

Compare (3.29) to another form of the energy equation (2.48) and <u>show</u> that

$$\int_1^2 Tds_i = u_2 - u_1 - q \qquad (3.30)$$

Does this result seem reasonable? To determine this, let us examine two extreme cases of the flow of a constant-density fluid. For the first case assume that the system is perfectly insulated. Since the integral of Tds_i is a positive quantity equation 3.30 shows that the losses (i.e., irreversible effects) will cause an increase in internal energy which means a temperature increase. Now consider an isothermal system. For this case how will the losses manifest themselves?

For the flow of a constant-density fluid, "losses" must appear in some combination of the above two forms. In either case mechanical energy has been degraded into a less useful form — thermal energy. Thus, when dealing with constant-density fluids we normally use a single loss term, and generally refer to it as a "head loss" or "friction loss," using the symbol h_ℓ or h_f in place of $\int Tds_i$. If you have studied fluid mechanics you have undoubtably used equation 3.29 in the form:

$$\frac{p_1}{\rho} + \frac{V_1^2}{2g_c} + \frac{g}{g_c}z_1 = \frac{p_2}{\rho} + \frac{V_2^2}{2g_c} + \frac{g}{g_c}z_2 + h_\ell + w_s \qquad (3.31)$$

How many restrictions and/or assumptions are embodied in equation 3.31?

Example 3.3 A turbine extracts 300 ft-lbf/lbm of water flowing. Frictional losses amount to $8V_p^2/2g_c$ where V_p is the velocity in a 2-ft-diameter pipe. Compute the power output of the turbine if it is 100% efficient and the available potential is 350 ft.

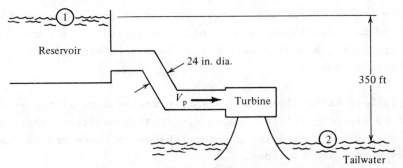

$$p_1 = p_{atmos} \qquad\qquad p_2 = p_{atmos} \qquad\qquad w_s = 300 \text{ ft-lbf/lbm}$$

$$V_1 \approx 0 \qquad\qquad V_2 \approx 0 \qquad\qquad h_\ell = 8V_p^2/2g_c$$

$$z_1 = 350 \text{ ft} \qquad\qquad z_2 = 0$$

Energy:
$$\frac{\not{p}}{\not\rho} + \frac{\not{V^2}}{\not{2g_c}} + \frac{g}{g_c}z_1 = \frac{\not{p}}{\not\rho} + \frac{\not{V^2}}{\not{2g_c}} + \frac{g}{\not{g_c}}z_2 + h_\ell + w_s$$

$$\frac{32.2}{32.2}(350) = \frac{8V_p^2}{2g_c} + 300$$

$$V_p^2 = \frac{2g_c(350-300)}{8} = 402.5$$

$$V_p = 20.1 \text{ ft/sec}$$

Flow rate: $\dot{m} = \rho AV = 62.4(\pi)20.1 = 3940 \text{ lbm/sec}$

Power: $HP = \dfrac{\dot{m}w_s}{550} = \dfrac{3940(300)}{550} = \underline{2150} \text{ HP}$

We can further restrict the flow to one in which no shaft work and no losses occur. In this case equation 3.31 simplifies

to
$$\frac{p_1}{\rho} + \frac{V_1^2}{2g_c} + \frac{g}{g_c}z_1 = \frac{p_2}{\rho} + \frac{V_2^2}{2g_c} + \frac{g}{g_c}z_2$$

or
$$\boxed{\frac{p}{\rho} + \frac{V^2}{2g_c} + \frac{g}{g_c}z = \text{const}} \qquad (3.32)$$

This is called "Bernoulli's equation" and it could also have been obtained by integrating Euler's equation (3.14) for a constant-density fluid. How many assumptions have been made to arrive at Bernoulli's equation?

Example 3.4 Water flows in a 6-inch-diameter duct with a velocity of 15 ft/sec. Within a short distance the duct converges to a 3-inch diameter. Find the pressure change if there are no losses between these two sections.

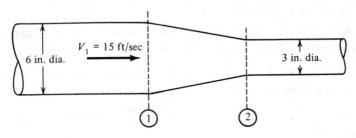

Bernoulli: $\quad \dfrac{p_1}{\rho} + \dfrac{V_1^2}{2g_c} + \dfrac{g}{g_c}z_1 = \dfrac{p_2}{\rho} + \dfrac{V_2^2}{2g_c} + \dfrac{g}{g_c}z_2$

$$p_1 - p_2 = \dfrac{\rho}{2g_c}(V_2^2 - V_1^2)$$

Continuity: $\quad \rho_1 A_1 V_1 = \rho_2 A_2 V_2$

$$V_2 = V_1 \dfrac{A_1}{A_2} = V_1\left(\dfrac{D_1}{D_2}\right)^2 = 15\left(\dfrac{6}{3}\right)^2 = 60 \text{ ft/sec}$$

Thus: $\quad p_1 - p_2 = \dfrac{62.4}{2(32.2)}\left[60^2 - 15^2\right] = 3270 \text{ lbf/ft}^2 = \underline{22.7} \text{ lbf/in}^2$

STAGNATION RELATIONS

We start by considering the property relation

$$Tds = du + p\,dv \tag{1.40}$$

If ρ = const, dv = 0, then

$$Tds = du \tag{3.33}$$

Note that for a process in which $ds = 0$, $du = 0$.

We also have, by definition,

$$c_v = \left(\dfrac{\partial u}{\partial T}\right)_v \tag{1.37}$$

But for a constant density fluid every process is one in which v = const. Thus, for these fluids we can drop the partial notation and write equation 1.37 as

$$c_v = \dfrac{du}{dT} \quad \text{or} \quad du = c_v dT \tag{3.34}$$

Note that for a process in which du = 0 , dT = 0 .

We now consider the stagnation process which by virtue of its definition is isentropic, or ds = 0.

From (3.33) we see that the internal energy does not change during the stagnation process.

$$u = u_t \qquad \text{for } \rho = \text{const} \qquad (3.35)$$

From (3.34) it must then be that the temperature also does not change during the stagnation process.

$$T = T_t \qquad \text{for } \rho = \text{const} \qquad (3.36)$$

Summarizing the above, we have shown that <u>for</u> a <u>constant-density fluid</u> the stagnation process is not only one of constant entropy but also one of constant temperature and internal energy. Let us continue and discover some other interesting relations.

From $\qquad\qquad\qquad h = u + pv \qquad\qquad (1.34)$

we have $\qquad\qquad dh = du + vdp + p\cancel{dv} \qquad (3.37)$

Let us integrate equation 3.37 between the static and stagnation states.

$$h_t - h = (u_t \cancel{-} u) + v(p_t - p) \qquad (3.38)$$

But we know that $\qquad h_t = h + \dfrac{v^2}{2g_c} + \dfrac{g}{g_c} z \qquad (3.17)$

Combining these last two equations yields:

$$\left[\cancel{h} + \frac{v^2}{2g_c} + \frac{g}{g_c} z\right] - \cancel{h} = v(p_t - p)$$

which becomes $\qquad \boxed{p_t = p + \dfrac{\rho v^2}{2g_c} + \rho\dfrac{g}{g_c} z} \qquad (3.39)$

This equation may also be familiar to those of you who have studied fluid mechanics. It is imperative to note that this relation between static and stagnation pressures <u>is</u> <u>only</u> <u>valid</u> <u>for</u> a <u>constant-density</u> <u>fluid</u>. In Section 4.5 we shall develop the corresponding relation for perfect gases.

Example 3.5 Water is flowing at a velocity of 20 m/s and has a pressure of 4 bars absolute. What is the total pressure?

$$p_t = p + \frac{\rho V^2}{2g_c} + \rho \frac{\cancel{g}}{\cancel{g_c}} z$$

$$p_t = 4 \times 10^5 + \frac{10^3 (20)^2}{2(1)} = 4 \times 10^5 + 2 \times 10^5$$

$$p_t = \underline{6 \times 10^5} \ N/m^2 \ \text{absolute}$$

3.8 MOMENTUM EQUATION

If we observe the motion of a given quantity of mass, Newton's Second Law tells us that its linear momentum will be changed in direct proportion to the applied forces. This is expressed with the following equation:

$$\Sigma \vec{F} = \frac{1}{g_c} \frac{d(\overrightarrow{momentum})}{dt} \tag{1.2}$$

We could write a similar expression relating torque and angular momentum but we shall confine our discussion to linear momentum.

Note that equation 1.2 is a vector relation and must be treated as such or we must carefully work with components of the equation. In nearly all fluid flow problems unbalanced forces exist and thus the momentum of the system being analyzed does not remain constant. Thus, we shall carefully avoid listing this as a "conservation law."

Again the question is, "What corresponding expression can we write for a control volume?" We note that the term on the right side of equation 1.2 is a material derivative and it must be transformed according to the relation developed in Section 2.4.

If we let N be the linear momentum of the system, then η represents the momentum per unit mass which is $\vec{V}$. Substitution into equation 2.22 yields:

$$\frac{d(\overrightarrow{Mom})}{dt} = \frac{\partial}{\partial t} \int_{cv} \vec{V} \rho dv + \int_{cs} \vec{V} \rho (\vec{V} \cdot \hat{n}) dA \tag{3.40}$$

and the transformed equation which is applicable to a control volume is:

$$\Sigma \vec{F} = \frac{1}{g_c} \frac{\partial}{\partial t} \int_{cv} \vec{V} \rho dv + \frac{1}{g_c} \int_{cs} \vec{V} \rho (\vec{V} \cdot \hat{n}) dA \tag{3.41}$$

This equation is usually called the momentum or momentum flux equation. The $\Sigma\vec{F}$ represents the summation of all forces on the control volume. What do the other terms represent? (See discussion after equation 2.22.)

In the solution of actual problems one normally works with the components of the momentum equation. In fact frequently only one component is required for the solution of a problem. The x-component of this equation would appear as:

$$\Sigma F_x = \frac{1}{g_c} \frac{\partial}{\partial t}\int_{cv} V_x\rho dv + \frac{1}{g_c}\int_{cs} V_x\rho(\vec{V}\cdot\hat{n})dA \qquad (3.42)$$

Note carefully how the last term is written.

In the event that one-dimensional flow exists the last integral in equation 3.41 is easy to evaluate as ρ and V are constant over any given cross section. If we choose the surface A perpendicular to the velocity then:

$$\int_{cs} \vec{V}\rho(\vec{V}\cdot\hat{n})dA = \Sigma\vec{V}\rho V\int dA = \Sigma\vec{V}\rho VA = \Sigma\dot{m}\vec{V} \qquad (3.43)$$

The summation is taken over all sections where fluid crosses the control surface and is positive where fluid leaves the control volume and negative where fluid enters the control volume.

If we now consider steady flow, the term involving the partial derivative with respect to time is zero. Thus, for steady one-dimensional flow the momentum equation for a control volume becomes:

$$\Sigma\vec{F} = \frac{1}{g_c} \Sigma\dot{m}\vec{V} \qquad (3.44)$$

If there is only one section where fluid enters and one section where fluid leaves the control volume we know (from continuity) that

$$\dot{m}_{in} = \dot{m}_{out} = \dot{m} \qquad (2.42)$$

and the momentum equation becomes:

$$\boxed{\Sigma\vec{F} = \frac{\dot{m}}{g_c} (\vec{V}_{out} - \vec{V}_{in})} \qquad (3.45)$$

This is the form of the equation for a finite control volume.

What assumptions have been fed into this equation? In using this relation one must be sure to

    a.   include all forces acting on the control volume, and
    b.   be extremely careful with the signs of all quantities.

Example 3.6   There is a steady one-dimensional flow of air through a 12-inch-diameter horizontal duct. At a section where the velocity is 460 ft/sec the pressure is 50 psia and the temperature is 550°R. At a downstream section the velocity is 880 ft/sec and the pressure is 23.9 psia. Determine the total wall shearing force between these sections.

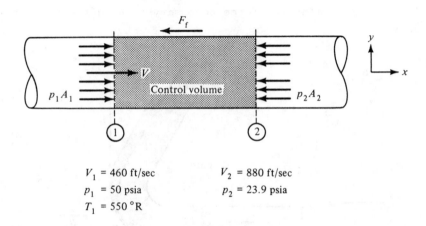

$$V_1 = 460 \text{ ft/sec} \qquad V_2 = 880 \text{ ft/sec}$$
$$p_1 = 50 \text{ psia} \qquad\quad p_2 = 23.9 \text{ psia}$$
$$T_1 = 550\,°\text{R}$$

We establish a coordinate system and indicate the forces on the control volume. Let F_f represent the frictional force of the duct on the gas. We write the x-component of equation 3.45.

$$F_x = \frac{\dot{m}}{g_c}\,(V_{out_x} - V_{in_x})$$

$$p_1 A_1 - p_2 A_2 - F_f = \frac{\dot{m}}{g_c}\,(V_2 - V_1) = \frac{\rho_1 A_1 V_1}{g_c}\,(V_2 - V_1)$$

Note that any force in the negative direction must include a minus sign. We divide by $A = A_1 = A_2$

$$p_1 - p_2 - \frac{F_f}{A} = \frac{\rho_1 V_1}{g_c}\,(V_2 - V_1)$$

$$\rho_1 = \frac{p_1}{RT_1} = \frac{50(144)}{53.3(550)} = 0.246 \text{ lbm/ft}^3$$

$$(50 - 23.9)144 - \frac{F_f}{A} = \frac{0.246(460)}{32.2}(880-460)$$

$$3758 - F_f/A = 1476$$

$$F_f = (3758 - 1476)\pi(0.5)^2 = \underline{1792} \text{ lbf}$$

Example 3.7 Water flowing at the rate of 0.05 m^3/s has a velocity of 40 m/s. The jet strikes a vane and is deflected 120°. Friction along the vane is negligible and the entire system is exposed to the atmosphere. Potential changes can also be neglected. Determine the force necessary to hold the vane stationary.

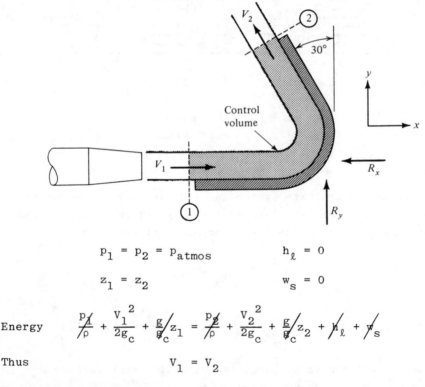

$$p_1 = p_2 = p_{atmos} \qquad\qquad h_\ell = 0$$

$$z_1 = z_2 \qquad\qquad w_s = 0$$

Energy $$\frac{\cancel{p_1}}{\cancel{\rho}} + \frac{V_1^2}{2g_c} + \frac{g}{\cancel{g_c}}\cancel{z_1} = \frac{\cancel{p_2}}{\cancel{\rho}} + \frac{V_2^2}{2g_c} + \frac{g}{\cancel{g_c}}\cancel{z_2} + \cancel{h_\ell} + \cancel{w_s}$$

Thus $$V_1 = V_2$$

We indicate the force components of the vane on the fluid as R_x and R_y and put them on the diagram in assumed directions. (If we have guessed wrong our answer will turn out to be negative.)

x-component $\quad \Sigma F_x = \dfrac{\dot{m}}{g_c} (V_{2x} - V_{1x})$

$$-R_x = \frac{\dot{m}}{g_c} \left[(-V_2 \sin 30) - V_1 \right] = \frac{\dot{m} V_1}{g_c} \left[-\sin 30 - 1 \right]$$

$$-R_x = \frac{10^3 (0.05)40}{(1)} (-0.5-1)$$

$$R_x = \underline{3000} \text{ newtons}$$

y-component $\quad \Sigma F_y = \dfrac{\dot{m}}{g_c} (V_{2y} - V_{1y})$

$$R_y = \frac{\dot{m}}{g_c} \left[(V_2 \cos 30) - 0 \right]$$

$$R_y = \frac{10^3 (0.05)40}{(1)} (0.866)$$

$$R_y = \underline{1732} \text{ newtons}$$

Note that the assumed directions for R_x and R_y were correct.

DIFFERENTIAL FORM OF MOMENTUM EQUATION

As a further example of the meticulous care that must be exercised when utilizing the momentum equation, we shall apply it to the differential control volume shown in Figure 3.6. Under conditions of steady one-dimensional flow, the properties of the fluid entering the control volume are designated as ρ, V, p, etc. Fluid leaves the control volume with slightly different properties as indicated by $\rho + d\rho$, $V + dV$, etc. The x-coordinate is chosen as positive in the direction of flow and the positive z-direction is opposite gravity. (Note that the x and z axes are not necessarily orthogonal.)

Now that the control volume has been identified we note all forces that act on it. The forces can be divided into two types:

a. Surface Forces - these act on the control surface and are either from normal or tangential stress components.
b. Body Forces - these act directly on the fluid within the control volume. Examples of these are gravity and electromagnetic forces. We shall limit our discussion to gravity forces.

Figure 3.6 Momentum analysis on
infinitesimal control volume

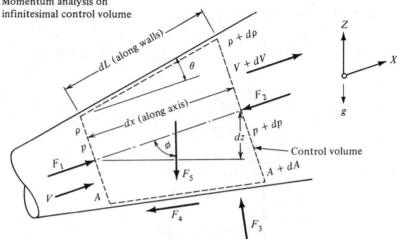

Thus we have: $F_1 \equiv$ Upstream pressure force

$F_2 \equiv$ Downstream pressure force

$F_3 \equiv$ Wall pressure force

$F_4 \equiv$ Wall friction force

$F_5 \equiv$ Gravity force

It should be mentioned that wall forces F_3 and F_4 are usually
lumped together into a single force called the "enclosure force"
for the reason that it is extremely difficult to account for them
separately in most finite control volumes. Fortunately, it is
the total enclosure force that is of significance in the solution
of these problems. However, in dealing with a differential control volume, it will be more instructive to separate each portion
of the enclosure force as we have indicated.

We write the x-component of the momentum equation for steady,
one-dimensional flow:

$$\Sigma F_x = \frac{\dot{m}}{g_c} (V_{out_x} - V_{in_x}) \qquad (3.46)$$

Now we proceed to evaluate the x-component of each force, taking
care to indicate whether it is in the positive or negative direction.

$$F_{1x} = F_1 = (\text{pressure})(\text{area})$$

$$F_{1x} = pA \qquad (3.47)$$

$$F_{2x} = -F_2 = -(\text{pressure})(\text{area})$$

H.O.T.

$$F_{2x} = -(p+dp)(A+dA) = -(pA + pdA + Adp + dpdA) \qquad (3.48)$$

Neglecting the higher order term, this becomes:

$$F_{2x} = -(pA + pdA + Adp) \qquad (3.49)$$

The wall pressure force can be obtained with a mean pressure value.

$$F_{3x} = F_3 \sin\theta = \left[(\text{mean pressure})(\text{wall area})\right]\sin\theta$$

but $\quad$ dA = (wall area) $\sin\theta$; and thus

$$F_{3x} = (p + \tfrac{dp}{2}) \, dA \qquad (3.50)$$

This same result could be obtained using principles of basic fluid mechanics which show that a component of the pressure force can be computed by considering the pressure distribution over the projected area. Expanding and neglecting the higher order term we have:

$$F_{3x} = pdA \qquad (3.51)$$

To compute the wall friction force we define

$\qquad \tau_w \equiv$ the mean shear stress along the wall

$\qquad P \equiv$ the mean wetted perimeter

$$F_{4x} = -F_4 \cos\theta = -\left[(\text{mean shear stress})(\text{wall area})\right]\cos\theta$$

$$F_{4x} = -\tau_w(PdL) \cos\theta \qquad (3.52)$$

but $\quad$ dx = dL $\cos\theta$, and thus

$$F_{4x} = -\tau_w Pdx \qquad (3.53)$$

For the body force we have

$$F_{5x} = -F_5 \cos\phi = -\left[(\text{volume})(\text{mean density})\tfrac{g}{g_c}\right]\cos\phi$$

$$F_{5x} = -\left[\left(A + \tfrac{dA}{2}\right) dx\right]\left[\rho + \tfrac{d\rho}{2}\right] \tfrac{g}{g_c} \cos\phi \qquad (3.54)$$

But $\quad$ dx $\cos\phi$ = dz, and thus

$$F_{5x} = -\left[A + \tfrac{dA}{2}\right]\left[\rho + \tfrac{d\rho}{2}\right] \tfrac{g}{g_c} dz \qquad (3.55)$$

Expand this and eliminate all the higher order terms to show that:

$$F_{5x} = -A\rho\tfrac{g}{g_c} dz \qquad (3.56)$$

Summarizing the above we have

$$\Sigma F_x = F_{1x} + F_{2x} + F_{3x} + F_{4x} + F_{5x}$$

$$= \cancel{pA} - (\cancel{pA} + \cancel{pdA} + Adp) + \cancel{pdA} - \tau_w Pdx - A\rho\frac{g}{g_c} dz$$

$$= -Adp - \tau_w Pdx - A\rho\frac{g}{g_c} dz \tag{3.57}$$

We now turn our attention to the right side of equation 3.46. Looking at Figure 3.6 we see that this is

$$\frac{\dot{m}}{g_c}(V_{out_x} - V_{in_x}) = \frac{\dot{m}}{g_c}\left[(V + dV) - V\right] = \frac{\dot{m}}{g_c} dV \tag{3.58}$$

Combining equations 3.57 and 3.58 yields the x-component of the momentum equation applied to a differential control volume:

$$\Sigma F_x = \frac{\dot{m}}{g_c}(V_{out_x} - V_{in_x}) \tag{3.46}$$

$$-Adp - \tau_w Pdx - A\rho\frac{g}{g_c} dz = \frac{\dot{m}}{g_c} dV = \frac{\rho A V dV}{g_c} \tag{3.59}$$

Equation 3.59 can be put into a more useful form by introducing the concepts of the "friction factor" and "equivalent diameter."

The friction factor (f) relates the average shear stress at the wall (τ_w) to the dynamic pressure in the following manner:

$$f \equiv \frac{4\tau_w}{\rho V^2/2g_c} \tag{3.60}$$

This is the "Darcy-Weisbach" friction factor and is the one we shall use in this book. Care should be taken when reading literature in this area since some authors use the "Fanning" friction factor which is only one-quarter as large due to omission of the factor of 4 in the definition.

Frequently, fluid flows through a non-circular cross-section such as a rectangular duct. In order to handle these problems an equivalent diameter has been devised which is defined as

$$D_e \equiv \frac{4A}{P} \tag{3.61}$$

where A ≡ cross-sectional area
       P ≡ perimeter of the enclosure wetted by the fluid.

Note that if equation 3.61 is applied to a circular duct completely filled with fluid, the equivalent diameter is the same as the actual diameter.

Use the definitions given for the friction factor and the equivalent diameter and <u>show</u> that equation 3.59 can be rearranged to

$$\frac{dp}{\rho} + f\frac{V^2}{2g_c}\frac{dx}{D_e} + \frac{g}{g_c}dz + \frac{VdV}{g_c} = 0 \qquad (3.62)$$

This is a very useful form of the momentum equation (written in the direction of flow) for steady one-dimensional flow through a differential control volume. The last term can be written in an alternate form to yield:

$$\frac{dp}{\rho} + f\frac{V^2}{2g_c}\frac{dx}{D_e} + \frac{g}{g_c}dz + \frac{dV^2}{2g_c} = 0 \qquad (3.63)$$

We shall use this equation in Unit 9 when we discuss flow through ducts with friction.

It might be instructive at this time to compare (3.63) with equation 3.13. Recall that (3.13) was derived from energy considerations whereas (3.63) was developed from momentum concepts. A comparison of this nature reinforces our division of entropy concept for it shows that

$$Tds_i = f\frac{V^2}{2g_c}\frac{dx}{D_e} \qquad (3.64)$$

3.9 SUMMARY

We have taken a new look at entropy changes by dividing them into two parts, that caused by heat transfer and that caused by irreversible effects. We then introduced the concept of a stagnation reference state. These two ideas permitted the energy equation to be written in alternate forms called "pressure-energy equations." Several interesting conclusions were drawn from these equations under appropriate assumptions.

Newton's Second Law was transformed into a form suitable for control volume analysis. Extreme care should be taken when the momentum equation is used. The following steps should be noted <u>in</u> addition to those listed in the summary for Unit 2:

1. Establish a coordinate system.
2. Indicate all forces acting on the fluid inside the control volume.
3. Be especially careful with the signs of vector quantities such as $\vec{F}$ and $\vec{V}$.

Some of the most frequently used equations that were developed in this unit are summarized below. Most are restricted to steady, one-dimensional flow; others involve additional assumptions. You should determine under what conditions each may be used.

a. Entropy Divison

$$ds = ds_e + ds_i = \frac{\delta q}{T} + ds_i \qquad\qquad (3.9) \ \& \ (3.10)$$

ds_e is positive or negative (depends on δq)

ds_i is always positive (irreversibilities)

b. Pressure-Energy Equation

$$\frac{dp}{\rho} + \frac{dV^2}{2g_c} + \frac{g}{g_c} dz + \delta w_s + Tds_i = 0 \qquad\qquad (3.13)$$

c. Stagnation Concept - (Depends on reference frame)

$$h_t = h + \frac{V^2}{2g_c} + \frac{g}{g_c} z \qquad (\text{neglect} \ z \ \text{for gas}) \qquad (3.17)$$

$$s_t = s$$

d. Energy Equation

$$h_{t1} + q = h_{t2} + w_s \qquad\qquad (3.19)$$

$$\delta q = \delta w_s + dh_t \qquad\qquad (3.20)$$

If $q = w_s = 0$, $h_t = \text{const}$

e. Stagnation Pressure-Energy Equation

$$\frac{dp_t}{\rho_t} + ds_e(T_t - T) + T_t ds_i + \delta w_s = 0 \qquad\qquad (3.25)$$

If $q = w_s = 0$, and Loss $= 0$, $p_t = \text{const}$.

f. Constant-Density Fluids

$$\frac{p_1}{\rho} + \frac{V_1^2}{2g_c} + \frac{g}{g_c}z_1 = \frac{p_2}{\rho} + \frac{V_2^2}{2g_c} + \frac{g}{g_c}z_2 + h_\ell + w_s \qquad (3.31)$$

$$u = u_t \qquad \text{and} \qquad T = T_t \qquad (3.35) \ \& \ (3.36)$$

$$p_t = p + \frac{\rho V^2}{2g_c} + \rho\frac{g}{g_c}z \qquad (3.39)$$

g. Second Law of Motion - Momentum Equation $\begin{cases} N = \overrightarrow{mom} \\ \eta = \overrightarrow{V} \end{cases}$

$$\Sigma\overrightarrow{F} = \frac{\partial}{\partial t}\int_{cv}\frac{\rho\overrightarrow{V}}{g_c}\,dv + \int_{cs}\frac{\rho\overrightarrow{V}}{g_c}(\overrightarrow{V}\cdot\hat{n})dA \qquad (3.41)$$

For steady, one-dimensional flow

$$\Sigma\overrightarrow{F} = \frac{\dot{m}}{g_c}(\overrightarrow{V}_{out} - \overrightarrow{V}_{in}) \qquad (3.45)$$

$$\frac{dp}{\rho} + f\frac{V^2}{2g_c}\frac{dx}{D_e} + \frac{g}{g_c}dz + \frac{dV^2}{2g_c} = 0 \qquad (3.63)$$

3.10 PROBLEMS

For those problems involving water you may use $\rho = 62.4$ lbm/ft^3 or 1000 kg/m^3, and the specific heat equals 1 Btu/lbm-$^\circ$R or 4187 J/kg-$^\circ$K.

1. Compare the pressure-energy equation (3.13) for the case of no external work with the differential form of the momentum equation (3.63). Does the result seem reasonable?

2. Consider steady flow of a perfect gas in a horizontal, insulated, frictionless duct. Start with the pressure-energy equation and show that

$$\frac{V^2}{2g_c} + \frac{\gamma}{(\gamma-1)}\frac{p}{\rho} = \text{const}$$

3. It is proposed to determine the flow rate through a pipe line from pressure measurements at two points of different cross-sectional areas. No energy transfers are involved ($q = w_s = 0$) and potential differences are negligible. Show

that for the steady, one-dimensional frictionless flow of an
incompressible fluid, the flow rate can be represented by

$$\dot{m} = A_1 A_2 \left[\frac{2\rho g_c (p_1 - p_2)}{A_1^{\ 2} - A_2^{\ 2}} \right]^{\frac{1}{2}}$$

4. Pressure taps in a low-speed wind tunnel reveal the differ-
    ence between stagnation and static pressure to be 0.5 psi.
    Calculate the test section air velocity under the assumption
    that the air density remains constant at 0.0765 lbm/ft^3.

5. Water flows through a duct of varying area. The difference
    in stagnation pressures between two sections is 4.5x10^5 N/m^2.
    (a) If the water remains at a constant temperature, how much
        heat will be transferred in this length of duct?
    (b) If the system is perfectly insulated against heat trans-
        fer, compute the temperature change of the water as it
        flows through the duct.

6. The following information is known about the steady flow of
    methane through a horizontal insulated duct:

Entering stagnation enthalpy	=	634 Btu/lbm
Leaving static enthalpy	=	532 Btu/lbm
Leaving static temperature	=	540°F
Leaving static pressure	=	50 psia

    (a) Determine the outlet velocity.
    (b) What is the stagnation temperature at the outlet?
    (c) Determine the stagnation pressure at the outlet.

7. Under what conditions would it be possible to have an adiabat-
    ic flow process with a real fluid (with friction) and have
    the stagnation pressures at inlet and outlet to the system be
    the same? Hint: Look at the stagnation pressure-energy equa-
    tion.

8. Simplify the stagnation pressure-energy equation (3.25) for
    the case of an incompressible fluid. Integrate the result
    and compare your answer to any other energy equation that you
    might use for an incompressible fluid (say equation 3.29).

9. An incompressible fluid (ρ = 55 lbm/ft^3) leaves the pipe with
    a velocity of 15 ft/sec.
    (a) Calculate the flow losses.

(b) Assume that all losses occur in the constant-area pipe
and find the pressure at the entrance to the pipe.

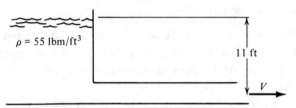

$\rho = 55\ \text{lbm/ft}^3$

11 ft

V

10. What Δz is required to produce a jet velocity (V_j) of 30
m/sec if the flow losses are $h_\ell = 15V_p^2/2g_c$?

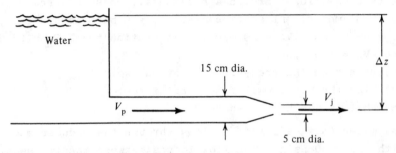

Water

15 cm dia.

Δz

V_p

V_j

5 cm dia.

11. Water flows in a 2-foot-diameter duct under the following
conditions: $p_1 = 55$ psia and $V_1 = 20$ ft/sec . At another
section 12 feet below the first the diameter is 1 foot and
the pressure $p_2 = 40$ psia . Compute the frictional losses
between these two sections and determine the direction of
flow.

12. Find the pipe diameter required to produce a flow rate of
50 kg/sec if the flow losses are $h_\ell = 6V^2/2g_c$.

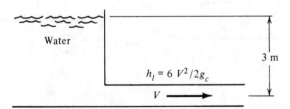

Water

3 m

$h_l = 6\,V^2/2g_c$

V

13. For a given mass we can relate the moment of the applied
force to the angular momentum by the following:

$$\Sigma \vec{M} = \frac{1}{g_c} \frac{d(\overline{\text{angular momentum}})}{dt}$$

(a) What is the angular momentum per unit mass?

(b) What form does the above equation take for the analysis of a control volume?

14. An incompressible fluid flows through a 10-inch-diameter, horizontal, constant-area pipe. At one section the pressure is 150 psia and 1000 feet downstream the pressure has dropped to 100 psia.

(a) Find the total frictional force exerted on the fluid by the pipe.

(b) Compute the average wall shear stress.

15. Methane gas flows through a horizontal, constant-area pipe of 15 cm diameter. At section one: p_1 = 6 bars absolute , T_1 = 66°C and V_1 = 30 m/sec. At section two: T_2 = 38°C and V_2 = 110 m/sec.

(a) Determine the pressure at section two.

(b) Find the total wall frictional force.

(c) What is the heat transfer?

16. Sea water (ρ = 64 lbm/ft^3) flows through the reducer shown with p_1 = 50 psig. The flow losses between the two sections amount to h_ℓ = 5.0 ft-lbf/lbm.

(a) Find V_2 and p_2 .

(b) Determine the force exerted by the reducer on the sea water between sections one and two.

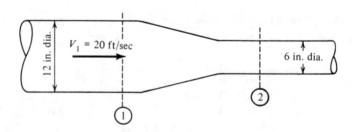

17. (a) Neglect all losses and compute the exit velocity from the tank shown below.

(b) If the opening is 4 inches in diameter, determine the mass flow rate.

(c) Compute the force tending to push the tank along the floor.

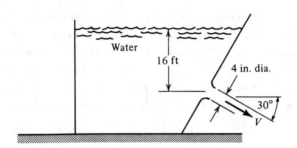

18. A jet of water with a velocity of 5 m/s has an area of 0.05 m^2. It strikes a 1-meter-thick concrete block at a point 2 m above the ground. After hitting the block the water drops straight to the ground. What minimum weight must the block have in order not to tip over?

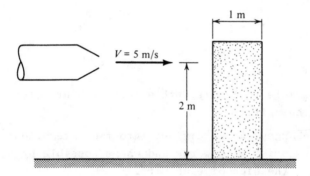

19. It is proposed to brake a racing car by opening an air-scoop to deflect the air as shown. You may assume that the density of the air remains approximately constant at the inlet conditions of 14.7 psia and 60°F. What inlet area is needed to provide a braking force of 2000 lbf when traveling at 300 mph?

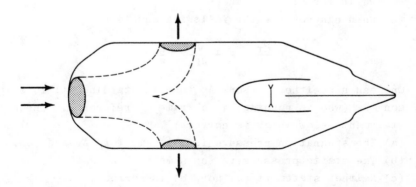

20. A fluid jet strikes a vane and is deflected through angle θ .
     For a given jet (fluid, area, and velocity are fixed) what
     deflection angle will cause the greatest x-component of force
     between the fluid and vane? You may assume an incompressible
     fluid and no friction along the vane. Set up the general
     problem and then differentiate to find a maximum.

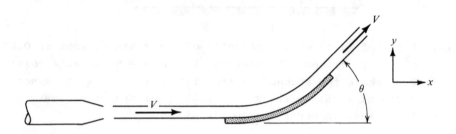

3.11 CHECK TEST

You should be able to complete this test without reference to
material in the unit.

1. Entropy changes can be divided into two categories. Define
    these categories with words and where possible by equations.
    Comment on the sign of each part.

2. Starting with the differential form of the energy equation,
    derive the pressure-energy equation.

3. (a) Define the stagnation process. Be careful to state all
        conditions.
    (b) Give a general equation for stagnation enthalpy that is
        valid for all substances.
    (c) When can you use the following equation?

$$\frac{p_t}{\rho} = \frac{p}{\rho} + \frac{V^2}{2g_c} + \frac{g}{g_c} z$$

4. One can use either person A (who is standing still) or per-
    son B (who is running) as a frame of reference. Check the
    statement below which is correct:
    (a) The stagnation pressure is the same for A and B .
    (b) The static pressure is the same for A and B .
    (c) Neither statement (a) nor (b) is correct.

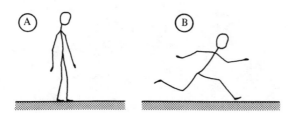

5. Consider the case of steady, one-dimensional flow with one stream in and one stream out of the control volume.

   (a) Under what conditions can we say that the stagnation enthalpy remains constant? (Can p_t vary under these conditions?)

   (b) If the conditions of part (a) are known to exist, what additional assumption is required before we can say that the stagnation pressure remains constant?

6. Under certain circumstances the momentum equation is sometimes written in the following form when used to analyze a control volume:

$$\Sigma \vec{F} = \frac{\dot{m}}{g_c} (\vec{V}_r - \vec{V}_s)$$

   (a) Which of the sections (r or s) represents the location where fluid enters the control volume?

   (b) What circumstances must exist before you can use the equation in this form?

7. Work problem #16 in Section 3.10.

UNIT 4

INTRODUCTION TO COMPRESSIBLE FLOW

4.1 INTRODUCTION

In the previous units we developed the fundamental relations
that are needed for the analysis of fluid flow. We have seen the
special form some of these take for the case of constant density
fluids. Our <u>main</u> interest now is in compressible fluids or gases.
We shall soon learn that it is not uncommon to encounter gases
which are traveling faster than the speed of sound. Furthermore,
their behavior when in this situation is quite different than
when traveling slower than the speed of sound.

Thus, we begin this unit by developing an expression for
sonic velocity through an arbitrary medium. The relation is simplified for the case of perfect gases. We then examine subsonic
and supersonic flows to gain some insight as to why their behavior is different.

Mach number is introduced as a key parameter and we find
that for the case of a perfect gas it is very simple to express

our basic equations and many supplementary relations in terms of
this new parameter. The unit will close with a discussion on the
significance of h-s and T-s diagrams and their importance in
visualizing flow problems.

4.2 OBJECTIVES

After successfully completing this unit you should be able to:

1. Explain how sound is propagated through any medium (solid,
    liquid or gas).

2. Define sonic velocity. State the basic differences between
    a shock wave and a sound wave.

3. Starting with the continuity and momentum equations for
    steady, one-dimensional flow, utilize a control volume analy-
    sis to derive the general expression for the velocity of an
    infinitesimal pressure disturbance in an arbitrary medium.
    (Optional.)

4. State the relations for:
    a. Speed of sound in an arbitrary medium.
    b. Speed of sound in a perfect gas.
    c. Mach number.

5. Discuss the propagation of signal waves from a moving body in
    a fluid and explain what is meant by "zone of action," "zone
    of silence," "Mach cone" and "Mach angle." Compare subsonic
    and supersonic flow in these respects.

6. Give an equation for the stagnation enthalpy (h_t) of a per-
    fect gas in terms of enthalpy (h), Mach number (M) and
    ratio of specific heats (γ).

7. Give an equation for the stagnation temperature (T_t) of a
    perfect gas in terms of temperature (T), Mach number (M)
    and ratio of specific heats (γ).

8. Give an equation for the stagnation pressure (p_t) of a per-
    fect gas in terms of pressure (p), Mach number (M) and
    ratio of specific heats (γ).

9. Demonstrate manipulative skills by developing simple rela-
    tions in terms of Mach number for a perfect gas, such as

$$p_t = p\left[1 + \frac{(\gamma-1)}{2} M^2\right]^{\frac{\gamma}{\gamma-1}}$$

10. Demonstrate the ability to utilize the above concepts in typical flow problems.

4.3 SONIC VELOCITY AND MACH NUMBER

We shall now examine the means by which disturbances pass through any elastic medium. A disturbance at a given point creates a region of compressed molecules which is passed along to its neighboring molecules, and in so doing creates a traveling wave. Waves come in various strengths, which are measured by the amplitude of the disturbance. The speed at which this disturbance is propagated through the medium is called the wave speed. This speed not only depends on the type of medium and its thermodynamic state but also is a function of the strength of the wave. The "stronger" the wave is the faster it moves.

If we are dealing with waves of large amplitude, which involve relatively large changes in pressure and density, we call these "shock waves". These will be studied in detail in Unit 6. If, on the other hand, we observe waves of very small amplitude, their speed is characteristic only of the medium and its state. These waves are of vital importance to us since sound waves fall into this category. Furthermore, the presence of an object in a medium can only be felt by the object's sending out infinitesimal waves which propagate at the characteristic "sonic velocity."

Let us hypothesize how we might form an infinitesimal pressure wave and then apply the fundamental concepts to determine the wave velocity. Consider a long, constant-area tube filled with fluid and having a piston at one end as shown in Figure 4.1. The fluid is initially at rest. At a certain instant the piston is given an incremental velocity dV to the left. The fluid particles immediately next to the piston are compressed a very small amount as they acquire the velocity of the piston.

As the piston (and these compressed particles) continue to move, the next group of fluid particles is compressed and the "wave front" is observed to propagate through the fluid at the characteristic "sonic" velocity of magnitude a. All particles between the wave front and the piston are moving with velocity dV to the left and have been compressed from ρ to $\rho + d\rho$ and have increased their pressure from p to $p + dp$. We next recognize that this is a difficult situation to analyze. Why? Because it is unsteady flow! (As you observe any

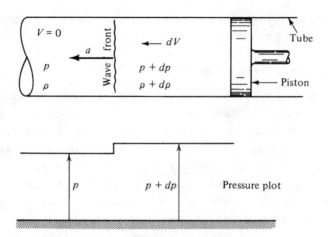

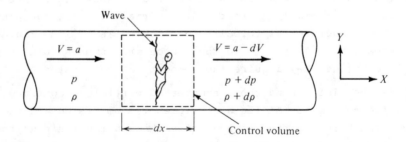

Figure 4.1 Initiation of infinitesimal pressure pulse

given point in the tube the properties change with time; e.g.,
pressure changes from p to p + dp as the wave front passes).
This difficulty can easily be solved by superimposing on the
entire flow field a velocity to the right of magnitude a . This
procedure changes the frame of reference to the wave front as it
now appears as a stationary wave. An alternate way of achieving
this same result is to jump on the wave front. Figure 4.2 shows
the problem that we now have. Note that changing the reference
frame in this manner does not in any way alter the actual (static)
thermodynamic properties of the fluid, although it will affect
the stagnation conditions.

Figure 4.2 Steady-flow picture corresponding to Figure 4.1

Since the wave front is extremely thin we can use a control
volume of infinitesimal thickness.

CONTINUITY

For steady, one-dimensional flow we have:

$$\dot{m} = \rho AV = \text{const} \qquad (2.30)$$

But A = const; thus $\qquad \rho V = \text{const} \qquad (4.1)$

Application of this to our problem yields:

$$\rho a = (\rho + d\rho)(a - dV)$$

Expanding $\qquad \cancel{\rho a} = \cancel{\rho a} - \rho dV + a d\rho - \overset{\text{H.O.T.}}{\cancel{d\rho dV}}$

Neglecting the higher order term and solving for dV we have:

$$dV = \frac{a d\rho}{\rho} \qquad (4.2)$$

MOMENTUM

Since the control volume has infinitesimal thickness we can neglect any shear stresses along the walls. We shall write the x-component of the momentum equation, taking forces and velocities as positive if to the right. For steady, one-dimensional flow we may write:

$$\Sigma F_x = \frac{\dot{m}}{g_c}(V_{out_x} - V_{in_x}) \qquad (3.46)$$

$$\cancel{\rho A} - (\cancel{\rho} + dp)A = \frac{\rho A a}{g_c}\left[(\cancel{a} - dV) - \cancel{a}\right]$$

$$A dp = \frac{\rho A a}{g_c} dV$$

Canceling the area and solving for dV we have:

$$dV = \frac{g_c dp}{\rho a} \qquad (4.3)$$

Equations 4.2 and 4.3 may now be combined to eliminate dV, with the result:

$$a^2 = g_c \frac{dp}{d\rho} \qquad (4.4)$$

However, the derivative $dp/d\rho$ is not unique. It depends entirely upon the process. Thus, it should really be written as a partial derivative with the appropriate subscript. But what subscript? What kind of a process are we dealing with?

Remember, we are analyzing an infinitesimal disturbance.
For this case we can assume negligible losses and heat transfer
as the wave passes through the fluid. Thus, the process is both
reversible and adiabatic, which means it is isentropic. (Why?)
After we have studied shock waves we shall prove that very weak
shock waves (i.e., small disturbances) approach an isentropic
process in the limit. Therefore, equation 4.4 should properly be
written as:

$$a^2 = g_c \left(\frac{\partial p}{\partial \rho} \right)_s$$ (4.5)

This can be expressed in an alternate form by introducing the
bulk or volume modulus of elasticity E_v . This is a relation
between volume or density changes which occur as a result of pressure fluctuations and is defined as:

$$E_v \equiv -v \left(\frac{\partial p}{\partial v} \right)_s \equiv \rho \left(\frac{\partial p}{\partial \rho} \right)_s$$ (4.6)

Thus, $$a^2 = g_c \left(\frac{E_v}{\rho} \right)$$ (4.7)

Equations 4.5 and 4.7 are equivalent general relations for sonic
velocity through any medium. The bulk modulus is normally used
in connection with liquids and solids. Table 4.1 gives some typical values of this modulus, the exact value depending on the temperature and pressure of the medium. For solids it also depends
on the type of loading. The reciprocal of the bulk modulus is
called the "compressibility." What is the sonic velocity in a
truly incompressible fluid? Hint: What is the value of $\left(\partial p / \partial \rho \right)_s$?

TABLE 4.1

Medium	Bulk Modulus (psi)
Oil	185,000 to 270,000
Water	300,000 to 400,000
Mercury	approx 4,000,000
Steel	approx 30,000,000

Equation 4.5 is normally used for gases and this can be greatly simplified for the case of a gas which obeys the perfect gas law. For an isentropic process we know that:

$$pv^\gamma = const \quad\quad or \quad\quad p = \rho^\gamma \, const \quad\quad (4.8)$$

Thus,
$$\left(\frac{\partial p}{\partial \rho}\right)_s = \gamma \, \rho^{\gamma-1} \, const$$

But from (4.8) the constant $= p/\rho^\gamma$

Therefore,
$$\left(\frac{\partial p}{\partial \rho}\right)_s = \gamma \, \rho^{\gamma-1} \, \frac{p}{\rho^\gamma} = \gamma \, \frac{p}{\rho} = \gamma RT$$

and from (4.5)
$$\boxed{a^2 = \gamma g_c RT} \quad\quad (4.9)$$

or
$$\boxed{a = \sqrt{\gamma g_c RT}} \quad\quad (4.10)$$

Notice that for perfect gases sonic velocity is a function of temperature <u>only</u>.

Example 4.1   Compute the sonic velocity in air at $70^\circ F$.

$$a^2 = \gamma g_c RT = (1.4)(32.2)(53.3)(460 + 70)$$

$$a = \underline{1128} \; ft/sec$$

Example 4.2   Sonic velocity through carbon dioxide is 275 m/s. What is the temperature in $^\circ K$?

$$a^2 = \gamma g_c RT$$

$$(275)^2 = (1.29)(1)(189)(T)$$

$$T = \underline{310.2} \; ^\circ K$$

Always keep in mind that, in general, sonic velocity is a property of the fluid and varies with the state of the fluid. <u>Only</u> for gases which can be treated as perfect is the sonic velocity a function of temperature alone.

MACH NUMBER

We define the Mach number as:

$$\boxed{M \equiv \frac{V}{a}} \quad\quad (4.11)$$

where   V ≡ the velocity of the medium

and   a ≡ sonic velocity through the medium.

It is important to realize that both V and a are computed
locally for conditions that actually exist at the same point.
If the velocity at one point in a flow system is twice that at
another point, we cannot say that the Mach number has doubled.
We must seek further information on the sonic velocity which has
probably also changed. (What property would we be interested in
if the fluid were a perfect gas?)

If the velocity is less than the local speed of sound, then
M is less than 1 and the flow is called subsonic. If the
velocity is greater than the local speed of sound, M is greater
than 1 and the flow is called supersonic. We shall soon see
that Mach number is the most important parameter in the analysis
of compressible flows.

4.4 WAVE PROPAGATION

Let us now examine a point disturbance that is at rest in a
fluid. Infinitesimal pressure pulses are continually being emitted and thus they travel through the medium at sonic velocity in
the form of spherical wave fronts. To simplify matters we shall
keep track of only those pulses that are emitted every second.
At the end of three seconds the picture will appear as shown in
Figure 4.3. Note that the wave fronts are concentric.

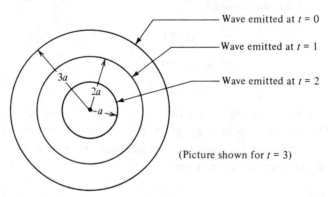

Wave emitted at $t = 0$

Wave emitted at $t = 1$

Wave emitted at $t = 2$

$3a$

$2a$

a

(Picture shown for $t = 3$)

Figure 4.3 Wave fronts from a stationary disturbance

Now consider a similar problem in which the disturbance is
no longer stationary. Assume that it is moving at a speed less
than sonic velocity, say a/2 . Figure 4.4 shows such a situa-

tion at the end of three seconds. Note that the wave fronts are no longer concentric. Furthermore, the wave that was emitted at t = o is always in front of the disturbance itself. Therefore, any person, object, or fluid particle located upstream will feel the wave fronts pass by and know that the disturbance is coming.

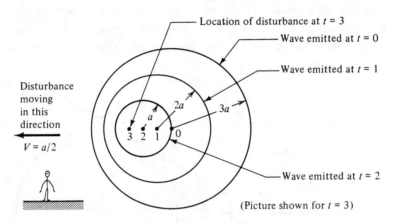

Figure 4.4 Wave fronts from subsonic disturbance

Next, let the disturbance move at exactly sonic velocity. Figure 4.5 shows this case and you will note that all wave fronts coalesce on the left side and move along with the disturbance. After a period of time this wave front would approximate a plane indicated by the dotted line. In this case, no region upstream is forewarned of the disturbance as the disturbance arrives at the same time as the wave front.

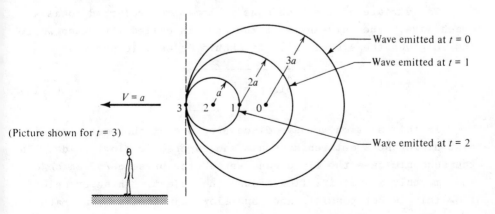

Figure 4.5 Wave fronts from sonic disturbance

The only other case to consider is that of a disturbance
moving at velocities greater than the speed of sound. Figure 4.6
shows a point disturbance moving at Mach number = 2 (twice sonic
velocity). The wave fronts have coalesced to form a cone with
the disturbance at the apex. This is called a "Mach cone." The
region inside the cone is called the "zone of action" since it
feels the presence of the waves. The outer region is called the
"zone of silence" as this entire region is unaware of the disturbance.

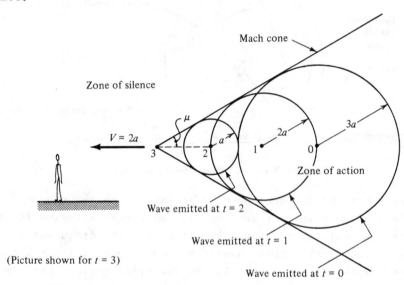

Figure 4.6 Wave fronts from supersonic disturbance

The surface of the Mach cone is sometimes referred to as a
"Mach wave," the half-angle at the apex is called the "Mach angle"
and is given the symbol μ . It should be easy to see that

$$\text{Sin } \mu = \frac{a}{V} = \frac{1}{M} \tag{4.12}$$

In this section we have discovered one of the most significant differences between subsonic and supersonic flow fields. In
the subsonic case the fluid can "sense" the presence of an object
and smoothly adjust its flow around the object. In supersonic
flow this is not possible and thus flow adjustments occur rather
abruptly in the form of shock or expansion waves. We shall study
these in great detail in Units 6 through 8.

4.5 EQUATIONS FOR A PERFECT GAS IN TERMS OF MACH NUMBER

The previous section has shown that supersonic and subsonic flows have totally different characteristics. This suggests that it would be instructive to use Mach number as a parameter in our basic equations. This can be done very easily for the flow of a perfect gas since in this case we have a simple equation of state and an explicit expression for sonic velocity. Development of some of the more important relations follow.

CONTINUITY

For steady, one-dimensional flow we have

$$\dot{m} = \rho A V = const \qquad (2.30)$$

From the perfect gas equation of state

$$\rho = \frac{p}{RT} \qquad (1.13)$$

and from the definition of Mach number

$$V = Ma \qquad (4.11)$$

Also recall the expression for sonic velocity in a perfect gas

$$a = \sqrt{\gamma g_c RT} \qquad (4.10)$$

Substitution of (1.13), (4.11) and (4.10) into equation 2.30

yields:
$$\rho A V = \frac{p}{RT} AM \sqrt{\gamma g_c RT} = pAM \sqrt{\frac{\gamma g_c}{RT}}$$

Thus, for steady, one-dimensional flow of a perfect gas, the continuity equation becomes:

$$\boxed{\dot{m} = pAM \sqrt{\frac{\gamma g_c}{RT}} = const} \qquad (4.13)$$

STAGNATION RELATIONS

For gases we eliminate the potential term and write

$$h_t = h + \frac{V^2}{2g_c} \qquad (3.18)$$

Knowing $$V^2 = M^2 a^2$$ (from 4.11)

and $$a^2 = \gamma g_c RT$$ (4.9)

we have $$h_t = h + \frac{M^2 \gamma g_c RT}{2 g_c} = h + \frac{M^2 \gamma RT}{2}$$ (4.14)

From equations 1.49 and 1.50 we can write the specific heat at constant pressure in terms of γ and R. Show that

$$c_p = \frac{\gamma R}{\gamma - 1}$$ (4.15)

Combining (4.15) and (4.14) we have

$$h_t = h + M^2 \frac{(\gamma - 1)}{2} c_p T$$ (4.16)

But for a gas we can say

$$h = c_p T$$ (1.48)

Thus, $$h_t = h + M^2 \frac{(\gamma - 1)}{2} h$$

or $$\boxed{h_t = h\left[1 + \frac{(\gamma - 1)}{2} M^2\right]}$$ (4.17)

Using $h = c_p T$ and $h_t = c_p T_t$, this can be written as:

$$\boxed{T_t = T\left[1 + \frac{(\gamma - 1)}{2} M^2\right]}$$ (4.18)

Equations 4.17 and 4.18 are used frequently. Memorize them!

Now, the stagnation process is isentropic. Thus, γ can be used as the exponent n in equation 1.57, and between any two points on the same isentropic we have

$$\frac{p_2}{p_1} = \left(\frac{T_2}{T_1}\right)^{\frac{\gamma}{\gamma - 1}}$$ (4.19)

Let point 1 refer to the static conditions and point 2 the stagnation conditions. Then combining (4.19) and (4.18) produces:

$$\frac{p_t}{p} = \left(\frac{T_t}{T}\right)^{\frac{\gamma}{\gamma - 1}} = \left[1 + \frac{(\gamma - 1)}{2} M^2\right]^{\frac{\gamma}{\gamma - 1}}$$ (4.20)

or

$$p_t = p\left[1 + \frac{(\gamma-1)}{2} M^2\right]^{\frac{\gamma}{\gamma-1}}$$ (4.21)

This expression for total pressure is important. <u>Learn it</u>!

Example 4.3 Air flows with a velocity of 800 ft/sec and has a pressure of 30 psia and temperature of 600°R. Determine the stagnation pressure.

$$a = (\gamma g_c RT)^{\frac{1}{2}} = \left[(1.4)(32.2)(53.3)(600)\right]^{\frac{1}{2}} = 1201 \text{ ft/sec}$$

$$M = v/a = 800/1201 = 0.666$$

$$p_t = p\left[1 + \frac{(\gamma-1)}{2} M^2\right]^{\frac{\gamma}{\gamma-1}} = 30\left[1 + \frac{(1.4-1)}{2}(0.666)^2\right]^{\frac{1.4}{1.4-1}}$$

$$p_t = 30\left[1 + 0.0887\right]^{3.5} = 30(1.346) = \underline{40.4} \text{ psia}$$

Example 4.4 Hydrogen has a static temperature of 25°C and a stagnation temperature of 250°C. What is the Mach number?

$$T_t = T\left[1 + \frac{\gamma-1}{2} M^2\right]$$

$$(250 + 273) = (25 + 273)\left[1 + \frac{1.41-1}{2} M^2\right]$$

$$523 = 298\left[1 + 0.205 M^2\right]$$

$$M^2 = 3.683 \quad \text{and} \quad M = \underline{1.92}$$

STAGNATION PRESSURE-ENERGY EQUATION

For steady, one-dimensional flow we have:

$$\frac{dp_t}{\rho_t} + ds_e(T_t - T) + T_t ds_i + \delta w_s = 0$$ (3.25)

For a perfect gas $\quad\quad p_t = \rho_t RT_t$ (4.22)

Substitute for the stagnation density and <u>show</u> that equation 3.25 can be written as

$$\frac{dp_t}{p_t} + \frac{ds_e}{R}\left(1 - \frac{T}{T_t}\right) + \frac{ds_i}{R} + \frac{\delta w_s}{RT_t} = 0$$ (4.23)

A large number of problems are adiabatic and involve no shaft work. In this case ds_e and δw_s are zero.

$$\frac{dp_t}{p_t} + \frac{ds_i}{R} = 0 \tag{4.24}$$

This can be integrated between two points in the flow system to give:

$$\ln \frac{p_{t2}}{p_{t1}} + \frac{s_{i2} - s_{i1}}{R} = 0 \tag{4.25}$$

But since $ds_e = 0$, $ds_i = ds$, and we really do not need to continue writing the subscript i under the entropy. Thus,

$$\ln \frac{p_{t2}}{p_{t1}} = -\frac{(s_2 - s_1)}{R} \tag{4.26}$$

Taking the anti-log this becomes:

$$\frac{p_{t2}}{p_{t1}} = e^{-\frac{(s_2 - s_1)}{R}} \tag{4.27}$$

or

$$\boxed{\frac{p_{t2}}{p_{t1}} = e^{-\Delta s/R}} \tag{4.28}$$

Watch your units when you use this equation! Total pressures must be absolute, and $\Delta s/R$ must be dimensionless. For this case of adiabatic, no-work flow, Δs will always be positive. (Why?) Thus, p_{t2} will always be less than p_{t1}. Only for the limiting case of no losses will the stagnation pressure remain constant.

This confirms previous knowledge gained from the stagnation pressure-energy equation; that for the case of an adiabatic, no-work system without flow losses p_t = const for any fluid. Thus, the stagnation pressure is seen to be a very important parameter which in many systems reflects the flow losses. Be careful to note, however, that the specific relation in equation 4.28 is only applicable to perfect gases, and even then only under certain flow conditions. What are these conditions?

Summarizing the above: for steady, one-dimensional flow we have:

$$\delta q = \delta w_s + dh_t \qquad (3.20)$$

Note that equation 3.20 is valid even if flow losses are present.

If　$\delta q = \delta w_s = 0$,　　　　　then　$h_t = const$

If in addition to the above no losses occur, that is

If　$\delta q = \delta w_s = ds_i = 0$,　　　then　$p_t = const$

Example 4.5　Oxygen flows in a constant-area, horizontal, insulated duct. Conditions at section one are: $p_1 = 50$ psia , $T_1 = 600^{\circ}R$, $V_1 = 2860$ ft/sec. At a downstream section the temperature is $T_2 = 1048^{\circ}R$.

　　(a) Determine M_1 and T_{t1} .
　　(b) Find V_2 and p_2 .
　　(c) What is the entropy change between the two sections?

(a) $a_1 = (\gamma g_c R T_1)^{\frac{1}{2}} = \left[(1.4)(32.2)(48.3)(600)\right]^{\frac{1}{2}} = 1143$ ft/sec

$M_1 = V_1/a_1 = 2860/1143 = \underline{2.50}$

$T_{t1} = T_1\left[1 + \frac{\gamma-1}{2} M_1^{2}\right] = 600\left[1 + \frac{1.4-1}{2}(2.5)^2\right] = \underline{1350^{\circ}R}$

(b) Energy　　　　　　　　　$h_{t1} + \cancel{q} = h_{t2} + \cancel{w}_s$

$$h_{t1} = h_{t2}$$

and since this is a perfect gas, then $T_{t1} = T_{t2}$

$$T_{t2} = T_2\left[1 + \frac{\gamma-1}{2} M_2^{2}\right]$$

$1350 = 1048\left[1 + \frac{1.4-1}{2} M_2^{2}\right]$　　　and　　$M_2 = \underline{1.20}$

$V_2 = M_2 a_2 = 1.20\left[(1.4)(32.2)(48.3)(1048)\right]^{\frac{1}{2}} = \underline{1813}$ ft/sec

Continuity　　　　　　　$\dot{m} = \rho_1 A_1 V_1 = \rho_2 A_2 V_2$

but　　　　　　$A_1 = A_2$,　　　and　　$\rho = p/RT$

Thus　　　　　　　　　　$\dfrac{p_1 V_1}{T_1} = \dfrac{p_2 V_2}{T_2}$

$$p_2 = \frac{V_1}{V_2} \frac{T_2}{T_1} p_1 = \frac{(2860)}{(1813)} \frac{(1048)}{(600)} (50) = \underline{137.8} \text{ psia}$$

(c) To obtain the entropy change we need p_{t1} and p_{t2}.

$$p_{t1} = p_1 \left[1 + \frac{\gamma-1}{2} M_1^2\right]^{\frac{\gamma}{\gamma-1}} = 50\left[1 + \frac{1.4-1}{2} (2.5)^2\right]^{\frac{1.4}{1.4-1}} = 854 \text{ psia}$$

Similarly, $p_{t2} = 334$ psia

$$e^{-\Delta s/R} = \frac{p_{t2}}{p_{t1}} = \frac{334}{854} = 0.391$$

$$\Delta s/R = \ln{(1/0.391)} = 0.939$$

$$\Delta s = (0.939)(48.3)/(778) = \underline{0.0583} \text{ Btu/lbm-}^{\circ}\text{R}$$

4.6 *h–s* AND *T–s* DIAGRAMS

Every problem should be approached with a simple sketch of the physical system and also a thermodynamic state diagram. Since the losses affect the entropy changes (through ds_i) one generally uses either an h-s or T-s diagram. In the case of perfect gases enthalpy is a function of temperature only and, therefore, the T-s and h-s diagrams are identical.

Consider a steady, one-dimensional flow of a perfect gas. Let us assume no heat transfer and no external work. From the energy equation

$$h_{t1} + \cancel{q} = h_{t2} + \cancel{w}_s \tag{3.19}$$

the stagnation enthalpy remains constant, and since it is a perfect gas the total temperature is also constant. This is represented by the solid horizontal line in Figure 4.7. Two particular sections in the system have been indicated by 1 and 2. The actual process that takes place between these points is indicated on the T-s diagram.

Notice that although the stagnation conditions do not actually exist in the system they are also shown on the diagram for reference. The distance between the static and stagnation points is indicative of the velocity that exists at that location (since gravity has been neglected). It can also be clearly seen that if there is a Δs_{1-2} then $p_{t2} < p_{t1}$ and the relationship between stagnation pressure and flow losses is again verified.

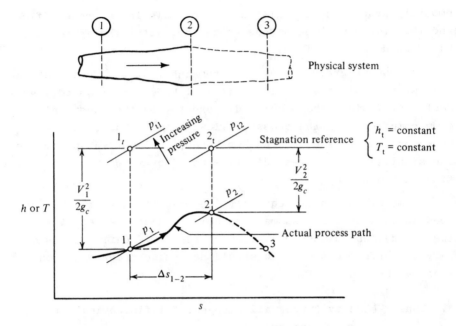

Figure 4.7 Diagram showing stagnation reference states

It is interesting to hypothesize a third section that just happens to be at the same enthalpy (and temperature) as the first. What else do these points have in common? The same velocity? Obviously! How about sonic velocity? (Recall for gases that this is a function of temperature only.) This means that points 1 and 3 would also have the same Mach number (something which is not immediately obvious). One can now imagine that someplace on this diagram there is a horizontal line that represents the locus of points having a Mach number of unity. Between this line and the stagnation line lie all points in the subsonic regime. Below this line lie all points in the supersonic regime. These conclusions are based on certain assumptions. What are they?

4.7 SUMMARY

In general, waves propagate at a speed which depends on the medium, its thermodynamic state, and the strength of the wave. However, infinitesimal disturbances travel at a speed determined only by the medium and its state. Sound waves fall into this latter category. A discussion of wave propagation and sonic velocity brought out a basic difference between subsonic and supersonic flows. If subsonic, the flow can "sense" objects and flow

smoothly around them. This is not possible in supersonic flow
and this topic will be discussed further after the appropriate
background has been laid.

As you progress through the remainder of this book and analyze specific flow situations, it will become increasingly evi-
dent that fluids behave quite differently in the supersonic regime compared with the subsonic flow regime. Thus, it will not
be surprising to see Mach number become an important parameter.
The significance of T-s diagrams as a key to problem visualiza-
tion should not be overlooked.

Some of the most frequently used equations that were developed in this unit are summarized below. Most are restricted to
the steady, one-dimensional flow of any fluid while others apply
only to perfect gases. You should determine under what conditions
each may be used.

a. Sonic Velocity (propagation speed of infinitesimal pressure
        pulses)

$$a^2 = g_c\left(\frac{\partial p}{\partial \rho}\right)_s = g_c\frac{E_v}{\rho} \qquad (4.5) \text{ \& } (4.7)$$

$$M = \frac{V}{a} \quad \text{(all at the same location)} \quad (4.11)$$

$$\sin \mu = \frac{1}{M} \qquad (4.12)$$

b. Special Relations for a Perfect Gas

$$a^2 = \gamma g_c RT \qquad (4.9)$$

$$h_t = h\left[1 + \frac{(\gamma-1)}{2} M^2\right] \qquad (4.17)$$

$$T_t = T\left[1 + \frac{(\gamma-1)}{2} M^2\right] \qquad (4.18)$$

$$p_t = p\left[1 + \frac{(\gamma-1)}{2} M^2\right]^{\frac{\gamma}{\gamma-1}} \qquad (4.21)$$

$$\frac{dp_t}{p_t} + \frac{ds_e}{R}\left(1 - \frac{T}{T_t}\right) + \frac{ds_i}{R} + \frac{\delta w_s}{RT_t} = 0 \qquad (4.23)$$

$$\frac{p_{t2}}{p_{t1}} = e^{-\Delta s/R} \qquad \text{for } Q = W = 0 \qquad (4.28)$$

4.8 PROBLEMS

1. Compute and compare sonic velocity in air, hydrogen, water and mercury. Assume normal room temperature and pressure.

2. Start with the relation for stagnation pressure which is valid for a perfect gas:

$$p_t = p\left[1 + \frac{(\gamma-1)}{2}M^2\right]^{\frac{\gamma}{\gamma-1}}$$

Expand the right side in a binomial series and evaluate the result for small (but not zero) Mach numbers. Show that your answer can be written as:

$$p_t = p + \frac{\rho V^2}{2g_c} + \text{H.O.T.}$$

Remember, the higher order terms are only negligible for very small Mach numbers. (See problem number 3.)

3. Measurement of air flow shows the static and stagnation pressures to be 30 and 32 psig, respectively. (Note that these are gage pressures.) Assume p_{amb} = 14.7 psia and the temperature is 120°F.
   (a) Find the flow velocity using equation 4.21.
   (b) Now assume that the air is incompressible and calculate the velocity using equation 3.39.
   (c) Repeat parts (a) and (b) for static and stagnation pressures of 30 and 80 psig, respectively.
   (d) Can you reach any conclusions concerning when a gas may be treated as a constant-density fluid?

4. If $\gamma = 1.2$ and the fluid is a perfect gas, what Mach number will give a temperature ratio of T/T_t = 0.909? What will the ratio of p/p_t be for this flow?

5. Carbon dioxide with a temperature of 335°K and a pressure of 1.4×10^5 N/m^2 is flowing with a velocity of 200 m/s.
   (a) Determine the sonic velocity and Mach number.
   (b) Determine the stagnation density.

6. The temperature of argon is 100°F, the pressure 42 psia, and the velocity 2264 ft/sec. Calculate the Mach number and stagnation pressure.

7. Helium flows in a duct with a temperature of 50°C, a pressure of 2.0 bars absolute, and a total pressure of 5.3 bars absolute. Determine the velocity in the duct.

8. An airplane flies 600 mph at an altitude of 16,500 ft where the temperature is $0°F$ and the pressure 1124 psfa. What temperature and pressure might you expect on the nose of the airplane?

9. Air flows at $M = 1.35$ and has a stagnation enthalpy of 4.5×10^5 J/kg. The stagnation pressure is 3.8×10^5 N/m^2. Determine the static conditions (pressure, temperature, and velocity).

10. A large chamber contains a perfect gas under conditions p_1, T_1, h_1, etc. If the gas is allowed to flow from the chamber (with $q = w_s = 0$) show that the velocity cannot be greater than

$$V_{max} = a_1 \left[\frac{2}{\gamma - 1} \right]^{\frac{1}{2}}$$

If the velocity is the maximum, what is the Mach number?

11. Air flows steadily in an adiabatic duct where no shaft work is involved. At one section the total pressure is 50 psia and at another section it is 67.3 psia. In which direction is the fluid flowing and what is the entropy change between these two sections?

12. Methane gas flows in an adiabatic, no-work system with negligible change in potential. At one section $p_1 = 14$ bars abs., $T_1 = 500°K$ and $V_1 = 125$ m/s. At a downstream section $M_2 = 0.8$.

    (a) Determine T_2 and V_2.
    (b) Find p_2 assuming there are no friction losses.
    (c) What is the area ratio A_2/A_1 ?

13. Air flows through a constant area insulated passage. Entering conditions are $T_1 = 520°R$, $p_1 = 50$ psia and $M_1 = 0.45$. At a point downstream the Mach number is found to be unity.
    (a) Solve for T_2 and p_2.
    (b) What is the entropy change between these two sections?
    (c) Determine the wall frictional force if the duct is 1 foot in diameter.

14. Carbon dioxide flows in a horizontal adiabatic, no-work system. Pressure and temperature at section one are 7 atmospheres and $600°K$. At a downstream section $p_2 = 4$ atmospheres, $T_2 = 550°K$ and the Mach number is $M_2 = 0.90$.

(a) Compute the velocity at the upstream location.

(b) What is the entropy change?

(c) Determine the area ratio A_2/A_1 .

15. Oxygen with T_{t1} = $1000^{\circ}R$, p_{t1} = 100 psia and M_1 = 0.2 enters a device with a cross-sectional area A_1 = 1 ft^2 . There is no heat transfer, work transfer, or losses as the gas passes through the device, and expands to 14.7 psia.

(a) Compute ρ_1 , V_1 and $\dot{m}$.

(b) Compute M_2 , T_2 , V_2 , ρ_2 and A_2 .

(c) What force does the fluid exert on the device?

16. Consider steady, one-dimensional, constant-area, horizontal, isothermal flow of a perfect gas with no shaft work. The duct has a cross-sectional area A and perimeter P . Let τ_w be the shear stress at the wall.

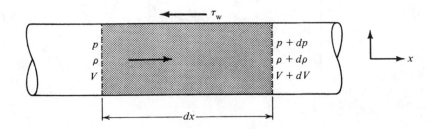

(a) Apply momentum concepts (equation 3.45) and show that

$$-dp + f \frac{dx}{D_e} \frac{\rho V^2}{2g_c} = \frac{\rho V dV}{g_c}$$

(b) From the concept of continuity and the equation of state show that

$$\frac{d\rho}{\rho} = \frac{dp}{p} = -\frac{dV}{V}$$

(c) Combine the results of (a) and (b) to show that

$$\frac{d\rho}{\rho} = \left[\frac{\gamma M^2}{2(\gamma M^2 - 1)} \right] \frac{f dx}{D_e}$$

4.9 CHECK TEST

You should be able to complete this test without reference to material in the unit.

1.   (a) Define Mach number and Mach angle.
     (b) Give an expression that represents sonic velocity in an
         arbitrary fluid.
     (c) Give the relation used to compute sonic velocity in a
         perfect gas.

2.   Consider the steady, one-dimensional flow of a perfect gas
     with heat transfer. The T-s diagram below shows both
     static and stagnation points at two locations in the system.
     It is known that A = B .
     (a) Is heat transferred into or out of the system?
     (b) Is $M_2 > M_1$, $M_2 = M_1$, or $M_2 < M_1$?

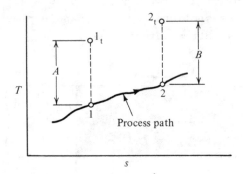

3.   Mark the following as true or false.
     (a) Changing the frame of reference (or superposition of a
         velocity onto an existing flow) does not change the
         static enthalpy.
     (b) Shock waves travel at sonic velocity through a medium.
     (c) In general, one can say that flow losses will show up
         as a decrease in stagnation enthalpy.
     (d) The stagnation process is one of constant entropy.
     (e) A Mach cone does not exist for subsonic flow.

4.   Cite the conditions that are necessary for the stagnation
     temperature to remain constant in a flow system.

5.   For steady flow of a perfect gas the continuity equation can
     be written as:

$$\dot{m} = f(p, M, T, \gamma, A, R, g_c) = \text{const}$$

     Determine the precise function.

6.   Work problem #13 in Section 4.8.

UNIT 5

VARYING-AREA ADIABATIC FLOW

5.1 INTRODUCTION

Area changes, friction, and heat transfer are the most important factors which affect the properties in a flow system. Although some situations may involve the simultaneous effects of
all three of these factors, the majority of engineering problems
are <u>predominantly</u> influenced by only one of these variables.
Thus, it is more than academic interest which leads to the separate study of each of the above mentioned effects. In this manner it is possible to consider only the controlling factor and
develop a simple solution which is within the realm of acceptable
engineering accuracy.

In this unit we shall consider the general problem of
varying-area flow under the assumptions of no heat transfer (adiabatic) and no shaft work. We shall first consider the flow of
an arbitrary fluid without losses and determine how its properties are affected by area changes. The case of a perfect gas

will then be considered and simple working equations will be developed to aid in the solution of problems with or without flow
losses. The latter case (isentropic flow) lends itself to the
construction of tables which are used throughout the remainder
of the book. The unit will close with a brief discussion of the
various ways in which nozzle and diffuser performance can be represented.

5.2 OBJECTIVES

After sucessfully completing this unit you should be able to:

1. Simplify the basic equations for continuity and energy to
    relate differential changes in density, pressure and velocity
    to Mach number and a differential change in area for steady,
    one-dimensional flow through a varying-area passage with no
    losses. (Optional.)

2. Show graphically how pressure, density, velocity, and area
    vary in steady, one-dimensional, isentropic flow as Mach num-
    ber ranges from zero to supersonic values.

3. Compare the function of a nozzle and a diffuser. Sketch
    physical devices that perform as each for subsonic and super-
    sonic flow.

4. Derive the working equations for a perfect gas relating prop-
    erty ratios between two points, in adiabatic no-work flow, as
    a function of Mach number (M), ratio of specific heats (γ)
    and change in entropy (Δs).

5. Define the * reference condition and the properties asso-
    ciated with it. (i.e., A^*, p^*, T^*, ρ^*, etc.)

6. Express the loss (Δs_i) as a function of stagnation pres-
    sures (p_t) <u>or</u> reference areas (A^*) between two points in
    the flow. Under what conditions are these relations true?

7. State and interpret the relation between stagnation pressure
    (p_t) and the reference area (A^*) for a process between two
    points in adiabatic no-work flow.

8. Explain how a converging nozzle performs with various re-
    ceiver pressures. Do the same for the <u>isentropic</u> performance
    of a converging-diverging nozzle.

9. State what is meant by the first and third critical modes of
    nozzle operation. Given the area ratio of a converging-

diverging nozzle, determine the operating pressure ratios which cause operation at the first and third critical points.

10. With the aid of an h-s diagram give a suitable definition for both nozzle efficiency and diffuser efficiency.

11. Describe what is meant by a "choked" flow passage.

12. Demonstrate the ability to utilize the adiabatic and isentropic flow relations and the isentropic tables to solve typical flow problems.

5.3 GENERAL FLUID—NO LOSSES

We shall first consider the general behavior of an arbitrary fluid. In order to isolate the effects of area change we make the following

Assumptions: Steady, one-dimensional flow

Adiabatic	$(\delta q = 0, \ ds_e = 0)$
No shaft work	$(\delta w_s = 0)$
Neglect potential	$(dz = 0)$
No losses	$(ds_i = 0)$

Our objective will be to obtain relations which indicate the variation of fluid properties with area changes <u>and</u> Mach number. In this manner we can distinguish the important differences between subsonic and supersonic behavior. We start with the energy equation.

$$\delta q = \delta w_s + dh + \frac{dV^2}{2g_c} + \frac{g}{g_c} dz \qquad (2.53)$$

But
$$\delta q = \delta w_s = 0$$

and
$$dz = 0$$

which leaves
$$0 = dh + \frac{dV^2}{2g_c} \qquad (5.1)$$

or
$$dh = - \frac{VdV}{g_c} \qquad (5.2)$$

We now introduce the property relation

$$Tds = dh - \frac{dp}{\rho} \qquad (1.41)$$

Since our flow situation has been assumed to be adiabatic $(ds_e = 0)$ and to contain no losses $(ds_i = 0)$, it is also isentropic (ds = 0). Thus, equation 1.41 becomes:

$$dh = \frac{dp}{\rho}$$

(5.3)

We equate equations 5.2 and 5.3 to obtain

$$-\frac{VdV}{g_c} = \frac{dp}{\rho}$$

or

$$dV = -\frac{g_c d_p}{\rho V}$$

(5.4)

We introduce this into equation 2.32 and the differential form of the continuity equation becomes:

$$\frac{d\rho}{\rho} + \frac{dA}{A} - \frac{g_c dp}{\rho V^2} = 0$$

(5.5)

Solve this for dp/ρ and show that

$$\frac{dp}{\rho} = \frac{V^2}{g_c}\left[\frac{d\rho}{\rho} + \frac{dA}{A}\right]$$

(5.6)

Recall the definition of sonic velocity:

$$a^2 = g_c\left(\frac{\partial p}{\partial \rho}\right)_s$$

(4.5)

Since our flow is isentropic, we may drop the subscript and change the partial derivative to an ordinary derivative

$$a^2 = g_c \frac{dp}{d\rho}$$

(5.7)

This permits equation 5.7 to be rearranged to

$$dp = \frac{a^2}{g_c} d\rho$$

(5.8)

Substituting this expression for dp into equation 5.6 yields

$$\frac{d\rho}{\rho} = \frac{V^2}{a^2}\left[\frac{d\rho}{\rho} + \frac{dA}{A}\right]$$

(5.9)

Introduce the definition of Mach number

$$M^2 = \frac{V^2}{a^2}$$

(4.11)

and combine the terms in dρ/ρ to obtain the following relation between density and area changes:

$$\frac{d\rho}{\rho} = \left[\frac{M^2}{1-M^2}\right]\frac{dA}{A} \qquad (5.10)$$

If we now substitute equation 5.10 into the differential form of the continuity equation (2.32) we can obtain a relation between velocity and area changes. <u>Show</u> that

$$\frac{dV}{V} = -\left[\frac{1}{1-M^2}\right]\frac{dA}{A} \qquad (5.11)$$

Now equation 5.4 can be divided by V to yield

$$\frac{dV}{V} = -\frac{g_c dp}{\rho V^2} \qquad (5.12)$$

If we equate (5.11) and (5.12) we can obtain a relation between pressure and area changes. <u>Show</u> that

$$dp = \frac{\rho V^2}{g_c}\left[\frac{1}{1-M^2}\right]\frac{dA}{A} \qquad (5.13)$$

For convenience, we collect the three important relations which will be referred to in the analysis that follows:

$$dp = \frac{\rho V^2}{g_c}\left[\frac{1}{1-M^2}\right]\frac{dA}{A} \qquad (5.13)$$

$$\frac{d\rho}{\rho} = \left[\frac{M^2}{1-M^2}\right]\frac{dA}{A} \qquad (5.10)$$

$$\frac{dV}{V} = -\left[\frac{1}{1-M^2}\right]\frac{dA}{A} \qquad (5.11)$$

Let us consider what is happening as fluid flows through a variable-area duct. For simplicity we shall <u>assume</u> <u>that</u> <u>the</u> <u>pressure</u> <u>is</u> <u>always</u> <u>decreasing</u>. Thus, dp is negative. From equation 5.13 you see that if $M < 1$, dA must be negative indicating that the area is decreasing; whereas if $M > 1$, dA must be positive and the area is increasing.

Now continue to assume that the pressure is decreasing. Knowing the area variation you can now consider equation 5.10. Fill in the following blanks with the words "increasing" or "decreasing." If $M < 1$ (and dA is _____) then $d\rho$ must be

_____. If M > 1 (and dA is _____) then dρ must be

_____.

   Looking at equation 5.11 reveals that if M < 1 (and dA
is _____) then dV must be _____ meaning that velocity is
_____, whereas if M > 1 (and dA is _____) then dV
must be _____ and velocity is _____.

   We summarize the above by saying that as the pressure decreases the following variations occur.

		Subsonic (M < 1)	Supersonic (M > 1)
Area	A	decreases	increases
Density	ρ	decreases	decreases
Velocity	V	increases	increases

A similar chart could easily be made for the situation where pressure increases but it is probably more convenient to express the
above in an alternate graphical form as shown in Figure 5.1.

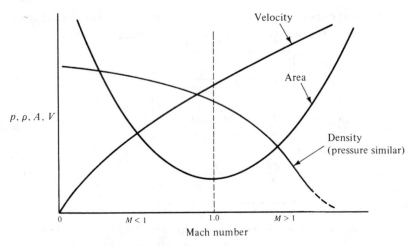

Figure 5.1 Property variation with area change

   The appropriate shape of these curves can be easily visualized if one combines equations 5.10 and 5.11 to eliminate the
term dA/A with the following result:

$$\frac{d\rho}{\rho} = - M^2 \frac{dV}{V} \tag{5.14}$$

From this equation we see that at low Mach numbers density variations will be quite small, whereas at high Mach numbers the density changes very rapidly. (Eventually, as V becomes very

large, and ρ becomes very small, then small density changes
occur once again.) This means that the density is nearly constant in the low subsonic regime (dρ ≈ 0) and the velocity
changes compensate for area changes. (See the differential form
of the continuity equation 2.32.) At a Mach number equal to unity we reach a situation where density changes and velocity changes
compensate for one another and thus no change in area is required
(dA = 0). As we move on into the supersonic area the density
decreases so rapidly that the accompanying velocity change cannot
accommodate the flow and thus the area must increase. We now
recognize another aspect of flow behavior which is exactly the
opposite in subsonic and supersonic flow. Consider the operation
of devices such as nozzles and diffusers.

A <u>nozzle</u> is a device that converts enthalpy (or pressure
energy for the case of an incompressible fluid) into kinetic energy. From Figure 5.1 we see that an increase in velocity is
accompanied by either an increase or decrease in area, depending
upon the Mach number. Figure 5.2 shows what these devices look
like in the subsonic and supersonic flow regimes.

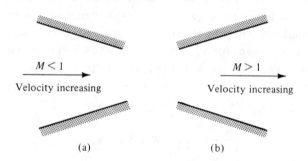

M < 1   *M* > 1
⟶   ⟶
Velocity increasing   Velocity increasing

(a)   (b)

Figure 5.2 Nozzle configurations

A <u>diffuser</u> is a device which converts kinetic energy into
enthalpy (or pressure energy for the case of an incompressible
fluid). Figure 5.3 shows what these devices look like in the
subsonic and supersonic regimes. Thus, we see that the same
piece of equipment can operate as either a nozzle or a diffuser
depending upon the flow regime.

Notice that a device is called a nozzle or a diffuser because of <u>what it does</u> and not what it looks like. Further consideration of Figures 5.1 and 5.2 leads to some interesting conclusions. If one attached a converging section (see Figure 5.2a)
to a high-pressure supply, one could never attain a flow greater

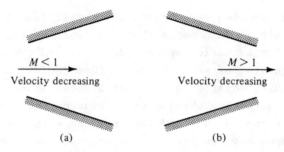

$$M < 1$$
Velocity decreasing

$$M > 1$$
Velocity decreasing

(a)   (b)

Figure 5.3 Diffuser configurations

than Mach 1 regardless of the pressure differential available.
On the other hand, if we made a converging-diverging device (combination of Figures 5.2a and 5.2b) we see a means of accelerating
the fluid into the supersonic regime, providing the proper pressure differential exists. Specific examples of these cases will
be discussed later in this unit.

5.4 PERFECT GAS—WITH LOSSES

Now that we understand the general effects of area change
in a flow system we will develop some specific working equations
for the case of a perfect gas. The term "working equations" will
be used throughout this book to indicate relations between properties at arbitrary sections of a flow system written in terms of
Mach numbers, specific heat ratio, and a loss indicator such as
Δs_i . An example of this for the system shown in Figure 5.4
would be

$$\frac{p_2}{p_1} = f(M_1, M_2, \gamma, \Delta s_i) \qquad (5.15)$$

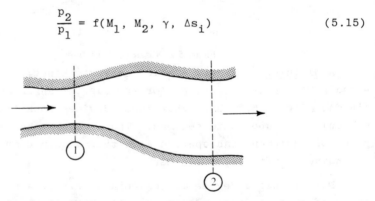

Figure 5.4 Varying area flow system

We begin by feeding the following assumptions into our fundamental concepts of state, continuity, and energy.

Assumptions: Steady, one-dimensional flow
                Adiabatic
                No shaft work
                Perfect gas
                Neglect potential

STATE - We have the perfect gas equation of state

$$p = \rho RT \tag{1.13}$$

CONTINUITY

$$\dot{m} = \rho AV = \text{const} \tag{2.30}$$

$$\rho_1 A_1 V_1 = \rho_2 A_2 V_2 \tag{5.16}$$

We first seek the area ratio

$$\frac{A_2}{A_1} = \frac{\rho_1 V_1}{\rho_2 V_2} \tag{5.17}$$

We substitute for the densities using the equation of state (1.13) and for velocities from the definition of Mach number (4.11)

$$\frac{A_2}{A_1} = \left[\frac{p_1}{RT_1}\right]\left[\frac{RT_2}{p_2}\right]\frac{M_1 a_1}{M_2 a_2} = \frac{p_1 T_2 M_1 a_1}{p_2 T_1 M_2 a_2} \tag{5.18}$$

Introduce the expression for the sonic velocity of a perfect gas:

$$a = \sqrt{\gamma g_c RT} \tag{4.10}$$

and <u>show</u> that equation 5.18 becomes

$$\frac{A_2}{A_1} = \frac{p_1 M_1}{p_2 M_2}\left(\frac{T_2}{T_1}\right)^{\frac{1}{2}} \tag{5.19}$$

We must now find a means to express the pressure and temperature ratios in terms of M_1, M_2, γ and Δs .

ENERGY - We start with

$$h_{t1} + q = h_{t2} + w_s \tag{3.19}$$

For an adiabatic and no-work process this shows that

$$h_{t1} = h_{t2} \tag{5.20}$$

However, we can go further than this since we know that for a perfect gas enthalpy is a function of temperature <u>only.</u>
Thus,

$$T_{t1} = T_{t2} \tag{5.21}$$

Recall from Unit 4 that we developed a general relationship between static and stagnation temperatures for a perfect gas as

$$T_t = T\left[1 + \frac{(\gamma-1)}{2} M^2\right] \tag{4.18}$$

Hence, equation 5.21 can be written as

$$T_1\left[1 + \frac{(\gamma-1)}{2} M_1^2\right] = T_2\left[1 + \frac{(\gamma-1)}{2} M_2^2\right] \tag{5.22}$$

or
$$\frac{T_2}{T_1} = \frac{1 + \frac{(\gamma-1)}{2} M_1^2}{1 + \frac{(\gamma-1)}{2} M_2^2} \tag{5.23}$$

which is the ratio desired for equation 5.19. Note that no subscripts have been put on the specific heat ratio γ which means
we are assuming $\gamma_1 = \gamma_2$. This might be questioned since the
specific heats c_p and c_v are known to vary somewhat with temperature. However, one can observe that they vary in a similar
manner and consequently their ratio (γ) does not exhibit much
variation except over large temperature ranges. Thus, the assumption of constant γ generally leads to acceptable engineering accuracy.

Recall again from Unit 4 that we also developed a general
relationship between static and stagnation pressures for a perfect gas.

$$p_t = p\left[1 + \frac{(\gamma-1)}{2} M^2\right]^{\frac{\gamma}{\gamma-1}} \tag{4.21}$$

Furthermore, the stagnation pressure-energy equation was easily
integrated for the case of a perfect gas in adiabatic, no-work
flow to yield:

$$\frac{p_{t2}}{p_{t1}} = e^{-\Delta s/R} \tag{4.28}$$

If we introduce equation 4.21 into 4.28 we have:

$$\frac{p_{t2}}{p_{t1}} = \frac{p_2}{p_1}\left[\frac{1 + \frac{(\gamma-1)}{2} M_2^2}{1 + \frac{(\gamma-1)}{2} M_1^2}\right]^{\frac{\gamma}{\gamma-1}} = e^{-\Delta s/R} \tag{5.24}$$

Rearrange this to obtain the desired ratio

$$\frac{p_1}{p_2} = \left[\frac{1 + \frac{(\gamma-1)}{2} M_2^2}{1 + \frac{(\gamma-1)}{2} M_1^2}\right]^{\frac{\gamma}{\gamma-1}} e^{+\Delta s/R} \tag{5.25}$$

We now have the desired information to accomplish the original objective. Direct substitution of equations 5.23 and 5.25 into 5.19 yields:

$$\frac{A_2}{A_1} = \left\{\left[\frac{1 + \frac{(\gamma-1)}{2} M_2^2}{1 + \frac{(\gamma-1)}{2} M_1^2}\right]^{\frac{\gamma}{\gamma-1}} e^{\Delta s/R}\right\} \frac{M_1}{M_2}\left[\frac{1 + \frac{(\gamma-1)}{2} M_1^2}{1 + \frac{(\gamma-1)}{2} M_2^2}\right]^{\frac{1}{2}} \tag{5.26}$$

<u>Show</u> that this can be simplified to

$$\boxed{\frac{A_2}{A_1} = \frac{M_1}{M_2}\left[\frac{1 + \frac{(\gamma-1)}{2} M_2^2}{1 + \frac{(\gamma-1)}{2} M_1^2}\right]^{\frac{\gamma+1}{2(\gamma-1)}} e^{\Delta s/R}} \tag{5.27}$$

Note that in order to obtain this equation we automatically discovered a number of other working equations which for convenience we summarize below.

$$T_{t1} = T_{t2} \tag{5.21}$$

$$\frac{p_{t2}}{p_{t1}} = e^{-\Delta s/R} \tag{4.28}$$

$$\frac{T_2}{T_1} = \frac{1 + \frac{(\gamma-1)}{2} M_1^2}{1 + \frac{(\gamma-1)}{2} M_2^2} \tag{5.23}$$

$$\frac{p_2}{p_1} = \left[\frac{1 + \frac{(\gamma-1)}{2} M_1^2}{1 + \frac{(\gamma-1)}{2} M_2^2}\right]^{\frac{\gamma}{\gamma-1}} e^{-\Delta s/R} \qquad \text{from (5.25)}$$

From equations 1.13, 5.23 and 5.25 you should also be able to show that

$$\frac{\rho_2}{\rho_1} = \left[\frac{1 + \frac{(\gamma-1)}{2} M_1^2}{1 + \frac{(\gamma-1)}{2} M_2^2}\right]^{\frac{1}{\gamma-1}} e^{-\Delta s/R} \tag{5.28}$$

Example 5.1   Air flows in an adiabatic duct without friction. At
one section the Mach number is 1.5 and farther downstream it has
increased to 2.8. Find the area ratio.

For a frictionless adiabatic system $\Delta s = 0$. We substitute
directly into equation 5.27.

$$\frac{A_2}{A_1} = \frac{1.5}{2.8} \left[\frac{1 + \frac{1.4-1}{2}(2.8)^2}{1 + \frac{1.4-1}{2}(1.5)^2} \right]^{\frac{1.4+1}{2(1.4-1)}} \quad (1) = \underline{2.98}$$

This problem is very simple since both Mach numbers are known.
The inverse problem (given A_1, A_2, M_1; find M_2) is not so
straightforward. We shall come back to this in Section 5.6 after
we develop a new concept.

5.5 THE * REFERENCE CONCEPT

In Section 3.5 the concept of a stagnation reference state
was introduced which by the nature of its definition turned out
to involve an isentropic process. Before going any further with
the working equations developed in Section 5.4 it will be convenient to introduce another reference conditon. We denote this
reference state with a superscript * and define it as "that
thermodynamic state which would exist if the fluid reached a Mach
number of unity by some particular process." The underlined
phrase is significant for there are many processes by which we
could reach Mach 1.0 from any given starting point and they would
each lead to a different thermodynamic state. Every time we
analyze a different flow phenomenon we will be considering different types of processes, and thus we will be dealing with a
different * reference state.

We first consider a * reference state reached under reversible, adiabatic conditions, i.e., by an isentropic process.
Every point in the flow system has its own * reference state,
just as it has its own stagnation reference state.

As an illustration, consider a system which involves the
flow of a perfect gas with no heat or work transfer. Figure 5.5
shows a T-s diagram indicating two points in such a flow system.
Above each point is shown its stagnation reference state and we
now add the isentropic * reference state that is associated

with each point. Not only is the stagnation line for the entire
system a horizontal line, but in this system all * reference
points will lie on a horizontal line (see discussion in Section
4.6). Is the flow subsonic or supersonic in the system depicted
in Figure 5.5?

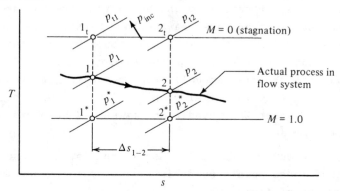

Figure 5.5 Isentropic * reference states

We shall now proceed to develop an extremely important relation. Keep in mind that the * reference states most likely
don't exist in the system — but with appropriate area changes
they could exist — and as such they represent legitimate section
locations to be used with any of the equations that we have previously developed (such as equations 5.23, 5.25, 5.27, etc.).
Specifically, let us consider:

$$\frac{A_2}{A_1} = \frac{M_1}{M_2} \left[\frac{1 + \frac{(\gamma-1)}{2} M_2^2}{1 + \frac{(\gamma-1)}{2} M_1^2} \right]^{\frac{\gamma+1}{2(\gamma-1)}} e^{\Delta s/R} \qquad (5.27)$$

In this equation, points 1 and 2 represent <u>any</u> two points
that could exist in a system (subject to the same assumptions
that led to the development of the relation). We now apply equation 5.27 between points 1^* and 2^* .

Thus: $\qquad A_1 \Rightarrow A_1^* \qquad\qquad M_1 \Rightarrow M_1^* \equiv 1$

$\qquad\qquad\quad A_2 \Rightarrow A_2^* \qquad\qquad M_2 \Rightarrow M_2^* \equiv 1$

and we have: $\quad \dfrac{A_2^*}{A_1^*} = \dfrac{1}{1} \left[\dfrac{1 + \frac{(\gamma-1)}{2} 1^2}{1 + \frac{(\gamma-1)}{2} 1^2} \right]^{\frac{\gamma+1}{2(\gamma-1)}} e^{\Delta s/R}$

or

$$\frac{A_2^{\,*}}{A_1^{\,*}} = e^{\Delta s/R}$$

(5.29)

Before going further, it might be instructive to check this relation to see if it appears reasonable. First, take the case of no losses where $\Delta s = 0$. Then equation 5.29 says that $A_1^{\,*} = A_2^{\,*}$. Check Figure 5.5 for the case of $\Delta s_{1-2} = 0$. Under these conditions the diagram collapses into a single isentropic line on which 1_t is identical with 2_t and 1^{*} is the same point as 2^{*}. Under this condition, it should be obvious that $A_1^{\,*}$ is the same as $A_2^{\,*}$.

Next, take the more general case where Δs_{1-2} is non-zero. Assuming that these points exist in a flow system, they must pass the same amount of fluid, or

$$\dot{m} = \rho_1^{\,*} A_1^{\,*} V_1^{\,*} = \rho_2^{\,*} A_2^{\,*} V_2^{\,*}$$

(5.30)

Recall from Section 4.6 that since these state points are on the same horizontal line

$$V_1^{\,*} = V_2^{\,*}$$

(5.31)

Similarly, we know that $T_1^{\,*} = T_2^{\,*}$, and from Figure 5.5 it is clear that $p_1^{\,*} > p_2^{\,*}$. Thus, from the equation of state we can easily determine that

$$\rho_2^{\,*} < \rho_1^{\,*}$$

(5.32)

Introduce equations 5.31 and 5.32 into 5.30 and show that for the case of $\Delta s_{1-2} > 0$

$$A_2^{\,*} > A_1^{\,*}$$

(5.33)

which agrees with equation 5.29.

We have previously developed a relation between the stagnation pressures (which involves the same assumptions as equation 5.29):

$$\frac{p_{t2}}{p_{t1}} = e^{-\Delta s/R}$$

(4.28)

Check Figure 5.5 to convince yourself that this equation also
appears to give reasonable answers for the special case of
$\Delta s = 0$ and for the general case of $\Delta s > 0$.

We now multiply equation 5.29 by equation 4.28.

$$\frac{A_2^*}{A_1^*} \frac{p_{t2}}{p_{t1}} = e^{\Delta s/R} \; e^{-\Delta s/R} = 1 \qquad (5.34)$$

or

$$\boxed{p_{t1} \, A_1^* = p_{t2} \, A_2^*} \qquad (5.35)$$

This is a most important relation that is frequently the key to a
problem solution. Learn equation 5.35 and the conditions under
which it applies.

5.6 ISENTROPIC TABLES

In Section 5.4 we considered the steady, one-dimensional
flow of a perfect gas under the conditions of no heat and work
transfer and negligible potential changes. Looking back over the
working equations that were developed reveals that many of them
do not include the loss term (Δs_i). In those where the loss
term does appear it takes the form of a simple multiplicative
factor such as $e^{\Delta s/R}$. This leads to the natural use of the
isentropic process as a standard for ideal performance with appropriate corrections made to account for losses when necessary. In
a number of cases we find that some actual processes are so efficient that they are very nearly isentropic and thus need no corrections.

If we simplify equation 5.27 for an isentropic process it
becomes:

$$\frac{A_2}{A_1} = \frac{M_1}{M_2} \left[\frac{1 + \frac{(\gamma-1)}{2} M_2^2}{1 + \frac{(\gamma-1)}{2} M_1^2} \right]^{\frac{\gamma+1}{2(\gamma-1)}} \qquad (5.36)$$

This is easy to solve for the area ratio if both Mach numbers are
known (see example 5.1), but let's consider a more typical problem. The physical situation is fixed, i.e., A_1 and A_2 are
known. The fluid (and thus γ) is known, and the Mach number at
one location (say M_1) is known. Our problem is to solve for the

Mach number (M_2) at the other location. Although this is not impossible it is messy and a lot of work.

We can simplify the solution by the introduction of the * reference state. Let point 2 be an arbitrary point in the flow system and let its isentropic * point be point 1.

Then $A_2 \Rightarrow A$ $M_2 \Rightarrow M$ (any value)

$\qquad\qquad A_1 \Rightarrow A^*$ $M_1 \Rightarrow 1$

and equation 5.36 becomes:

$$\frac{A}{A^*} = \frac{1}{M}\left[\frac{1 + \frac{(\gamma-1)}{2} M^2}{\frac{\gamma+1}{2}}\right]^{\frac{\gamma+1}{2(\gamma-1)}} = f(M, \gamma) \qquad (5.37)$$

We see that $A/A^* = f(M, \gamma)$ and we can easily construct a table giving values of A/A^* vs. M for a particular γ. The problem previously posed could then be solved as follows:

Given: γ, A_1, A_2, M_1, and flow is isentropic.

Find: M_2

We approach the solution by formulating the ratio A_2/A_2^* in terms of known quantities.

$$\frac{A_2}{A_2^*} = \frac{A_2}{A_1} \frac{A_1}{A_1^*} \frac{A_1^*}{A_2^*} \qquad (5.38)$$

Given ────────┘   │   └─┤ Evaluated by equation 5.29 and equals 1.0 if flow is isentropic

└─┤ A function of M_1 Look up in isentropic table

Thus, A_2/A_2^* can be calculated, and by entering the isentropic table with this value M_2 can be determined. A word of caution here! The value of A_2/A_2^* will be found in <u>two</u> places in the table, as we are really solving equation 5.36, or for the more general case equation 5.27, which is a quadratic for M_2. One value will be in the subsonic region and the other in the supersonic regime. You should have no difficulty determining which answer is correct when you consider the physical appearance of the system together with the concepts developed in Section 5.3.

Note that the general problem <u>with losses</u> can also be solved by the same technique as long as information is available concerning the loss. This could be given to us in the form of A_1^*/A_2^*, p_{t2}/p_{t1}, or possibly as Δs_{1-2}. All three of these represent equivalent ways of expressing the loss (through equations 4.28 and 5.29).

We now realize that the key to simplified problem solution is to have available a table of property ratios as a function of γ and <u>one</u> Mach number only. These are obtained by taking the equations developed in Section 5.4 and introducing a reference state, either the $*$ reference condition (reached by an isentropic process) or the stagnation reference condition (reached by an isentropic process). We proceed with equation 5.23.

$$\frac{T_2}{T_1} = \frac{1 + \frac{(\gamma-1)}{2} M_1^{\ 2}}{1 + \frac{(\gamma-1)}{2} M_2^{\ 2}} \qquad (5.23)$$

Let point 2 be any arbitrary point in the system and let its stagnation point be point 1.

Then $\qquad T_2 \Rightarrow T \qquad\qquad M_2 \Rightarrow M$ (any value)

$\qquad\qquad T_1 \Rightarrow T_t \qquad\qquad M_1 \Rightarrow 0$

and equation 5.23 becomes

$$\frac{T}{T_t} = \frac{1}{1 + \frac{(\gamma-1)}{2} M^2} = f(M, \gamma) \qquad (5.39)$$

Equation 5.25 can be treated in a similar fashion. In this case we let 1 be the arbitrary point and its stagnation point is taken as 2.

Then $\qquad p_1 \Rightarrow p \qquad\qquad M_1 \Rightarrow M$ (any value)

$\qquad\qquad p_2 \Rightarrow p_t \qquad\qquad M_2 \Rightarrow 0$

and when we remember that the stagnation process is isentropic, equation 5.25 becomes:

$$\frac{p}{p_t} = \left[\frac{1}{1 + \frac{(\gamma-1)}{2} M^2}\right]^{\frac{\gamma}{\gamma-1}} = f(M, \gamma) \qquad (5.40)$$

Equations 5.39 and 5.40 are not surprising as we have previously

developed these by other methods (see equations 4.18 and 4.21).
The tabulation of equation 5.40 may be used to solve problems in
the same manner as the area ratio. For example, assume we are

Given: γ, p_1, p_2, M_2, and Δs_{1-2}

Find: M_1

To solve this problem we seek the ratio p_1/p_{t1} in terms of
known ratios.

$$\frac{p_1}{p_{t1}} = \frac{p_1}{p_2}\frac{p_2}{p_{t2}}\frac{p_{t2}}{p_{t1}} \tag{5.41}$$

Given ⟶ ⎰Evaluated by equation 4.28
         ⎱as a function of Δs_{1-2}

⎰A function of M_2
⎱Look up in isentropic tables

After calculating the value of p_1/p_{t1} we enter the isentropic
tables and find M_1. Note that even though the flow from station 1 to 2 is not isentropic, the functions for p_1/p_{t1}
and p_2/p_{t2} are isentropic by definition; thus, the isentropic
tables can be used to solve this problem. The connection between
the two points is made through p_{t2}/p_{t1} which involves the entropy change.

We could continue to develop other isentropic relations as
functions of Mach number and γ. Apply the previous techniques
to equation 5.28 and show that:

$$\frac{\rho}{\rho_t} = \left[\frac{1}{1 + \frac{(\gamma-1)}{2}M^2}\right]^{\frac{1}{\gamma-1}} \tag{5.42}$$

Another interesting relationship is the product of equations 5.37
and 5.40.

$$\frac{A}{A^*}\frac{p}{p_t} = f(M, \gamma) \tag{5.43}$$

Determine what unique function of M and γ is represented in
equation 5.43. Since this is a $f(M, \gamma)$ it is normally also
listed in the isentropic tables and we shall see later that it
provides the only direct means of solving certain types of problems.

Values of these isentropic flow parameters have been calculated from equations 5.37, 5.39, 5.40, etc., and tabulated in the

Appendix. To convince yourself that there is nothing magical
about these tables you might want to check some of the numbers
found in them opposite a particular Mach number. In fact, as an
exercise in programming a digital computer you could work up your
own set of tables for values of γ other than 1.4, which is the
only one included in the Appendix.

As you read the following examples look up the numbers in
the isentropic table to convince yourself that you know how to
find them.

Example 5.2 You are now in a position to rework Example 5.1 with
a minimum of calculation. Recall that $M_1 = 1.5$ and $M_2 = 2.8$.

$$\frac{A_2}{A_1} = \frac{A_2}{A_2^*} \frac{A_2^*}{A_1^*} \frac{A_1^*}{A_1} = (3.5001)(1) \frac{1}{1.1762} = \underline{2.98}$$

The following picture and information are common to the next
three examples. We are given the steady, one-dimensional flow of
air $(\gamma = 1.4)$ which can be treated as a perfect gas. Assume
$Q = W_s = 0$ and negligible potential changes. $A_1 = 2.0 \text{ ft}^2$ and
$A_2 = 5.0 \text{ ft}^2$.

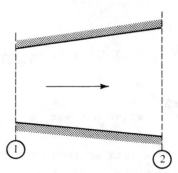

Example 5.3 Given that $M_1 = 1.0$ and $\Delta s_{1-2} = 0$
         Find the possible values of M_2

To determine conditions at section 2 we establish the ratio:

$$\frac{A_2}{A_2^*} = \frac{A_2}{A_1} \frac{A_1}{A_1^*} \frac{A_1^*}{A_2^*} = \frac{5}{2} (1.000)(1) = 2.5$$

Equal — since isentropic
From isentropic tables @ M = 1.0
From given physical configuration

Look up $A/A^* = 2.5$ in the isentropic tables and determine that $M_2 = 0.24$ or 2.44 .

Example 5.4   Given that $M_1 = 0.5$, $p_1 = 4$ bars and $\Delta s_{1-2} = 0$
Find M_2 and p_2

$$\frac{A_2}{A_2^*} = \frac{A_2}{A_1} \frac{A_1}{A_1^*} \frac{A_1^*}{A_2^*} = \frac{5}{2} (1.3398)(1) = 3.35$$

Thus $M_2 \approx \underline{0.175}$ (Why isn't it 2.75?)

$$p_2 = \frac{p_2}{p_{t2}} \frac{p_{t2}}{p_{t1}} \frac{p_{t1}}{p_1} p_1 = (0.9788)(1) \frac{1}{0.8430} (4) = \underline{4.64} \text{ bars}$$

Example 5.5   Given $M_1 = 1.5$, $T_1 = 70^\circ F$ and $\Delta s_{1-2} = 0$
Find M_2 and T_2

Find $\frac{A_2}{A_2^*} = ?$ (Thus, $M_2 \approx 2.62$)

Once M_2 is known we can find T_2 .

$$T_2 = \frac{T_2}{T_{t2}} \frac{T_{t2}}{T_{t1}} \frac{T_{t1}}{T_1} T_1 = (0.4214)(1) \frac{1}{(0.6897)} 530 = \underline{324}^\circ R$$

Why is $T_{t1} = T_{t2}$?
(Write an energy equation between 1 and 2).

Example 5.6   Oxygen flows into an insulated device with the following initial conditions: $p_1 = 20$ psia, $T_1 = 600^\circ R$, and
$V_1 = 2,960$ ft/sec . After a short distance the area has con-
verged from 6 ft^2 to 2.5 ft^2 . You may assume steady, one-
dimensional flow and a perfect gas. (See table in the Appendix
for gas properties.)

(a)   Find M_1, p_{t1}, T_{t1} and h_{t1} .
(b)   If there are losses such that $\Delta s_{1-2} = 0.005$ Btu/lbm-$^\circ R$
find M_2, p_2, and T_2.

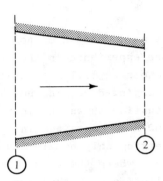

(a) First, we determine conditions at station 1.

$$a_1 = (\gamma g_c R T_1)^{\frac{1}{2}} = \left[(1.4)(32.2)(48.3)(600)\right]^{\frac{1}{2}} = 1143 \text{ ft/sec}$$

$$M_1 = V_1/a_1 = 2960/1143 = \underline{2.59}$$

$$p_{t1} = \frac{p_{t1}}{p_1} \, p_1 = \frac{1}{0.0509} \, (20) = \underline{393} \text{ psia}$$

$$T_{t1} = \frac{T_{t1}}{T_1} \, T_1 = \frac{1}{0.4271} \, (600) = \underline{1405}°R$$

$$h_{t1} = c_p T_{t1} = (0.218)(1405) = \underline{306} \text{ Btu/lbm}$$

(b) For a perfect gas with $q = w_s = 0$, $T_{t1} = T_{t2}$ (from an energy equation), and also from equation 5.29

$$\frac{A_1^{*}}{A_2^{*}} = e^{-\Delta s/R} = e^{-\frac{(0.005)(778)}{48.3}} = 0.9226$$

Thus, $\dfrac{A_2}{A_2^{*}} = \dfrac{A_2}{A_1} \dfrac{A_1}{A_1^{*}} \dfrac{A_1^{*}}{A_2^{*}} = \dfrac{2.5}{6} \, (2.8688)(0.9226) = 1.1028$

From the isentropic tables we find that $M_2 \approx$ _____ .
Why is the use of the isentropic tables legitimate here when
there are losses in the flow?

Continue and compute p_2 and T_2

$$p_2 =$$ $(p_2 \approx 117 \text{ psia})$

$$T_2 =$$ $(T_2 \approx 1017°R)$

Could you find the velocity at section 2?

5.7 NOZZLE OPERATION

We will now start a discussion of nozzle operation and at the same time gain more experience in the use of the isentropic tables. Two types of nozzles will be considered, a converging-only nozzle and a converging-diverging nozzle. We start by examining the physical situation shown in Figure 5.6. A source of air at 100 psia and 600°R is contained in a large tank. Connected to the tank is a converging-only nozzle and it exhausts into an extremely large receiver where the pressure can be regulated. We can neglect frictional effects as they are very small in a converging section.

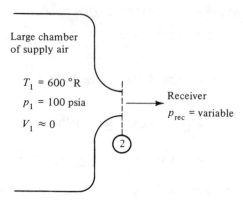

Large chamber
of supply air

$T_1 = 600\,°R$

$p_1 = 100$ psia

$V_1 \approx 0$

Receiver

$p_{rec} =$ variable

Figure 5.6 Converging nozzle

If the receiver pressure is set at 100 psia no flow results. Once the receiver pressure is lowered below 100 psia air will flow from the supply tank. Since the supply tank has a large cross section relative to the nozzle outlet area the velocities in the tank may be neglected. Thus, $T_1 \approx T_{t1}$ and $p_1 \approx p_{t1}$. There is no shaft work and we assume no heat transfer. We identify section 2 as the nozzle outlet.

ENERGY
$$h_{t1} + \cancel{q} = h_{t2} + \cancel{w}_s \qquad (3.19)$$

$$h_{t1} = h_{t2}$$

and, since we can treat this as a perfect gas,

$$T_{t1} = T_{t2}$$

It is important to recognize that the receiver pressure is controlling the flow. The velocity will increase and the pressure will decrease as we progress through the nozzle until the pressure at the nozzle outlet equals that of the receiver. This will always be true as long as the nozzle outlet can "sense" the receiver pressure. Can you think of a situation where pressure pulses from the receiver could not be "felt" inside the nozzle? (Recall Section 4.4.)

Let us assume that $p_{rec} = 80.2$ psia.

Then $$p_2 = p_{rec} = 80.2 \text{ psia}$$

and $$\frac{p_2}{p_{t2}} = \frac{p_2}{p_{t1}} \frac{p_{t1}}{p_{t2}} = \frac{80.2}{100}(1) = 0.802$$

Note that $p_{t1} = p_{t2}$ by equation 4.28 since we are neglecting friction.

From the isentropic tables corresponding to $p/p_t = 0.802$, we see

that $M_2 = 0.57$ and $T_2/T_{t2} = 0.939$

Thus, $$T_2 = 0.939(T_{t2}) = 0.939(600) = 563^{\circ}R$$

$$a_2^2 = (1.4)(32.2)(53.3)(563)$$

$$a_2 = 1163 \text{ ft/sec}$$

and $$V_2 = M_2 a_2 = 0.57(1163) = \underline{663} \text{ ft/sec}$$

Figure 5.7 shows this process on a T-s diagram as an isentropic expansion. If the pressure in the receiver were lowered further, the air would expand to this lower pressure and the Mach number and velocity would increase.

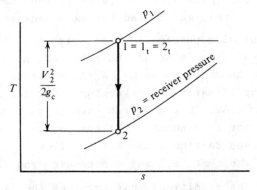

Figure 5.7 $T-s$ diagram for converging nozzle

Assume that the receiver pressure is lowered to 52.83 psia.

Show that $p_2/p_{t2} = 0.5283$

and thus $M_2 = 1.00$ with $V_2 = 1096$ ft/sec

Notice that the air velocity coming out of the nozzle is exactly
sonic. If we now drop the receiver pressure below this "critical
pressure" (52.83 psia) the nozzle has no way of adjusting to
these conditions. Why not? Assume that the nozzle outlet pressure could continue to drop along with the receiver. This would
mean $p_2/p_{t2} < 0.5283$ which corresponds to a supersonic velocity.
We know that if the flow is to go supersonic the area must reach
a minimum and then increase (see Section 5.3). Thus, for a converging-only nozzle, the flow is governed by the receiver pressure until sonic velocity is reached at the nozzle outlet and
further reduction of the receiver pressure will have no effect
on the flow conditions inside the nozzle. Under these conditions
the nozzle is said to be "choked" and the nozzle outlet pressure
remains at the "critical pressure." Expansion to the receiver
pressure takes place outside the nozzle.

In reviewing this example you should realize that there is
nothing magical about a receiver pressure of 52.83 psia. The
significant item is the ratio of the static to total pressure at
the exit plane, which for the case of no losses is the ratio of
the receiver pressure to the inlet pressure. With sonic velocity
at the exit this ratio is 0.5283.

Now let us examine a similar situation; only we shall deal
with a converging-diverging nozzle (sometimes called a DeLaval
nozzle) as shown in Figure 5.8. We identify the "throat" (or
section of minimum area) as 2 and the exit section as 3.

The distinguishing physical characteristic of this type of
nozzle is the area ratio, meaning the ratio of the exit area to
the throat area. Assume this to be $A_3/A_2 = 2.494$. Keep in mind
that the objective of making a converging-diverging nozzle is to
obtain supersonic flow. Let us first examine the "design operating condition" for this nozzle. If the nozzle is to operate as
desired we know (see Section 5.3) that the flow will be subsonic
from 1 to 2, sonic at 2, and supersonic from 2 to 3.

To discover the conditions that exist at the exit (under
design operation) we seek the ratio A_3/A_3^* .

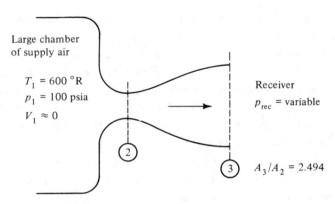

Figure 5.8 Converging–diverging nozzle

$$\frac{A_3}{A_3{}^*} = \frac{A_3}{A_2} \frac{A_2}{A_2{}^*} \frac{A_2{}^*}{A_3{}^*} = (2.494)(1)(1) = 2.494$$

Note that $A_2 = A_2{}^*$ since $M_2 = 1$, and $A_2{}^* = A_3{}^*$ by equation 5.29 as we are still assuming isentropic operation. We look for $A/A^* = 2.494$ in the <u>supersonic</u> section of the isentropic tables and see that

$$M_3 = 2.44, \quad p_3/p_{t3} = 0.0643 \quad \text{and} \quad T_3/T_{t3} = 0.4565$$

Thus, $\qquad p_3 = \frac{p_3}{p_{t3}} \frac{p_{t3}}{p_{t1}} p_{t1} = 0.0643(1)100 = \underline{6.43}$ psia

and to operate the nozzle at this "design condition" the receiver pressure <u>must be</u> at 6.43 psia. The pressure variation through the nozzle for this case is shown as curve "a" in Figure 5.9. This mode is sometimes referred to as "third critical." From the temperature ratio T_3/T_{t3} we can easily compute T_3, a_3 and then V_3 by the procedure shown previously.

One can also find $A/A^* = 2.494$ in the subsonic section of the isentropic tables. (Recall that these two answers come from the solution of a quadratic equation.) For this case

$$M_3 = 0.24, \quad p_3/p_{t3} = 0.9607 \quad \text{and} \quad T_3/T_{t3} = 0.9886$$

Thus, $\qquad p_3 = \frac{p_3}{p_{t3}} \frac{p_{t3}}{p_{t1}} p_{t1} = 0.9607(1)100 = \underline{96.07}$ psia

and to operate at this condition the receiver pressure <u>must be</u> at

96.07 psia. With this receiver pressure the flow is subsonic
from 1 to 2, sonic at 2, and <u>subsonic</u> again from 2 to 3.
The device is nowhere near its design condition and is really
operating as a "venture tube;" i.e., the converging section is
operating as a nozzle and the diverging section is operating as a
diffuser. The pressure variation through the nozzle for this
case is shown as curve "b" in Figure 5.9. This mode of operation
is frequently called "first critical."

Note that at both first and third critical the flow variations are identical from the inlet to the throat. Once the receiver pressure has been lowered to 96.07 psia Mach 1.0 exists
in the throat and the device is said to be "choked." <u>Any further</u>
<u>lowering of the receiver pressure will not change the flow rate</u>.

Again, realize that it is not the pressure in the receiver
by itself but rather the receiver pressure <u>relative</u> to the inlet
pressure that determines the mode of operation.

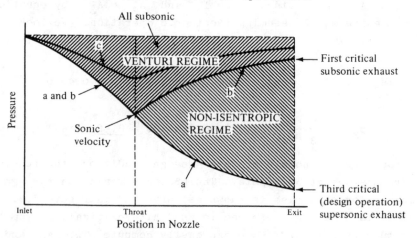

Figure 5.9 Pressure variation through converging–diverging nozzle

Example 5.7 A converging-diverging nozzle with an area ratio of
3.0 exhausts into a receiver where the pressure is 1 bar. The
nozzle is supplied by air at $22^\circ C$ from a large chamber. At what
pressure should the air in the chamber be in order for the nozzle to operate at its design condition (third critical)? What
will the outlet velocity be?

With reference to Figure 5.8, $A_3/A_2 = 3.0$

$$\frac{A_3}{A_3^*} = \frac{A_3}{A_2}\frac{A_2}{A_2^*}\frac{A_2^*}{A_3^*} = (3.0)(1)(1) = 3.0$$

From the isentropic tables:

$$M_3 = 2.64, \quad p_3/p_{t3} = 0.0471, \quad T_3/T_{t3} = 0.4177$$

$$p_1 = p_{t1} = \frac{p_{t1}}{p_{t3}} \frac{p_{t3}}{p_3} p_3 = (1) \frac{1}{0.0471} (1 \times 10^5) = \underline{21.2 \times 10^5} \text{ N/m}^2$$

$$T_3 = \frac{T_3}{T_{t3}} \frac{T_{t3}}{T_{t1}} T_{t1} = 0.4177(1)(22 + 273) = 123.2^\circ\text{K}$$

$$V_3 = M_3 a_3 = 2.64\left[(1.4)(1)(287)(123.2)\right]^{\frac{1}{2}} = \underline{587} \text{ m/s}$$

We have discussed only two specific operating conditions and one might ask what happens at other receiver pressures. We can state that the first and third criticals represent the only operating conditions that satisfy the following criteria:

(a) Mach one in the throat
(b) Isentropic flow throughout the nozzle
(c) Nozzle exit pressure equal to receiver pressure.

With receiver pressures above the first critical the "nozzle" operates as a venturi and we never reach sonic velocity in the throat. An example of this mode of operation is shown as curve "c" in Figure 5.9. The nozzle is no longer choked and the flow rate is less than the maximum. Conditions at the exit can be determined by the procedure previously shown for the converging-only nozzle. Then properties in the throat can be found if desired.

Operation between first and third critical is <u>not</u> isentropic. We shall learn later that under these conditions shocks will occur in either the diverging portion of the nozzle or after the exit. If the receiver pressure is below third critical the nozzle operates <u>internally</u> as though it were at the design condition but expansion waves occur <u>outside</u> the nozzle. These latter operating modes will be discussed in detail as soon as the appropriate background has been developed.

5.8 NOZZLE PERFORMANCE

We have seen that the isentropic operating conditions are very easy to determine. Friction losses can then be taken into account by one of several methods. Direct information on the entropy change could be given although this is usually not avail-

able. Sometimes equivalent information is provided in the form
of the stagnation pressure ratio. Normally, however, nozzle performance is indicated by an efficiency parameter which is defined
as follows:

$$\eta_n \equiv \frac{\text{Actual change in kinetic energy}}{\text{Ideal change in kinetic energy}}$$

or
$$\eta_n \equiv \frac{\Delta KE_{actual}}{\Delta KE_{ideal}} \qquad (5.44)$$

Since most nozzles involve negligible heat transfer (per unit
mass of fluid flowing) we have from

$$h_{t1} + \cancel{q} = h_{t2} + \cancel{w}_s \qquad (3.19)$$

$$h_{t1} = h_{t2} \qquad (5.45)$$

Thus,
$$h_1 + \frac{V_1^2}{2g_c} = h_2 + \frac{V_2^2}{2g_c} \qquad (5.46)$$

or
$$h_1 - h_2 = \frac{V_2^2 - V_1^2}{2g_c} \qquad (5.47)$$

Therefore, one normally sees the nozzle efficiency expressed as

$$\eta_n = \frac{\Delta h_{actual}}{\Delta h_{ideal}} \qquad (5.48)$$

With reference to Figure 5.10 this becomes

$$\eta_n = \frac{h_1 - h_2}{h_1 - h_{2s}} \qquad (5.49)$$

Since nozzle outlet velocities are quite large (relative to
the velocity at the inlet) one can normally neglect the inlet
velocity with little error. This is the case shown in Figure
5.10. Also note that the ideal process is assumed to take place
down to the actual available receiver pressure. This definition
of nozzle efficiency and its application appear quite reasonable
since a nozzle is subjected to fixed (inlet and outlet) operating
pressures and its purpose is to produce kinetic energy. The question is how well does it do this, and η_n not only answers the

question very quickly but permits a rapid determination of the actual outlet state.

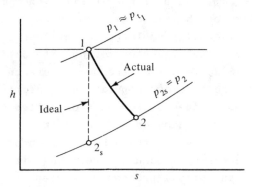

Figure 5.10 *h–s* diagram for a nozzle with losses

Example 5.8   Air at $800^\circ R$ and 80 psia feeds a converging-only nozzle having an efficiency of 96%.   The receiver pressure is 50 psia.   What is the actual nozzle outlet temperature?

Note that since p_{rec}/p_{inlet} = 50/80 = 0.625 > 0.528 , the nozzle will not be choked, flow will be subsonic at the exit and $p_2 = p_{rec}$ (see Figure 5.10).

$$\frac{p_{2s}}{p_{t2s}} = \frac{p_{2s}}{p_{t1}}\frac{p_{t1}}{p_{t2s}} = \frac{50}{80}(1) = 0.625$$

From tables; $M_{2s} \approx 0.85$ and $T_{2s}/T_{t2s} = 0.8737$

$$T_{2s} = \frac{T_{2s}}{T_{t2s}}\frac{T_{t2s}}{T_{t1}}T_{t1} = 0.8737(1)800 = 699^\circ R$$

$$\eta_n = \frac{T_1 - T_2}{T_1 - T_{2s}} ; \qquad 0.96 = \frac{800 - T_2}{800 - 699}$$

$$T_2 = \underline{703^\circ R}$$

Can you find the actual outlet velocity?

Another method of expressing nozzle performance is with a "velocity coefficient" which is defined as:

$$C_v \equiv \frac{\text{Actual Outlet Velocity}}{\text{Ideal Outlet Velocity}} \qquad (5.50)$$

Sometimes a "discharge coefficient" is used and is defined as:

$$C_d \equiv \frac{\text{Actual Mass Flow Rate}}{\text{Ideal Mass Flow Rate}} \qquad (5.51)$$

5.9 DIFFUSER PERFORMANCE

Although the common use of nozzle efficiency makes this parameter well understood by all engineers there is no single parameter that is universally employed for diffusers. Nearly a
dozen criteria have been suggested to indicate diffuser performance. (See page 392, Vol. 1 of Zucrow; reference no. 24.) Two
or three of these are the most popular, but unfortunately, even
these are sometimes defined differently or called by different
names. The following discussion refers to the h-s diagram
shown in Figure 5.11. Recall that the function of a diffuser is
to convert ·kinetic energy into pressure; thus, it is logical to
compare the ideal and actual processes between the same two enthalpy levels which represent the same kinetic energy change.

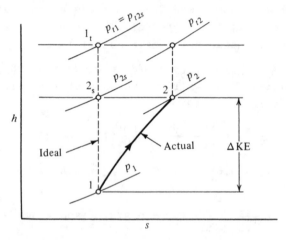

Figure 5.11 *h—s* diagram for a diffuser with losses

A suitable definition of diffuser efficiency is

$$\eta_d \equiv \frac{\text{Actual Pressure Rise}}{\text{Ideal Pressure Rise}} \qquad (5.52)$$

or

$$\eta_d \equiv \frac{p_2 - p_1}{p_{2s} - p_1} \qquad (5.53)$$

Another criterion is the stagnation pressure ratio which is
sometimes called the "total-pressure recovery factor" and designated as

$$\eta_r \equiv \frac{p_{t2}}{p_{t1}} \qquad (5.54)$$

This relation is directly related to the area ratio A_1^*/A_2^* or the entropy change Δs_{1-2} which we have previously shown to be equivalent loss indicators.

You are again warned to be extremely cautious in accepting any performance figure for a diffuser without also obtaining a precise definition of what is meant by the criterion.

Example 5.9 A steady flow of air at $650^\circ R$ and 30 psia enters a diffuser with a Mach number of 0.8. The total-pressure recovery factor $\eta_r = 0.95$. Determine the static pressure and temperature at the exit if M = 0.15 at that section.

With reference to Figure 5.11:

$$p_2 = \frac{p_2}{p_{t2}} \frac{p_{t2}}{p_{t1}} \frac{p_{t1}}{p_1} p_1 = 0.9844(0.95) \frac{1}{0.6560} (30) = \underline{42.8} \text{ psia}$$

$$T_2 = \frac{T_2}{T_{t2}} \frac{T_{t2}}{T_{t1}} \frac{T_{t1}}{T_1} T_1 = 0.9955(1) \frac{1}{0.8865} (650) = \underline{730}^\circ R$$

5.10 SUMMARY

We analyzed a general varying-area configuration and found that properties vary in a radically different manner depending upon whether the flow is subsonic or supersonic. The case of a perfect gas enabled the development of simple working equations for flow analysis. We then introduced the concept of a * reference state. The combination of the * and the stagnation reference states led to the development of isentropic tables which greatly aid problem solution. Deviations from isentropic flow can be handled by appropriate loss factors or efficiency criteria.

A large number of useful equations were developed; however, most of these are of the type that need not be memorized. Equations 5.10, 5.11, and 5.13 were used for the general analysis of varying-area flow and these are summarized in the middle of Section 5.3. The working equations that apply to a perfect gas are summarized at the end of Section 5.4 and are 4.28, 5.21, 5.23, 5.25, 5.27, and 5.28. Equations used as a basis for tables are numbered 5.37, 5.39, 5.40, 5.42, and 5.43 and are located in Section 5.6.

Those equations that are most frequently used are summarized below. You should be familiar with the conditions under which

each may be used. Go back and review the equations listed in
previous summaries, particularly those in Unit 4.

(a)   For Steady, One-dimensional Flow of a perfect gas when
      $Q = W = 0$

$$\frac{p_{t2}}{p_{t1}} = e^{-\Delta s/R} \qquad\qquad (4.28)$$

$$\frac{A_2^{\,*}}{A_1^{\,*}} = e^{\Delta s/R} \qquad\qquad (5.29)$$

$$p_{t1}A_1^{\,*} = p_{t2}A_2^{\,*} \qquad\qquad (5.35)$$

(b)   Nozzle Efficiency (between same pressures)

$$\eta_n \equiv \frac{\Delta KE_{actual}}{\Delta KE_{ideal}} = \frac{h_1 - h_2}{h_1 - h_{2s}} \qquad (5.44) \,\&\, (5.49)$$

(c)   Diffuser Efficiency (between same enthalpies)

$$\eta_d \equiv \frac{\text{Actual Pressure Rise}}{\text{Ideal Pressure Rise}} = \frac{p_2 - p_1}{p_{2s} - p_1} \qquad (5.52) \,\&\, (5.53)$$

5.11 PROBLEMS

1. The following information is common to each of parts (a) and
    (b). Nitrogen flows through a diverging section with
    $A_1 = 1.5$ ft^2 and $A_2 = 4.5$ ft^2 . You may assume steady,
    one-dimensional flow, $Q = W_s = 0$, negligible potential
    changes and no losses.
    (a) If $M_1 = 0.7$ and $p_1 = 70$ psia find M_2 and p_2 .
    (b) If $M_1 = 1.7$ and $T_1 = 95°F$ find M_2 and T_2 .

2. Air enters a converging section where $A_1 = 0.50$ m^2 . At a
    downstream section $A_2 = 0.25$ m^2 , $M_2 = 1.0$ and $\Delta s_{1-2} = 0$.
    It is known that $p_2 > p_1$. Find the initial Mach number
    (M_1) and the temperature ratio (T_2/T_1).

3. Oxygen flows into an insulated device with initial conditions
    as follows: $p_1 = 30$ psia , $T_1 = 750°R$ and $V_1 = 639$ ft/sec.
    The area changes from $A_1 = 6$ ft^2 to $A_2 = 5$ ft^2 .
    (a) Compute M_1 , p_{t1} , and T_{t1} .
    (b) Is this device a nozzle or diffuser?

(c) Determine M_2 , p_2 and T_2 if there are no losses.

4. Air flows with T_1 = 250°K , p_1 = 3 bars absolute , p_{t1} = 3.4 bars absolute and the cross-section area A_1 = 0.40 m^2 . The flow is isentropic to a point where A_2 = 0.30 m^2 . Determine the temperature at section 2.

5. The following information is known about the steady flow of air through an adiabatic system:
   At section 1; T_1 = 556°R , p_1 = 28.0 psia
   At section 2; T_2 = 70°F , T_{t2} = 109°F , p_2 = 18 psia
   (a) Find M_2 , V_2 and P_{t2} .
   (b) Determine M_1 , V_1 and P_{t1} .
   (c) Compute the area ratio A_2/A_1 .
   (d) Sketch a physical diagram of the system along with a T-s diagram.

6. Assuming the flow of a perfect gas in an adiabatic, no-work system, show that sonic velocity corresponding to the stagnation conditions (a_t) is related to sonic velocity where the Mach number is unity (a^*) by the following equation:

$$\frac{a^*}{a_t} = \left[\frac{2}{\gamma+1}\right]^{\frac{1}{2}}$$

7. Carbon monoxide flows through an adiabatic system. M_1 = 4.0 and p_{t1} = 45 psia. At a point downstream M_2 = 1.8 and p_2 = 7.0 psia.
   (a) Are there losses in this system? If so, compute Δs.
   (b) Determine the ratio of A_2/A_1 .

8. Two venturi meters are installed in a 30-cm-diameter duct that is insulated. The conditions are such that sonic flow exists at each throat; i.e., M_1 = M_4 = 1.0 . Although each venturi is isentropic, the connecting duct has friction and hence losses exist between sections 2 and 3. p_1 = 3 bars absolute and p_4 = 2.5 bars abs. If the diameter at section 1 is 15 cm and the fluid is air:
   (a) Compute Δs for the connecting duct.
   (b) Find the diameter at section 4.

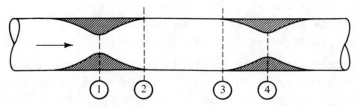

9. A smooth 3-inch-diameter hole is punched into the side of a
    large chamber where oxygen is stored at 500^OR and 150 psia.
    Assume frictionless flow.
    (a) Compute the initial mass flow rate from the chamber if
        the surrounding pressure is 15.0 psia.
    (b) What is the flow rate if the pressure of the surroundings
        is lowered to zero?
    (c) What is the flow rate if the chamber pressure is raised
        to 300 psia?

10. Nitrogen is stored in a large chamber under conditions of
    450^OK and 1.5×10^5 N/m^2. The gas leaves the chamber through
    a converging-only nozzle whose outlet area is 30 cm^2. The
    ambient room pressure is 1×10^5 N/m^2 and there are no losses.
    (a) What is the velocity of the nitrogen at the nozzle exit?
    (b) What is the mass flow rate?
    (c) What is the maximum flow rate that could be obtained by
        lowering the ambient pressure?

11. A converging-only nozzle has an efficiency of 96%. Air
    enters with negligible velocity at a pressure of 150 psia and
    a temperature of 750^OR. The receiver pressure is 100 psia.
    What are the actual outlet temperature, Mach number and veloc-
    ity?

12. A large chamber contains air at 80 psia and 600^OR. The air
    enters a converging-diverging nozzle which has an area ratio
    (exit to throat) of 3.0.
    (a) What pressure must exist in the receiver for the nozzle
        to operate at its first critical point?
    (b) What should the receiver pressure be for third critical
        (design point) operation?
    (c) If operating at third critical what are the density and
        velocity of the air at the nozzle exit plane?

13. Air enters a convergent-divergent nozzle at 20 bars absolute
    and 40^OC. At the end of the nozzle the pressure is 2.0 bars
    abs. Assume a frictionless adiabatic process. The throat
    area is 20 cm^2.
    (a) What is the area at the nozzle exit?
    (b) What is the mass flow rate in kg/sec?

14. A converging-diverging nozzle is designed to operate with an
    exit Mach number of M = 2.25 . It is fed by a large chamber

of oxygen at 15.0 psia and 600°R and exhausts into the room
at 14.7 psia. Assuming the losses to be negligible, compute
the velocity in the nozzle throat.

15. A converging-diverging nozzle discharges air into a receiver
where the static pressure is 15 psia. A one-square-foot duct
feeds the nozzle with air at 100 psia, 800°R and a velocity
such that the Mach number $M_1 = 0.3$. The exit area is such
that the pressure at the nozzle exit exactly matches the
receiver pressure. Assume steady, one-dimensional flow, perfect gas, etc. The nozzle is adiabatic and there are no
losses.
(a) Calculate the flow rate.
(b) Determine the throat area.
(c) Calculate the exit area.

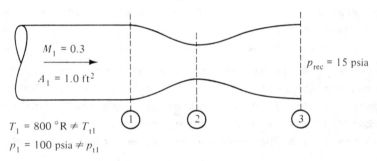

$T_1 = 800 \,°R \neq T_{t1}$
$p_1 = 100 \text{ psia} \neq p_{t1}$

16. Ten kilograms per second of air are flowing in an adiabatic
system. At one section the pressure is 2.0×10^5 N/m^2 , the
temperature is 650°C and the area is 50 cm^2 . At a downstream section $M_2 = 1.2$.
(a) Sketch the general shape of the system.
(b) Find A_2 if the flow is frictionless.
(c) Find A_2 if there is an entropy change between these two
sections of 42 J/kg-°K .

17. Carbon monoxide is expanded adiabatically from 100 psia,
540°F and negligible velocity through a converging-diverging
nozzle to a pressure of 20 psia.
(a) What is the ideal exit Mach number?
(b) If the actual exit Mach number is found to be $M = 1.6$,
what is the nozzle efficiency?
(c) What is the entropy change for the flow?
(d) Draw a T-s diagram showing the ideal and actual processes. Indicate pertinent temperatures, pressures, etc.

18. Air enters a converging-diverging nozzle with $T_1 = 22^\circ C$,
     $p_1 = 10$ bars absolute and $V_1 \approx 0$. The exit Mach number
     is 2.0, exit area is 0.25 m^2 and the nozzle efficiency is
     0.95.
     (a) What are the actual exit T , p and p_t ?
     (b) What is the ideal exit Mach number?
     (c) Assume that all the losses occur in the diverging portion
         of the nozzle and compute the throat area.
     (d) What is the mass flow rate?

19. A diffuser receives air at 500°R, 18 psia and a velocity of
     750 ft/sec. The diffuser has an efficiency of 90% (as de-
     fined by equation 5.53) and discharges the air with a veloci-
     ty of 150 ft/sec.
     (a) What is the pressure of the discharge air?
     (b) What is the total-pressure recovery factor as given by
         equation 5.54?
     (c) Determine the area ratio of the diffuser.

20. Consider the steady, one-dimensional flow of a perfect gas
     through a horizontal system with no shaft work. No frictional
     losses are involved but area changes and heat transfer effects
     provide a flow at constant temperature.
     (a) Start with the pressure energy equation and develop

$$\frac{p_2}{p_1} = e^{\frac{\gamma}{2}(M_1^2 - M_2^2)}$$

$$\frac{p_{t2}}{p_{t1}} = e^{\frac{\gamma}{2}(M_1^2 - M_2^2)} \left[\frac{1 + \frac{\gamma-1}{2} M_2^2}{1 + \frac{\gamma-1}{2} M_1^2} \right]^{\frac{\gamma}{\gamma-1}}$$

     (b) From the continuity equation show that

$$\frac{A_1}{A_2} = \frac{M_2}{M_1} e^{\frac{\gamma}{2}(M_1^2 - M_2^2)}$$

     (c) By letting M_1 be any Mach number and $M_2 = 1.0$, write
         the expression for A/A^* . Show that the section of
         minimum area occurs at $M = 1/\sqrt{\gamma}$.

21. Consider the steady, one-dimensional flow of a perfect gas
     through a horizontal system with no heat transfer or shaft
     work. Friction effects are present but area changes cause

the flow to be at constant Mach number.

(a) Recall the arguments of Section 4.6 and determine what other properties remain constant in this flow.

(b) Apply the concepts of continuity and momentum (equation 3.63) to show that

$$D_2 - D_1 = \frac{fM^2 \gamma}{4} (x_2 - x_1)$$

You may assume a circular duct and constant friction factor.

22. Write a computer program and construct a table of isentropic flow parameters for $\gamma \neq 1.4$. (Useful values might be $\gamma = 1.2$, 1.3 or 1.67 .) Use the following headings: M , p/p_t , T/T_t , ρ/ρ_t , A/A^* , $pA/p_t A^*$.

5.12 CHECK TEST

You should be able to complete this test without reference to material in the unit.

1. Define the $*$ reference condition.

2. In adiabatic, no-work flow, the losses can be expressed by three different parameters. List these parameters and show how they are related to one another.

3. In the T-s diagram point 1 represents a stagnation condition. Proceeding isentropically from 1 , the flow reaches a Mach number of unity at 1^* . Point 2 represents another stagnation condition in the same flow system. Assuming that the fluid is a perfect gas, locate the corresponding isentropic 2^* and prove that T_2^* is either greater than, equal to, or less than T_1^* .

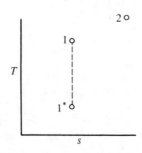

4. A supersonic nozzle is fed by a large chamber and produces
    Mach 3.0 at the exit. Sketch the following curves which show
    how properties vary through the nozzle as the Mach number in-
    creases from zero to 3.0. (No particular scale.)

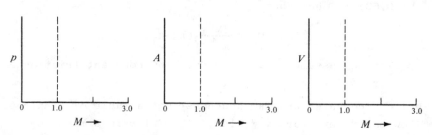

5. Give a suitable definition for nozzle efficiency in terms of
    enthalpies. Sketch an h–s diagram to identify your state
    points.

6. Air flows steadily with no losses through a converging-
    diverging nozzle with an area ratio of 1.50 . Conditions in
    the supply chamber are $T = 500^{\circ}R$ and $p = 150$ psia.
    (a) To what pressure must the receiver be lowered in order to
        choke the flow?
    (b) If the nozzle is choked, determine the density and veloci-
        ty at the throat.
    (c) If the receiver is at the pressure determined in part (a)
        and the diverging portion of the nozzle is removed, what
        will the exit Mach number be?

7. For steady, one-dimensional flow of a perfect gas in an adia-
    batic, no-work system, derive the working relation between
    the temperatures at two locations.

$$T_2/T_1 = f(M_1, M_2, \gamma)$$

8. Work problem #19 in Section 5.11.

UNIT 6

STANDING
NORMAL SHOCKS

6.1 INTRODUCTION

Up to this point we have considered only continuous flows; i.e., flow systems in which state changes occur continuously and thus the processes can easily be identified and plotted. Recall from Section 4.3 that _infinitesimal_ pressure disturbances are called sound waves and these travel at a characteristic velocity which is determined by the medium and its thermodynamic state. For the next two units we shall turn our attention to some _finite_ pressure disturbances which are frequently encountered. Although incorporating large changes in fluid properties, the region occupied by these disturbances is extremely small. Typical thicknesses are on the order of a few mean free molecular paths and thus they appear as _discontinuities_ in the flow and are called "shock waves."

Due to the complex interactions involved, an analysis of the changes within the shock wave is beyond the scope of this book.

Thus, we shall deal only with the properties that exist on each
side of the discontinuity. We first consider a standing normal
shock, that is, a stationary wave front which is perpendicular to
the direction of flow. We will discover that this phenomenon is
found only when supersonic flow exists and that it is basically a
form of a compression process. We will apply the basic concepts
of gas dynamics to analyze a shock wave in an arbitrary fluid
and then develop working equations for a perfect gas. This procedure leads naturally to the compilation of tables which greatly
simplify problem solution. The unit will close with a discussion
of shocks found in the diverging portion of supersonic nozzles.

6.2 OBJECTIVES

After sucessfully completing this unit you should be able to:

1. List the assumptions used to analyze a standing normal shock.

2. Start with the continuity, energy and momentum equations for
    steady, one-dimensional flow, and utilize a control volume
    analysis to derive the relations between properties on each
    side of a standing normal shock for an arbitrary fluid.

3. Derive the working equations for a perfect gas relating prop-
    erty ratios on each side of a standing normal shock as a func-
    tion of Mach number (M) and specific heat ratio (γ).

4. Given the working equations for a perfect gas, show that a
    unique relationship must exist between the Mach numbers before
    and after a standing normal shock.

5. Explain how normal-shock tables may be developed which give
    property ratios across the shock in terms of only the Mach
    number before the shock.

6. Sketch a normal shock "process" on a T-s diagram indicating
    as many pertinent features as possible — such as static and
    total pressures, static and total temperatures, and veloci-
    ties. Indicate each of the preceding before and after the
    shock.

7. Explain why an "expansion shock" cannot exist.

8. State what is meant by the second critical mode of nozzle
    operation. Given the area ratio of a converging-diverging
    nozzle, determine the operating pressure ratio which causes
    operation at the second critical point.

9. Describe how a converging-diverging nozzle operates between first and second critical points.

10. Demonstrate the ability to solve typical standing normal-shock problems by use of the tables and equations.

6.3 SHOCK ANALYSIS –GENERAL FLUID

Figure 6.1 shows a standing normal shock in a section of varying area. We first establish a control volume which includes the shock region and an infinitesimal amount of fluid on each side of the shock. In this manner we deal only with the changes which occur across the shock. It is important to recognize that, since the shock wave is so thin (about 10^{-6} meters), a control volume chosen in the above manner is extremely thin in the x-direction. This permits the following simplifications to be made without introducing error in the analysis:

a. The area on both sides of the shock may be considered to be the same.

b. There is negligible surface in contact with the wall and thus frictional effects may be omitted.

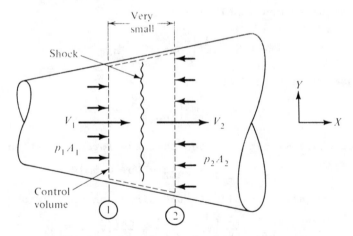

Figure 6.1 Control volume for shock analysis

We begin to apply the basic concepts of continuity, energy, and momentum under the following

Assumptions: Steady, one-dimensional flow

Adiabatic $\qquad (\delta q = 0, \ ds_e = 0)$

No shaft work $\qquad (\delta w_s = 0)$

Neglect potential $\qquad (dz = 0)$

Constant area $\qquad\qquad (A_1 = A_2)$
Neglect wall shear

CONTINUITY - $\qquad\qquad \dot{m} = \rho A V = \text{const}$ $\qquad\qquad$ (2.30)

$$\rho_1 A_1 V_1 = \rho_2 A_2 V_2 \qquad\qquad (6.1)$$

But since the area is constant

$$\boxed{\rho_1 V_1 = \rho_2 V_2} \qquad\qquad (6.2)$$

ENERGY - We start with

$$h_{t1} + q = h_{t2} + w_s \qquad\qquad (3.19)$$

For adiabatic and no work this becomes

$$h_{t1} = h_{t2} \qquad\qquad (6.3)$$

or $\qquad\qquad \boxed{h_1 + \dfrac{V_1^{\,2}}{2g_c} = h_2 + \dfrac{V_2^{\,2}}{2g_c}} \qquad\qquad$ (6.4)

MOMENTUM - The x-component of the momentum equation for steady one-dimensional flow is

$$\Sigma F_x = \frac{\dot{m}}{g_c}\left[V_{out_x} - V_{in_x}\right] \qquad\qquad (3.46)$$

which when applied to Figure 6.1 becomes

$$\Sigma F_x = \frac{\dot{m}}{g_c}\left[V_{2x} - V_{1x}\right] \qquad\qquad (6.5)$$

From Figure 6.1 we can also see that the force summation is

$$\Sigma F_x = p_1 A_1 - p_2 A_2 = \left(p_1 - p_2\right)A \qquad\qquad (6.6)$$

Thus, the momentum equation in the direction of flow becomes

$$\left(p_1 - p_2\right)A = \frac{\dot{m}}{g_c}\left[V_2 - V_1\right] = \frac{\rho A V}{g_c}\left[V_2 - V_1\right] \qquad\qquad (6.7)$$

With $\dot{m}$ written as $\rho A V$ we can cancel the area from both sides. Now the ρV remaining can be written as either $\rho_1 V_1$ or $\rho_2 V_2$ (see equation 6.2) and equation 6.7 becomes

$$p_1 - p_2 = \frac{\rho_2 V_2^{\,2} - \rho_1 V_1^{\,2}}{g_c} \qquad\qquad (6.8)$$

or

$$p_1 + \frac{\rho_1 V_1^2}{g_c} = p_2 + \frac{\rho_2 V_2^2}{g_c}$$ (6.9)

For the general case of an arbitrary fluid we have arrived at <u>three</u> governing equations; 6.2, 6.4, and 6.9. A typical problem would be: knowing the fluid and the conditions before the shock, predict the conditions that would exist after the shock. The unknown parameters are then <u>four</u> in number (ρ_2, p_2, h_2, V_2) which requires additional information for a problem solution. The missing information is supplied in the form of property relations for the fluid involved. For the general fluid (not a perfect gas) this leads to iterative type solutions but with modern computers these can be handled quite easily.

6.4 WORKING EQUATIONS FOR PERFECT GAS

In the previous section we have seen that a typical normal shock problem has four unknowns which can be found through the use of the three governing equations (from continuity, energy and momentum concepts) plus additional information on property relations. For the case of a perfect gas this additional information is supplied in the form of an equation of state and the assumption of constant specific heats. We now proceed to develop working equations in terms of Mach numbers and the specific heat ratio.

CONTINUITY - We start with the continuity equation developed in the previous section

$$\rho_1 V_1 = \rho_2 V_2$$ (6.2)

Substitute for the density from the perfect gas equation of state

$$p = \rho RT$$ (1.13)

and for the velocity from equations 4.10 and 4.11

$$V = Ma = M\sqrt{\gamma g_c RT}$$ (6.10)

<u>Show</u> that the continuity equation can now be written as:

$$\frac{p_1 M_1}{\sqrt{T_1}} = \frac{p_2 M_2}{\sqrt{T_2}}$$ (6.11)

ENERGY - From the previous section we have

$$h_{t1} = h_{t2} \qquad (6.3)$$

But since we are now restricted to a perfect gas for which enthalpy is a function of temperature only, we can say that

$$T_{t1} = T_{t2} \qquad (6.12)$$

Recall from Unit 4 that for a perfect gas with constant specific heats

$$T_t = T\left[1 + \frac{(\gamma-1)}{2} M^2\right] \qquad (4.18)$$

Hence, the energy equation across a standing normal shock can be written as

$$T_1\left[1 + \frac{(\gamma-1)}{2} M_1^2\right] = T_2\left[1 + \frac{(\gamma-1)}{2} M_2^2\right] \qquad (6.13)$$

MOMENTUM - The momentum equation in the direction of flow was seen to be

$$p_1 + \frac{\rho_1 V_1^2}{g_c} = p_2 + \frac{\rho_2 V_2^2}{g_c} \qquad (6.9)$$

Substitutions are made for the density from the equation of state (1.13) and for the velocity from equation 6.10.

$$p_1 + \left[\frac{p_1}{RT_1}\right]\frac{M_1^2 \gamma g_c RT_1}{g_c} = p_2 + \left[\frac{p_2}{RT_2}\right]\frac{M_2^2 \gamma g_c RT_2}{g_c} \qquad (6.14)$$

and the momentum equation becomes

$$p_1\left[1 + \gamma M_1^2\right] = p_2\left[1 + \gamma M_2^2\right] \qquad (6.15)$$

The governing equations for a standing normal shock have now been simplified for a perfect gas and for convenience are summarized below:

$$\frac{p_1 M_1}{\sqrt{T_1}} = \frac{p_2 M_2}{\sqrt{T_2}} \qquad (6.11)$$

$$T_1\left[1 + \frac{(\gamma-1)}{2} M_1^2\right] = T_2\left[1 + \frac{(\gamma-1)}{2} M_2^2\right] \qquad (6.13)$$

$$p_1\left[1 + \gamma M_1^2\right] = p_2\left[1 + \gamma M_2^2\right] \qquad (6.15)$$

There are seven variables involved in these equations:

$$\gamma, \ p_1, \ M_1, \ T_1, \ p_2, \ M_2, \ T_2$$

Once the gas is identified γ is known, and a given state preceding the shock fixes p_1, M_1, and T_1 . Thus, the three equations above (6.11, 6.13, and 6.15) are sufficient to solve for the unknowns after the shock, p_2, M_2, and T_2 .

Rather than struggle through the details of the solution for every shock problem that we encounter, let's solve it once and for all right now. We proceed to combine the above equations and derive an expression for M_2 in terms of the given information. First we rewrite equation 6.11 as

$$\frac{p_1 M_1}{p_2 M_2} = \sqrt{\frac{T_1}{T_2}} \tag{6.16}$$

equation 6.13 as

$$\sqrt{\frac{T_1}{T_2}} = \left[\frac{1 + \frac{(\gamma-1)}{2} M_2^{\ 2}}{1 + \frac{(\gamma-1)}{2} M_1^{\ 2}} \right]^{\frac{1}{2}} \tag{6.17}$$

and equation 6.15 as

$$\frac{p_1}{p_2} = \frac{1 + \gamma M_2^{\ 2}}{1 + \gamma M_1^{\ 2}} \tag{6.18}$$

We then substitute equations 6.17 and 6.18 into equation 6.16 which yields

$$\left[\frac{1 + \gamma M_2^{\ 2}}{1 + \gamma M_1^{\ 2}} \right] \frac{M_1}{M_2} = \left[\frac{1 + \frac{(\gamma-1)}{2} M_2^{\ 2}}{1 + \frac{(\gamma-1)}{2} M_1^{\ 2}} \right]^{\frac{1}{2}} \tag{6.19}$$

At this point notice that M_2 is a function of only M_1 and γ . A trivial solution of this is seen to be $M_1 = M_2$ which represents the degenerate case of no shock. To solve the nontrivial case we square equation 6.19, cross multiply, and arrange the result as a quadratic in $M_2^{\ 2}$.

$$A(M_2^{\ 2})^2 + BM_2^{\ 2} + C = 0 \tag{6.20}$$

where A, B, and C are functions of M_1 and γ . Only if you have considerable time and energy should you attempt to carry

out the tedious algebra required to show that the solution of
this quadratic is

$$M_2^{\ 2} = \frac{M_1^{\ 2} + \frac{2}{(\gamma-1)}}{\left(\frac{2\gamma}{\gamma-1}\right)M_1^{\ 2} - 1} \tag{6.21}$$

For our typical shock problem the Mach number after the
shock is computed with the aid of equation 6.21 and then T_2
and p_2 can easily be found from equations 6.13 and 6.15. To
complete the picture the total pressures p_{t1} and p_{t2} can be
computed in the usual manner. It turns out that if M_1 is
supersonic M_2 will always be subsonic and a typical problem is
shown on the T-s diagram in Figure 6.2.

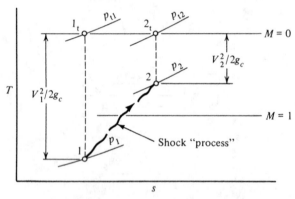

Figure 6.2 *T–s* diagram for typical normal shock

The end points 1 and 2 (before and after the shock) are
well defined states but the changes that occur within the shock
do not follow an equilibrium process in the usual thermodynamic
sense. For this reason the shock "process" is usually shown by
a dotted or wiggly line. Note that when points 1 and 2 are
located on the T-s diagram it can immediately be seen that an
entropy change is involved in the shock "process". This will be
discussed in greater detail in the next section.

Example 6.1 Helium is flowing at a Mach number of 1.80 and
enters a normal shock. Determine the pressure ratio across the
shock.

We use equation 6.21 to find the Mach number after the shock and
6.15 to obtain the pressure ratio.

$$M_2^{\,2} = \frac{M_1^{\,2} + \frac{2}{\gamma-1}}{\left(\frac{2\gamma}{\gamma-1}\right)M_1^{\,2}-1} = \frac{(1.8)^2 + \frac{2}{1.67-1}}{\left(\frac{2\times1.67}{1.67-1}\right)(1.8)^2-1} = 0.411, \quad M_2 = \underline{0.641}$$

$$\frac{p_2}{p_1} = \frac{1 + \gamma M_1^{\,2}}{1 + \gamma M_2^{\,2}} = \frac{1 + (1.67)(1.8)^2}{1 + (1.67)(0.411)} = \underline{3.80}$$

6.5 NORMAL-SHOCK TABLES

We have found that for any given fluid with a specific set of conditions entering a normal shock there is one and only one set of conditions that can result after the shock. An iterative type solution results for a fluid which cannot be treated as a perfect gas, whereas the case of the perfect gas produces an explicit solution. The latter case opens the door to further simplifications since equation 6.21 yields the exit Mach number M_2 for any given inlet Mach number M_1 and we can now eliminate M_2 from all previous equations.

For example, equation 6.13 can be solved for the temperature ratio

$$\frac{T_2}{T_1} = \frac{1 + \frac{(\gamma-1)}{2} M_1^{\,2}}{1 + \frac{(\gamma-1)}{2} M_2^{\,2}} \tag{6.22}$$

If we now eliminate M_2 by the use of equation 6.21 the result will be

$$\frac{T_2}{T_1} = \frac{\left[1 + \frac{(\gamma-1)}{2} M_1^{\,2}\right]\left[\frac{2\gamma}{(\gamma-1)} M_1^{\,2} - 1\right]}{\frac{(\gamma+1)^2}{2(\gamma-1)} M_1^{\,2}} \tag{6.23}$$

Similarly equation 6.15 can be solved for the pressure ratio

$$\frac{p_2}{p_1} = \frac{1 + \gamma M_1^{\,2}}{1 + \gamma M_2^{\,2}} \tag{6.24}$$

and elimination of M_2 through the use of equation 6.21 will produce

$$\frac{p_2}{p_1} = \frac{2\gamma}{(\gamma+1)} M_1^{\,2} - \left(\frac{\gamma-1}{\gamma+1}\right) \tag{6.25}$$

If you are very persistent (and in need of algebraic exercise) you might carry out the development of equations 6.23 and 6.25.

Also, these can be combined to form the density ratio

$$\frac{\rho_2}{\rho_1} = \frac{(\gamma+1)\ M_1^2}{(\gamma-1)\ M_1^2 + 2} \tag{6.26}$$

Other interesting ratios can be developed, each as a function of only M_1 and γ. For instance

since
$$p_t = p\left[1 + \frac{(\gamma-1)}{2}\ M^2\right]^{\frac{\gamma}{\gamma-1}} \tag{4.21}$$

we may write
$$\frac{p_{t2}}{p_{t1}} = \frac{p_2}{p_1}\left[\frac{1 + \frac{(\gamma-1)}{2}\ M_2^2}{1 + \frac{(\gamma-1)}{2}\ M_1^2}\right]^{\frac{\gamma}{\gamma-1}} \tag{6.27}$$

The ratio p_2/p_1 can be eliminated by equation 6.25 with the following result:

$$\frac{p_{t2}}{p_{t1}} = \left[\frac{\frac{(\gamma+1)}{2}\ M_1^2}{1 + \frac{(\gamma-1)}{2}\ M_1^2}\right]^{\frac{\gamma}{\gamma-1}}\left[\frac{2\gamma}{(\gamma+1)}\ M_1^2 - \frac{(\gamma-1)}{(\gamma+1)}\right]^{\frac{1}{1-\gamma}} \tag{6.28}$$

Equation 6.28 is extremely important since the stagnation pressure ratio is related to the entropy change through equation 4.28.

$$\frac{p_{t2}}{p_{t1}} = e^{-\Delta s/R} \tag{4.28}$$

In fact we could combine equations 4.28 and 6.28 to obtain an explicit relation for Δs as a function of M_1 and γ.

Note that for a given fluid (γ known) equations 6.23, 6.25, 6.26, and 6.28 express property ratios as a function of the entering Mach number only. This suggests that we could easily construct a table giving values of M_2, T_2/T_1, p_2/p_1, ρ_2/ρ_1, p_{t2}/p_{t1}, etc., versus M_1 for a particular γ. Such a table of normal-shock parameters may be found in the Appendix. These tables greatly aid problem solution as the following example shows.

Example 6.2 Fluid is air and can be treated as a perfect gas. If the conditions before the shock are: $M_1 = 2.0$, $p_1 = 20$ psia, $T_1 = 500°R$; determine the conditions after the shock and the entropy change across the shock.

First we compute p_{t1} with the aid of the isentropic tables.

$$p_{t1} = \frac{p_{t1}}{p_1} \, p_1 = \frac{1}{0.1278} \, (20) = \underline{156.5} \text{ psia}$$

Now from the normal-shock tables opposite $M_1 = 2.0$ we find:

$M_2 = 0.57735$, $p_2/p_1 = 4.5000$, $T_2/T_1 = 1.6875$, $p_{t2}/p_{t1} = 0.72087$

Thus,
$$p_2 = \frac{p_2}{p_1} \, p_1 = 4.5 \, (20) = \underline{90} \text{ psia}$$

$$T_2 = \frac{T_2}{T_1} \, T_1 = 1.6875 \, (500) = \underline{844}^{\circ}R$$

$$p_{t2} = \frac{p_{t2}}{p_{t1}} \, p_{t1} = 0.72087 \, (156.5) = \underline{112.8} \text{ psia}$$

or p_{t2} can be computed with the aid of the isentropic tables:

$$p_{t2} = \frac{p_{t2}}{p_2} \, p_2 = \frac{1}{0.7978} \, (90) = \underline{112.8} \text{ psia}$$

To compute the entropy change we use equation 4.28.

$$\frac{p_{t2}}{p_{t1}} = 0.72087 = e^{-\Delta s/R}$$

$$\Delta s/R = 0.3273$$

$$\Delta s = \frac{(0.3273)(53.3)}{778} = \underline{0.0224} \text{ Btu/lbm-}^{\circ}R$$

It is interesting to note that as far as the governing equations are concerned the previous problem could be completely reversed. The fundamental relations of continuity (6.11), energy (6.13), and momentum (6.15) would be completely satisfied if we changed the problem to $M_1 = 0.577$, $p_1 = 90$ psia, $T_1 = 844^{\circ}R$ with the resulting $M_2 = 2.0$, $p_2 = 20$ psia, and $T_2 = 500^{\circ}R$ (which would represent an "expansion shock"). However, in this latter case the entropy change would be negative which clearly violates the Second Law of Thermodynamics for an adiabatic system.

The above example and discussion clearly show that the shock phenomenon is a one-way process (i.e., irreversible). It is always a compression shock and for a normal shock the flow is always supersonic before the shock and subsonic after the shock. One can note from the tables that as M_1 increases, the pressure, temperature, and density ratios increase, indicating a stronger shock (or compression). One can also note that as M_1 increases p_{t2}/p_{t1} decreases, which means that the entropy change increases. Thus, as the strength of the shock increases the losses also increase.

Example 6.3   Air has a temperature and pressure of $300^{\circ}K$ and 2 bars absolute respectively. It is flowing with a velocity of 868 m/s and enters a normal shock. Determine the density before and after the shock.

$$\rho_1 = \frac{p_1}{RT_1} = \frac{2 \times 10^5}{(287)(300)} = \underline{2.32} \text{ kg/m}^3$$

$$a_1 = \left[\gamma g_c RT\right]^{\frac{1}{2}} = \left[(1.4)(1)(287)(300)\right]^{\frac{1}{2}} = 347 \text{ m/s}$$

$$M_1 = V_1/a_1 = 868/347 = 2.50$$

From the shock tables we obtain

$$\frac{\rho_2}{\rho_1} = \frac{p_2}{p_1}\frac{T_1}{T_2} = (7.125)\frac{1}{(2.1375)} = 3.333$$

$$\rho_2 = 3.3333 \, \rho_1 = (3.3333)(2.32) = \underline{7.73} \text{ kg/m}^3$$

Example 6.4   Oxygen enters the converging section shown and a normal shock occurs at the exit. The entering Mach number is 2.8 and the area ratio $A_1/A_2 = 1.7$. Compute the overall static temperature ratio T_3/T_1 . Neglect all frictional losses.

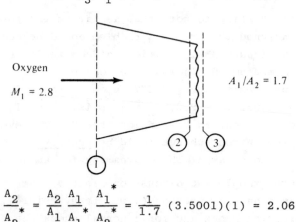

$$\frac{A_2}{A_2^*} = \frac{A_2}{A_1}\frac{A_1}{A_1^*}\frac{A_1^*}{A_2^*} = \frac{1}{1.7}(3.5001)(1) = 2.06$$

Thus, $M_2 \approx 2.23$ and from the shock tables we get
$M_3 = 0.5431$ and $T_3/T_2 = 1.8835$

$$\frac{T_3}{T_1} = \frac{T_3}{T_2}\frac{T_2}{T_{t2}}\frac{T_{t2}}{T_{t1}}\frac{T_{t1}}{T_1} = (1.8835)(0.5014)(1)\frac{1}{0.3894} = \underline{2.43}$$

We can also develop a relation for the velocity change across a standing normal shock. Starting with the basic continuity equation

$$\rho_1 V_1 = \rho_2 V_2 \tag{6.2}$$

we introduce the density relation from 6.26

$$\frac{V_2}{V_1} = \frac{\rho_1}{\rho_2} = \frac{(\gamma-1)M_1^2 + 2}{(\gamma+1)M_1^2} \tag{6.29}$$

and subtract one from each side.

$$\frac{V_2 - V_1}{V_1} = \frac{(\gamma-1)M_1^2 + 2 - (\gamma+1)M_1^2}{(\gamma+1)M_1^2} \tag{6.30}$$

$$\frac{V_2 - V_1}{M_1 a_1} = \frac{2(1 - M_1^2)}{(\gamma+1)M_1^2} \tag{6.31}$$

or

$$\boxed{\frac{V_1 - V_2}{a_1} = \frac{2}{\gamma+1}\left[\frac{M_1^2 - 1}{M_1}\right]} \tag{6.32}$$

This is another parameter which is a function of M_1 and γ and thus may be added to our shock tables. Its usefulness for solving certain types of problems will become apparent in Unit 7.

6.6 SHOCKS IN NOZZLES

In Section 5.7 we discussed the isentropic operation of a converging-diverging nozzle. Remember that this type of nozzle is physically distinguished by its "area ratio" or ratio of the exit area to the throat area. Furthermore its flow conditions are determined by the "operating pressure ratio" or ratio of the receiver pressure to the inlet stagnation pressure. We identified two critical pressure ratios which were significant. For any pressure ratio above the first critical the nozzle is not choked and has subsonic flow throughout (typical venturi opera-

tion). The first critical represents flow which is subsonic in
both the convergent and divergent sections but is choked with
a Mach number of 1.0 in the throat. The third critical represents operation at the design condition with supersonic flow in
the entire diverging section. It is also choked with Mach 1.0
in the throat. The first and third criticals represent the only
operating points that have (a) isentropic flow throughout, (b)
Mach number of one at the throat, and (c) exit pressure equal
to receiver pressure.

Remember that, with subsonic flow at the exit, the exit pressure <u>must</u> equal the receiver pressure. Imposing a pressure ratio
slightly below that of first critical presents a problem in that
there is no way that <u>isentropic</u> flow can meet the boundary condition of pressure equilibrium at the exit. However, there is nothing to prevent a <u>non-isentropic</u> flow adjustment from occuring
within the nozzle. This internal adjustment takes the form of
a standing normal shock which we now know involves an entropy
change.

As the pressure ratio is lowered below the first critical
point, a normal shock forms just downstream of the throat. The
remainder of the "nozzle" is now acting as a diffuser since after
the shock the flow is subsonic and the area is increasing. The
shock will locate itself in a position such that the pressure
changes that occur ahead of the shock, across the shock, and downstream of the shock will produce a pressure that <u>exactly</u> <u>matches</u>
<u>the</u> <u>outlet</u> <u>pressure</u>. In other words, <u>the</u> <u>operating</u> <u>pressure</u> <u>ratio</u>
<u>determines</u> <u>the</u> <u>location</u> <u>and</u> <u>strength</u> <u>of</u> <u>the</u> <u>shock</u>. An example of
this mode of operation is shown in Figure 6.3. As the pressure
ratio is lowered further, the shock continues to move toward the
exit. When the shock is located at the exit plane this condition
is referred to as the "second critical point."

If the operating pressure ratio is between second and third
critical, then a compression takes place outside the nozzle.
This is called "overexpansion" (i.e., the flow has been expanded
too far within the nozzle). If the receiver pressure is below
third critical, then an expansion takes place outside the nozzle.
This condition is called "underexpansion." We shall investigate
these conditions later in the next two units after the appropriate background has been covered.

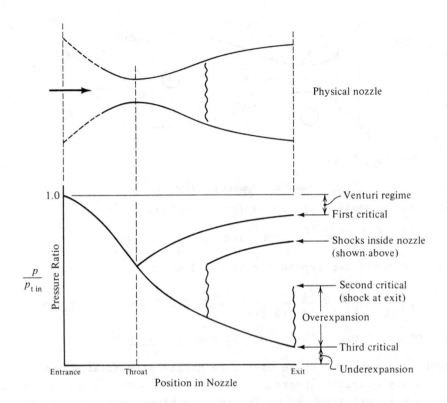

Figure 6.3 Operating modes for DeLaval nozzle

For the present we shall proceed to investigate the operational regime between first and second critical. Let us work with the same nozzle and inlet conditions that we used in Section 5.7. The nozzle has an area ratio of 2.494 and is fed by air at 100 psia and 600°R from a large tank. Thus, the inlet conditions are essentially stagnation. For these fixed inlet conditions we previously found that a receiver pressure of 96.07 psia (an operating pressure ratio of 0.9607) identifies the first critical point and a receiver pressure of 6.426 psia (an operating pressure ratio of 0.06426) exists at the third critical point.

What receiver pressure do we need to operate at the second critical point? Figure 6.4 shows such a condition and you should recognize that the entire nozzle up to the shock is operating at its design or third critical condition.

From the isentropic tables at $A/A^* = 2.494$ we have

$$M_3 = 2.44 \quad \text{and} \quad p_3/p_{t3} = 0.06426$$

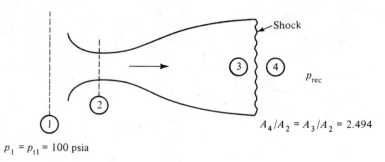

$p_1 = p_{t1} = 100$ psia

Figure 6.4 Operation at second critical

From the normal-shock tables for $M_3 = 2.44$ we have

$$M_4 = 0.5189 \quad \text{and} \quad p_4/p_3 = 6.7792$$

and the operating pressure ratio will be

$$\frac{p_{rec}}{p_{t1}} = \frac{p_4}{p_{t1}} = \frac{p_4}{p_3} \frac{p_3}{p_{t3}} \frac{p_{t3}}{p_{t1}} = 6.7792(0.06426)(1) = \underline{0.436}$$

or for $p_1 = p_{t1} = 100$ psia, $p_4 = p_{rec} = \underline{43.6}$ psia

Thus, for our converging-diverging nozzle with an area ratio of 2.494 any operating pressure ratio between 0.9607 and 0.436 will cause a normal shock to be located someplace in the diverging portion of the nozzle.

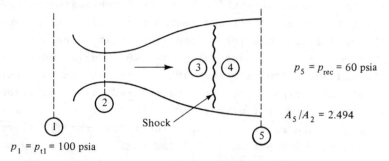

$p_1 = p_{t1} = 100$ psia

Figure 6.5 DeLaval nozzle with normal shock in diverging section

Suppose we are given that the operating pressure ratio is 0.60. The logical question to ask is, "Where is the shock?" This situation is shown in Figure 6.5. We must take advantage of the only two available pieces of information and from these construct a solution.

We know that $\dfrac{A_5}{A_2} = 2.494$ and $\dfrac{p_5}{p_{t1}} = 0.60$

Also we may assume that all losses occur across the shock and we know that $M_2 = 1.0$. It might also be helpful to visualize the flow on a T-s diagram and this is shown in Figure 6.6.

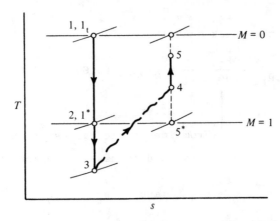

Figure 6.6 *T–s* diagram for DeLaval nozzle with normal shock (for physical picture see Figure 6.5)

Since there are no losses up to the shock we know that

$$A_2 = A_1^{*}$$

Thus,

$$\frac{A_5}{A_2} \frac{p_5}{p_{t1}} = \frac{A_5}{A_1^{*}} \frac{p_5}{p_{t1}} \tag{6.33}$$

We also know from equation 5.35 that for the case of adiabatic, no-work flow of a perfect gas

$$A_1^{*} \, p_{t1} = A_5^{*} \, p_{t5} \tag{6.34}$$

Thus,

$$\frac{A_5 \, p_5}{(A_1^{*} \, p_{t1})} = \frac{A_5 \, p_5}{(A_5^{*} \, p_{t5})}$$

In summary:

$$\frac{A_5}{A_2} \frac{p_5}{p_{t1}} = \frac{A_5}{A_1^{*}} \frac{p_5}{p_{t1}} = \frac{A_5}{A_5^{*}} \frac{p_5}{p_{t5}} \tag{6.35}$$

$$\underbrace{\qquad}_{\text{known}}$$

$$(2.494)(0.6) = 1.4964$$

Note that we have manipulated the known information into an expression with all similar station subscripts. In Section 5.6 we showed with equation 5.43 that the ratio Ap/A^*p_t is a simple function of M and γ and thus is listed in the isentropic tables. A check in the tables shows that the exit Mach number is $M_5 \approx 0.38$.

To locate the shock seek the ratio

$$\frac{p_{t5}}{p_{t1}} = \frac{p_{t5}}{p_5}\frac{p_5}{p_{t1}} = \frac{1}{0.9052}(0.6) = 0.664$$

Given

From isentropic tables at $M = 0.38$

And since all the loss is assumed to take place across the shock we have

$$p_{t5} = p_{t4} \quad \text{and} \quad p_{t1} = p_{t3}$$

Thus,

$$\frac{p_{t4}}{p_{t3}} = \frac{p_{t5}}{p_{t1}} = 0.664$$

Knowing the total pressure ratio across the shock we can determine from the normal-shock tables that $M_3 \approx 2.12$ and then from the isentropic tables we note that this Mach number will occur at an area ratio of about $A_3/A_3^* = A_3/A_2 = 1.869$. More accurate answers could be obtained by interpolating within the tables.

We see that if we are given a physical converging-diverging nozzle (area ratio is known) and an operating pressure ratio between first and second critical, it is a simple matter to determine the position and strength of the normal shock which is located in the diverging section.

Example 6.5 A converging-diverging nozzle has an area ratio of 3.50. At off-design conditions the exit Mach number is observed to be 0.3. What operating pressure ratio would cause this situation?

Using the section numbering system of Figure 6.5,

for $M_5 = 0.3$ we have $\dfrac{p_5 A_5}{p_{t5} A_5^*} = 1.9119$

$$\frac{p_5}{p_{t1}} = \frac{p_5 A_5}{p_{t5}A_5^*}\left(\frac{p_{t5}A_5^*}{p_{t1}A_1^*}\right)\frac{A_1^*}{A_2}\frac{A_2}{A_5} = (1.9119)(1)(1)\frac{1}{3.50} = \underline{0.546}$$

Could you now find the shock location and Mach number?

Example 6.6   Air enters a converging-diverging nozzle which has
an overall area ratio of 1.76.   A normal shock occurs at a section where the area is 1.19 times that of the throat.   Neglect
all friction losses and find the operating pressure ratio.
Again, we use the numbering system shown in Figure 6.5.
From the isentropic tables at A_3/A_2 = 1.19, M_3 = 1.52.
From the shock tables, M_4 = 0.6941, p_{t4}/p_{t3} = 0.9233

$$\frac{A_5}{A_5^*} = \frac{A_5}{A_2}\frac{A_2}{A_4}\frac{A_4}{A_4^*}\frac{A_4^*}{A_5^*} = (1.76)\,\frac{1}{1.19}\,(1.0988)(1) = 1.625$$

Thus, $M_5 \approx 0.389$

$$\frac{p_5}{p_{t1}} = \frac{p_5}{p_{t5}}\frac{p_{t5}}{p_{t4}}\frac{p_{t4}}{p_{t3}}\frac{p_{t3}}{p_{t1}} = (0.9007)(1)(0.9233)(1) = \underline{0.832}$$

6.7 SUPERSONIC WIND TUNNEL OPERATION

To provide a test section with supersonic flow requires a
converging-diverging nozzle.   To operate economically, the nozzle-
test-section combination must be followed by a diffusing section
which also must be converging-diverging.   This configuration presents some interesting problems in flow analysis.   Starting up
such a wind tunnel is another example of nozzle operation at
pressure ratios above the second critical point.   Figure 6.7
shows a typical tunnel in its most <u>unfavorable</u> operating condition which occurs at start-up.   A brief analysis of the situation
follows.

As the exhauster is started this reduces the pressure and
produces a flow through the tunnel.   At first the flow is subsonic throughout but at increased power settings the exhauster
reduces pressures still further and causes increased flow rates
until the nozzle throat (section 2) becomes choked.   At this
point the nozzle is operating at its first critical condition.
As power is further increased a normal shock is formed just downstream of the throat and, if the tunnel pressure is continuously
decreased, the shock will move down the diverging portion of the
nozzle and rapidly pass through the test section and into the
diffuser.   Figure 6.8 shows this general running condition which
is called the most <u>favorable</u> condition.

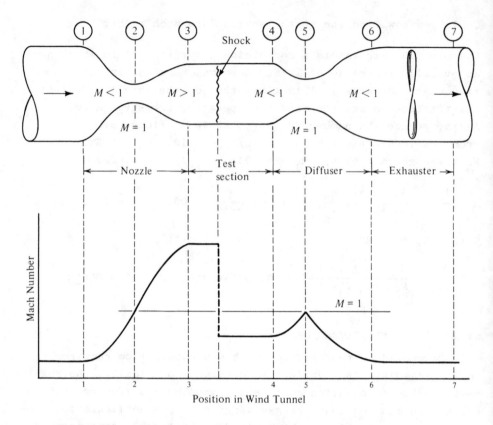

Figure 6.7 Supersonic tunnel at start-up (with associated Mach number variation)

We return to Figure 6.7 which shows the shock located in the test section. The variation of Mach number throughout the flow system is also shown for this case. This is called the most unfavorable condition because the shock occurs at the highest possible Mach number and thus the losses are greatest. We might also point out that the diffuser throat (section 5) must be sized for this condition. Let us see how this is done.

Recall the relation $p_t A^* = $ constant

Thus,
$$p_{t2} A_2^* = p_{t5} A_5^*$$

But since Mach one exists at both sections 2 and 5 (during start-up)

then
$$A_2 = A_2^* \quad \text{and} \quad A_5 = A_5^*$$

Hence
$$p_{t2} A_2 = p_{t5} A_5 \qquad (6.36)$$

Due to the shock losses (and other friction losses) we know that

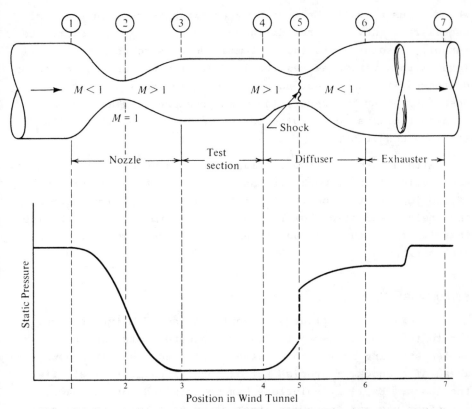

Figure 6.8 Supersonic tunnel in running condition (with associated pressure variation)

$p_{t5} < p_{t2}$ and, therefore, A_5 must be greater than A_2. Knowing the test section design Mach number fixes the shock strength in this unfavorable condition and A_5 is easily determined from equation 6.36. Keep in mind that this represents a <u>minimum</u> area for the diffuser throat. If it is made any smaller than this the tunnel could never be started; i.e., we could never get the shock into the test section. (In fact if A_5 is made too small, the flow will choke first in this throat and never get a chance to reach sonic conditions in section 2.)

Once the shock has passed into the diffuser throat, knowing that $A_5 > A_2$, we realize that the tunnel can never run with sonic velocity at section 5. Thus, to operate as a diffuser there must be a shock at this point as shown in Figure 6.8. We have also shown the pressure variation through the tunnel for this running condition.

To keep the losses during running a minimum, the shock in the diffuser should occur at the lowest possible Mach number,

which means a small throat. However, we have seen that it is
necessary to have a large diffuser throat in order to start the
tunnel. A solution to this dilemma would be to construct a diffuser with a variable-area throat. After start-up A_5 could be
decreased with a corresponding decrease in shock strength and
operating power. However, the power required for any installation must always be computed on the basis of the unfavorable
start-up condition.

Although the supersonic wind tunnel is used mainly for aero-
nautically oriented work, its operation serves to solidify many
of the important concepts of variable area flow, normal shocks,
and their associated flow losses. Equally important is the fact
that it begins to focus our attention on some practical design
applications.

6.8 SUMMARY

We examined stationary discontinuities of a type perpendicular to the flow. These are finite pressure disturbances and are
called standing normal shock waves. If conditions are known
ahead of a shock, a precise set of conditions must exist after
the shock. Explicit solutions can be obtained for the case of
a perfect gas and these lend themselves to tabulation for various
specific heat ratios.

Shocks are only found in supersonic flow and the flow is
always subsonic after a normal shock. The shock wave is a type
of compression process, although a rather inefficient one since
relatively large losses are involved in the process. Shocks provide a means of flow adjustment to meet imposed pressure conditions in supersonic flow.

As in the previous unit, most of the equations in this unit
need not be memorized. However, you should be completely familiar with the fundamental relations which apply to all fluids
across a normal shock. These are equations 6.2, 6.4, and 6.9.
Essentially, these say that the end points of a shock have three
things in common:

> (a) the same mass flow per unit area.
> (b) the same stagnation enthalpy.
> (c) the same value of $p + \rho V^2/g_c$

The working equations that apply to perfect gases are summarized in the middle of Section 6.4 and are equations 6.11, 6.13, and 6.15.

You should also be familiar with the various ratios which have been tabulated in the Appendix. Just knowing what kind of information you have available is frequently very helpful in setting up a problem solution.

6.9 PROBLEMS

Unless indicated otherwise you may assume that there is no friction in any of the following flow systems; thus, the only losses are those generated by shocks.

1. A standing normal shock occurs in air which is flowing at a Mach number of 1.8.
   (a) What are the pressure, temperature and density ratios across the shock?
   (b) Compute the entropy change for the air as it passes through the shock.
   (c) Repeat part (b) for flows at $M = 2.8$ and 3.8.

2. The difference between the total and static pressure before a shock is 75 psi. What is the maximum static pressure that can exist at this point ahead of the shock? The gas is oxygen.

3. In an arbitrary perfect gas, the Mach number before a shock is infinite.
   (a) Determine a general expression for the Mach number after the shock. What is the value of this expression for $\gamma = 1.4$?
   (b) Determine general expressions for the ratios p_2/p_1, T_2/T_1, ρ_2/ρ_1, p_{t2}/p_{t1}. Do these agree with the values shown in the Appendix for $\gamma = 1.4$?

4. It is known that sonic velocity exists in each throat of the system shown. The entropy change for the air is 0.062 Btu/lbm-$^{\circ}$R. Negligible friction exists in the duct. Determine the area ratios A_3/A_1 and A_2/A_1.

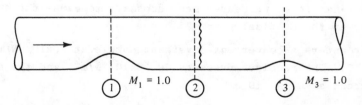

$M_1 = 1.0$ $M_3 = 1.0$

5. Air flows in the system shown below. It is known that the
    Mach number after the shock is $M_3 = 0.52$. Considering p_1
    and p_2 , it is also known that one of these pressures is
    twice the other.
    (a) Compute the Mach number at section 1.
    (b) What is the area ratio A_1/A_2 ?

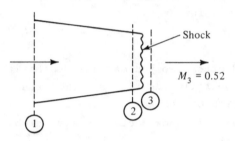

6. A shock stands at the inlet to the system shown. The free-
    stream Mach number is $M_1 = 2.90$, fluid is nitrogen,
    $A_2 = 0.25$ m^2 and $A_3 = 0.20$ m^2 . Find the outlet Mach num-
    ber and the temperature ratio T_3/T_1 .

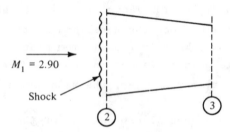

7. A converging-diverging nozzle is designed to produce a Mach
    number of 2.5 with air.
    (a) What operating pressure ratio ($p_{rec}/p_{t\ inlet}$) will cause
        this nozzle to operate at the first, second, and third
        critical points?
    (b) If the inlet stagnation pressure is 150 psia, what re-
        ceiver pressures represent operation at these critical
        points?
    (c) Suppose the receiver pressure were fixed at 15 psia.
        What inlet pressures are necessary to cause the operation
        at the critical points?

8. Air enters a convergent-divergent nozzle at 20x10^5 N/m^2 and
    40°C. The receiver pressure is 2x10^5 N/m^2 and the nozzle
    throat area is 10 cm^2 .

(a) What should the exit area be for the above design conditions (i.e., to operate at third critical)?

(b) With the nozzle area fixed at the value determined in (a)
and the inlet pressure held at 20×10^5 N/m^2 , what receiver pressure would cause a shock to stand at the exit?

(c) What receiver pressure would place the shock at the
throat?

9. In the sketch below $M_1 = 3.0$ and $A_1 = 2.0$ ft^2 . If the
fluid is carbon monoxide and the shock occurs at an area of
1.8 ft^2 , what is the minimum area that section 4 can have?

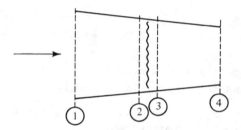

10. A converging-diverging nozzle has an area ratio of 7.8 but
is not being operated at its design pressure ratio. Consequently, a normal shock is found in the diverging section at
an area which is twice that of the throat. The fluid is
oxygen.

(a) Find the Mach number at the exit and the operating pressure ratio.

(b) What is the entropy change through the nozzle if there
is negligible friction?

11. The diverging section of a supersonic nozzle is formed from
the frustrum of a cone. When operating at its third critical
point with nitrogen the exit Mach number is 2.6. Compute the
operating pressure ratio which will locate a normal shock as
shown below.

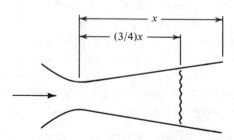

12. A converging-diverging nozzle receives air from a tank at
    100 psia and 600°R. The pressure is 28.0 psia immediately
    preceding a plane shock which is located in the diverging
    section. The Mach number at the exit is 0.5 and the flow
    rate is 10 lbm/sec. Determine:
    (a) The throat area.
    (b) Area at which the shock is located.
    (c) The outlet pressure required to operate the nozzle in the
        above manner.
    (d) The outlet area.
    (e) The design Mach number.

13. Air enters a device with a Mach number of M_1 = 2.0 and
    leaves with M_2 = 0.25 . The ratio of exit to inlet area is
    A_2/A_1 = 3.0 .
    (a) Find the static pressure ratio p_2/p_1 .
    (b) Determine the stagnation pressure ratio p_{t2}/p_{t1} .

14. Oxygen, with p_t = 95.5 psia , enters a diverging section
    where the area is 3.0 ft^2 . At the outlet the area is
    4.5 ft^2 , the Mach number is 0.43 and the static pressure
    is 75.3 psia. Determine the possible values of Mach number
    that could exist at the inlet.

15. A converging-diverging nozzle has an area ratio of 3.0 .
    The stagnation pressure at the inlet is 8.0 bars and the
    receiver pressure is 3.5 bars. Assume γ = 1.4 .
    (a) Compute the critical operating pressure ratios for the
        nozzle and show that a shock is located within the di-
        verging section.
    (b) Compute the Mach number at the outlet.
    (c) Compute the shock location (area) and the Mach number
        before the shock.

16. Nitrogen flows through a converging-diverging nozzle which
    was designed to operate at a Mach number of 3.0 . If it is
    subjected to an operating pressure ratio of 0.5 ,
    (a) Determine the Mach number at the exit.
    (b) What is the entropy change in the nozzle?
    (c) Compute the area ratio at the shock location.
    (d) What value of the operating pressure ratio would be re-
        quired to move the shock to the exit?

17. Consider the wind tunnel shown in Figures 6.7 and 6.8. Atmo-

spheric air enters the system with a pressure and temperature
of 14.7 psia and 80°F, respectively, and has negligible velocity at section 1. The test section has a cross-sectional
area of 1 ft² and operates at a Mach number of 2.5. You may
assume that the diffuser reduces the velocity to approximately
zero and that final exhaust is to the atmosphere with negligible velocity. The system is fully insulated and there are
no friction losses. Find

(a) The throat area of the nozzle.
(b) The mass flow rate.
(c) The minimum possible throat area of the diffuser.
(d) The total pressure entering exhauster at start-up
    (Figure 6.7).
(e) The total pressure entering exhauster when running
    (Figure 6.8).
(f) The HP required for exhauster (based on an isentropic
    compression).

6.10 CHECK TEST

You should be able to complete this test without reference to
material in the unit.

1. Starting with the continuity, energy and momentum equations
   in a form suitable for steady, one-dimensional flow, analyze
   a standing normal shock in an arbitrary fluid. Then simplify
   your results for the case of a perfect gas.

2. Fill in the following blanks with "increases," "decreases,"
   or "remains constant." Across a standing normal shock the
   (a) temperature   _____
   (b) stagnation pressure   _____
   (c) velocity   _____
   (d) density   _____

3. Consider a converging-diverging nozzle with an area ratio of
   3.0 and assume operation with a perfect gas ($\gamma = 1.4$).
   Determine the operating pressure ratios which would cause
   operation at the first, second, and third critical points.

4. Sketch a T-s diagram for a standing normal shock in a perfect gas. Indicate static and total pressures, static and
   total temperatures, and velocities (both before and after
   the shock).

5. Nitrogen flows in an insulated, variable-area system with friction. The area ratio is $A_2/A_1 = 2.0$ and the static pressure ratio is $p_2/p_1 = 0.20$. The Mach number at section 2 is $M_2 = 3.0$.

    (a) What is the Mach number at section 1?

    (b) Is the gas flowing from 1 to 2 or from 2 to 1?

6. A large chamber contains air at 100 psia and 600°R. A converging-diverging nozzle with an area ratio equal to 2.50 is connected to the chamber and the receiver pressure is 60 psia.

    (a) Determine the outlet Mach number and velocity.

    (b) Find the Δs across the shock.

    (c) Draw a T-s diagram for the flow through the nozzle.

UNIT 7

MOVING AND
OBLIQUE SHOCKS

7.1 INTRODUCTION

In Section 4.3 we superimposed a uniform velocity on a traveling sound wave so that we could obtain a standing wave and analyze it by the use of steady flow equations. We will use precisely the same technique in this unit to compare standing and
moving normal shocks. Recall that velocity superposition does
not affect the _static_ thermodynamic state of a fluid but it does
change the _stagnation_ conditions. (See Section 3.5.)

We will then superimpose a velocity tangential to a standing
normal shock and find that this results in the formation of an
oblique shock; i.e., one in which the wave front is at an angle
of other than 90° to the approaching flow. The case of an oblique
shock in a perfect gas will then be analyzed in detail and, as
you might suspect, these results lend themselves to the construction of tables and charts which greatly aid problem solution.
The unit will close on a discussion of a number of places where

oblique shocks can be found along with an investigation of the
boundary conditions that control the shock formation.

7.2 OBJECTIVES

After successfully completing this unit you should be able to:

1. Identify the properties that remain constant and the properties that change when a uniform velocity is superimposed on
   another flow field.

2. Describe how moving normal shocks can be analyzed with the
   relations developed for standing normal shocks.

3. Explain how an oblique shock can be described by the superposition of a normal shock and another flow field.

4. Sketch an oblique shock and define the "shock angle" and
   "deflection angle."

5. Analyze an oblique shock in a perfect gas and develop the
   relation among shock angle, deflection angle, and entering
   Mach number. (Optional.)

6. Describe the general results of an oblique shock analysis in
   terms of a diagram such as shock angle versus inlet Mach number for various deflection angles.

7. Distinguish between weak and strong shocks. Know when each
   might result.

8. Describe the conditions which cause a detached shock to form.

9. State what operating conditions will cause an oblique shock
   to form at a supersonic nozzle exit.

10. Demonstrate the ability to solve typical problems involving
    moving normal shocks or oblique shocks by use of the appro-
    priate equations and tables or charts.

7.3 NORMAL VELOCITY SUPERPOSITION
MOVING NORMAL SHOCKS

Let us consider a plane shock wave that is moving into a
stationary fluid such as shown in Figure 7.1. Such a wave could
be found traveling down a shock tube or it could have originated
from an explosive device in open air. In the latter case the
shock travels out from the explosion point in the form of a spherical wave front. However, very quickly the radius of curvature

becomes so large that it may be treated as a planar wave front with little error. A typical problem might be to determine the conditions that exist after the passage of the shock front assuming we know the original conditions and the speed of the shock wave.

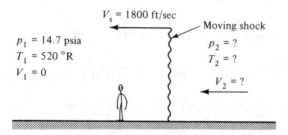

Figure 7.1 Moving normal shock with ground as reference

In Figure 7.1 we are on the ground viewing a normal shock which is moving to the left at a speed V_s into standard sea level air. This is an <u>unsteady</u> picture and we seek a means to make this fit the analysis made in the previous unit. To do this we superimpose on the entire flow field a velocity of V_s to the right. An alternate way of accomplishing the same effect is to get on the shock wave and go for a ride as shown in Figure 7.2. By either method the result is to change the frame of reference to the shock wave and thus it appears to be a standing normal shock.

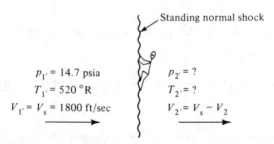

Figure 7.2 Moving shock transformed into stationary shock

The shock was given as moving at 1800 ft/sec into air at 14.7 psia and 520°R. We can now proceed to solve the problem represented in Figure 7.2 by the methods developed in Unit 6.

$$a_1' = \sqrt{\gamma g_c R T_1'} = \sqrt{(1.4)(32.2)(53.3)(520)} = 1118 \text{ ft/sec}$$

$$M_1' = V_1'/a_1' = 1800/1118 = 1.61$$

From the normal-shock tables we find:

$$M_2' = 0.6655, \quad p_2'/p_1' = 2.8575, \quad T_2'/T_1' = 1.3949$$

Thus: $p_2' = \dfrac{p_2'}{p_1'} p_1' = 2.8575(14.7) = \underline{42.0} \text{ psia} = p_2$

$T_2' = \dfrac{T_2'}{T_1'} T_1' = 1.3949(520) = \underline{725}^O R = T_2$

$a_2' = \sqrt{\gamma g_c R T_2'} = \sqrt{(1.4)(32.2)(53.3)(725)} = 1320 \text{ ft/sec} = a_2$

$V_2' = M_2' a_2' = (0.6655)(1320) = 878 \text{ ft/sec}$

$V_2 = V_S - V_2' = 1800 - 878 = \underline{922} \text{ ft/sec}$

Therefore, after the shock passes (referring now to Figure 7.1) the pressure and temperature will be 42 psia and 725°R, respectively, and the air will have acquired a velocity of 922 ft/sec to the left. It will be interesting to compute and compare the stagnation pressures in each case. Notice that they are completely different because of the change in reference that has taken place.

For Figure 7.1 $p_{t1} = p_1 = \underline{14.7} \text{ psia}$

$M_2 = V_2/a_2 = 922/1320 = 0.698$

$p_{t2} = \dfrac{p_{t2}}{p_2} p_2 = \dfrac{1}{0.7222} (42) = \underline{58.2} \text{ psia}$

For Figure 7.2 $p_{t1}' = \dfrac{p_{t1}'}{p_1'} p_1' = \dfrac{1}{0.2318} (14.7) = \underline{63.4} \text{ psia}$

$p_{t2}' = \dfrac{p_{t2}'}{p_2'} p_2' = \dfrac{1}{0.7430} (42) = \underline{56.5} \text{ psia}$

For the steady flow picture $p_{t2}' < p_{t1}'$ as expected. However, note that this decrease in stagnation pressure does not occur for the unsteady case. You might compute the stagnation temperatures on each side of the shock for the unsteady and steady flow cases. Would you expect $T_{t2} = T_{t1}$? How about T_{t1}' and T_{t2}' ?

Another type of moving shock is illustrated in Figure 7.3 where air is flowing through a duct under known conditions and a valve is suddenly closed. The fluid is compressed as it is quickly brought to rest. This results in a shock wave propagating back through the duct as shown. In this case the problem is not only to determine the conditions that exist after passage of the shock but also to predict the speed of the shock wave.

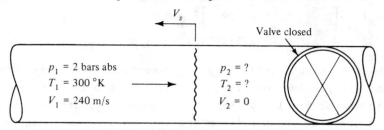

Figure 7.3 Moving normal shock in duct

Our procedure is exactly the same as before. We hop on the shock wave and with this new frame of reference we have the standing normal shock problem shown in Figure 7.4. (We have merely superimposed the velocity V_s on the entire flow field.)

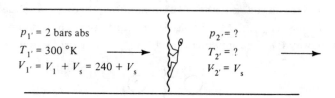

Figure 7.4 Moving shock transformed into stationary shock

The solution of this problem, however, is not so straightforward as the previous example for the reason that the velocity of the shock wave is unknown. Since V_s is unknown, V_1' is unknown and M_1' cannot be calculated. We could approach this as a trial and error problem but a direct solution is available to us. Recall the relation for the velocity difference across a normal shock which was developed in the previous unit (equation 6.32). Applied to Figure 7.4 this becomes

$$\frac{V_1' - V_2'}{a_1'} = \frac{2}{\gamma+1}\left[\frac{M_1'^2 - 1}{M_1'}\right] \tag{7.1}$$

$$a_1' = \left[\gamma g_c R T_1'\right]^{\frac{1}{2}} = \left[(1.4)(1)(287)(300)\right]^{\frac{1}{2}} = 347 \text{ m/s}$$

$$(V_1'-V_2')/a_1' = 240/347 = \underline{0.6916}$$

From the normal-shock tables we see that $M_1' \approx 1.5$

and $M_2' = 0.7011$, $T_2'/T_1' = 1.3202$, $p_2'/p_1' = 2.4583$

$\quad\quad p_2' = (2.4583)(2) = \underline{4.92}$ bars abs $= p_2$

$\quad\quad T_2' = (1.3202)(300) = \underline{396}^\circ K = T_2$

$\quad\quad a_2' = \left[(1.4)(1)(287)(396)\right]^{\frac{1}{2}} = 399$ m/s

$\quad\quad V_2' = M_2' a_2' = (0.7011)(399) = \underline{280}$ m/s $= V_s$

Do not forget that the <u>static</u> temperatures and pressures obtained in problem solutions of this type are the desired answers to the original problem but the velocities and Mach numbers for the standing shock problem are <u>not</u> the same as those in the original moving shock problem.

7.4 TANGENTIAL VELOCITY SUPERPOSITION OBLIQUE SHOCKS

We now consider the standing normal shock shown in Figure 7.5. To emphasize the fact that these velocities are normal to the shock front we label them V_{1n} and V_{2n}. Recall that the velocity is decreased as the fluid passes through a shock wave and thus $V_{1n} > V_{2n}$. Also remember that for this type of shock V_{1n} must always be supersonic and V_{2n} is always subsonic.

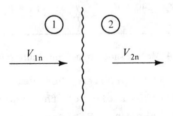

Figure 7.5 Standing normal shock

Now let us superimpose on the entire flow field a velocity of magnitude V_t which is perpendicular to V_{1n} and V_{2n}. This is equivalent to running <u>along</u> the shock front at a speed of V_t. The resulting picture is shown in Figure 7.6. As before, we realize that velocity superposition does not affect the static states of the fluid. What does change?

We normally would view this picture in a slightly different manner. If we concentrate on the total velocity (rather than its components) we see the flow as illustrated in Figure 7.7 and immediately notice several things.

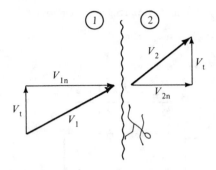

Figure 7.6 Standing normal shock plus tangential velocity

(1) The shock is no longer normal to the approaching flow; hence, it is called an <u>oblique shock</u>.

(2) The flow has been deflected <u>away</u> from the normal.

(3) V_1 must still be supersonic.

(4) V_2 could be supersonic (if V_t is large enough).

We define the "shock angle" θ as the acute angle between the approaching flow (V_1) and the shock front. The "deflection angle" δ is the angle through which the flow has been deflected.

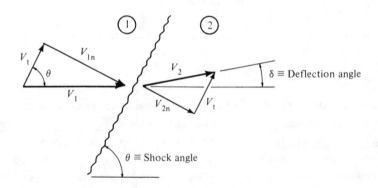

Figure 7.7 Oblique shock with angle definitions

Viewing the oblique shock in the above manner (as a combination of a normal shock and a tangential velocity) permits one to use the normal-shock equations and tables to solve oblique-shock problems for perfect gases providing proper care is taken.

$$V_{1n} = V_1 \sin \theta \qquad (7.2)$$

Since sonic velocity is a function of temperature only

$$a_{1n} = a_1 \tag{7.3}$$

Dividing (7.2) by (7.3) we have

$$\frac{V_{1n}}{a_{1n}} = \frac{V_1 \sin \theta}{a_1} \tag{7.4}$$

or

$$\boxed{M_{1n} = M_1 \sin \theta} \tag{7.5}$$

Thus, if we know the approaching Mach number (M_1) and the shock angle (θ) the normal-shock tables can be utilized by using the "normal Mach number" (M_{1n}). This procedure can be used to obtain static temperature and pressure changes across the shock, since these are unaltered by the superposition of V_t on the original normal-shock picture.

Let us now investigate the range of possible shock angles that may exist for a given Mach number. We know that for a shock to exist

$$M_{1n} \geq 1 \tag{7.6}$$

Thus,

$$M_1 \sin \theta \geq 1 \tag{7.7}$$

and the minimum θ will occur when $M_1 \sin \theta = 1$

or

$$\theta_{min} = \sin^{-1} \frac{1}{M_1} \tag{7.8}$$

Recall that this is the same expression that was developed for the Mach angle μ . Hence, the Mach angle is the minimum possible shock angle. Note that this is a limiting condition and really no shock exists since for this case $M_{1n} = 1.0$. For this reason these are called "Mach waves" or "Mach lines" rather than shock waves. The maximum value that θ can achieve is obviously 90°. This is another limiting condition and represents our familiar normal shock.

Notice that as the shock angle θ decreases from 90° to the Mach angle μ, M_{1n} decreases from M_1 to 1 . Since the strength of a shock is dependent upon the normal Mach number we have the means to produce a shock of any strength equal to or less than the normal shock. Do you see any possible application of this information for the case of a converging-diverging nozzle with an operating pressure ratio someplace between second and third critical? We shall return to this thought in Section 7.8.

The following example is presented in order to obtain a better understanding of the correlation between oblique and normal shocks. Figure 7.8 shows the given information and we proceed to compute the conditions after the shock.

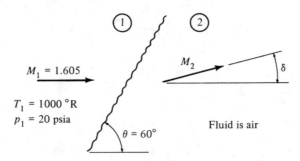

Figure 7.8 Example oblique shock problem

$$a_1 = \left[\gamma g_c R T_1\right]^{\frac{1}{2}} = \left[(1.4)(32.2)(53.3)(1000)\right]^{\frac{1}{2}} = 1550 \text{ ft/sec}$$

$$V_1 = M_1 a_1 = 1.605(1550) = 2488 \text{ ft/sec}$$

$$M_{1n} = M_1 \sin \theta = 1.605 \sin 60^{\circ} = 1.39$$

$$V_{1n} = M_{1n} a_1 = 1.39(1550) = 2155 \text{ ft/sec}$$

$$V_t = V_1 \cos \theta = 2488 \cos 60^{\circ} = 1244 \text{ ft/sec}$$

Using information from the normal-shock tables at $M_{1n} = 1.39$ we find $M_{2n} = 0.7440$, $T_2/T_1 = 1.2483$, $p_2/p_1 = 2.0875$ and $p_{t2}/p_{t1} = 0.9607$. Remember that the static temperatures and pressures are the same whether we are talking about the normal shock or the oblique shock.

$$p_2 = (p_2/p_1)(p_1) = 2.0875(20) = 41.7 \text{ psia}$$

$$T_2 = (T_2/T_1)(T_1) = 1.2483(1000) = 1248^{\circ}R$$

$$a_2 = \left[\gamma g_c R T_2\right]^{\frac{1}{2}} = \left[(1.4)(32.2)(53.3)(1248)\right]^{\frac{1}{2}} = 1732 \text{ ft/sec}$$

$$V_{2n} = M_{2n} a_2 = 0.7440(1732) = 1289 \text{ ft/sec}$$

$$V_{2t} = V_{1t} = V_t = 1244 \text{ ft/sec}$$

$$V_2 = \left[(V_{2n})^2 + (V_{2t})^2\right]^{\frac{1}{2}} = \left[(1289)^2 + (1244)^2\right]^{\frac{1}{2}} = 1791 \text{ ft/sec}$$

$$M_2 = V_2/a_2 \qquad\qquad = 1791/1732 \qquad\qquad = 1.034$$

Note that, although the <u>normal component</u> is subsonic after the shock, the velocity after the shock is supersonic in this case.

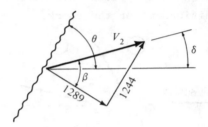

We now calculate the deflection angle.

$$\tan \beta = \frac{1244}{1289} = 0.9651, \quad \beta = 44^{\circ}$$

$$90 - \theta = \beta - \delta$$

Thus, $\delta = \theta - 90 + \beta = 60 - 90 + 44 = 14^{\circ}$

Once δ is known, an alternate calculation for M_2 would be

$$\boxed{M_2 = \frac{M_{2n}}{\sin(\theta-\delta)}}$$

$$M_2 = \frac{0.7440}{\sin(60-14)} = 1.034$$

Let us also compute the stagnation pressures.

$$p_{t1} = \frac{p_{t1}}{p_1} \, p_1 = \frac{1}{0.2335} \, (20) = 85.7 \text{ psia}$$

$$p_{t2} = \frac{p_{t2}}{p_2} \, p_2 = \frac{1}{0.5075} \, (41.7) = 82.2 \text{ psia}$$

If we looked at the normal-shock problem and computed stagnation pressures on the basis of the <u>normal</u> Mach numbers, we would have:

$$p_{t1n} = \left(\frac{p_{t1}}{p_1}\right)_n \, p_1 = \frac{1}{0.3187} \, (20) = 62.8 \text{ psia}$$

$$p_{t2n} = \left(\frac{p_{t2}}{p_2}\right)_n \, p_2 = \frac{1}{0.6925} \, (41.7) = 60.2 \text{ psia}$$

Calculate the stagnation temperatures and show that for the actual oblique-shock problem $T_t = 1515°R$ and for the normal-shock problem $T_t = 1386°R$. All of these static and stagnation pressures and temperatures are shown in the T-s diagram of Figure 7.9. This clearly shows the effect of superimposing the tangential velocity on top of the normal-shock problem with the corresponding change in stagnation reference. It is interesting to note that the <u>ratio</u> of stagnation pressures is the same whether figured from the oblique shock problem or the normal shock problem.

$$\frac{p_{t2}}{p_{t1}} = \frac{82.2}{85.7} = 0.959 \qquad \frac{p_{t2n}}{p_{t1n}} = \frac{60.2}{62.8} = 0.959$$

Is this a coincidence? No! Remember that the stagnation pressure <u>ratio</u> is a measure of the loss across the shock. Superposition of a tangential velocity onto a normal shock does not affect the actual shock process, so the losses remain the same. Thus, although one cannot use the stagnation pressures from the normal-shock problem, one can use the stagnation pressure <u>ratio</u> (which is listed in the tables). Be careful! These conclusions do <u>not</u> apply to the moving normal shock which was discussed in the previous section.

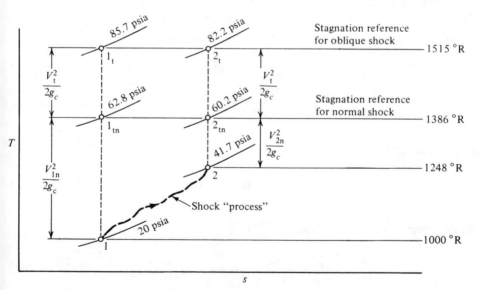

Figure 7.9 *T–s* diagram for oblique shock
(showing the included normal shock)

7.5 OBLIQUE-SHOCK ANALYSIS—PERFECT GAS

In the previous section we saw how an oblique shock could be viewed as a combination of a normal shock and a tangential velocity. If the initial conditions and the shock angle are known the problem can be solved through careful application of the normal shock tables. Frequently, however, the shock angle is not known and thus we seek a new approach to the problem. The oblique shock with its components and angles is shown again in Figure 7.10.

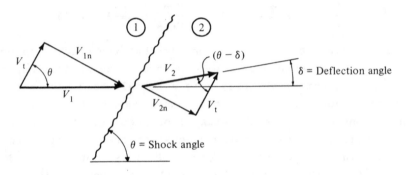

Figure 7.10 Oblique shock

Our objective will be to relate the deflection angle (δ) to the shock angle (θ) and the entering Mach number. We start by applying the continuity equation to a unit area at the shock.

$$\rho_1 V_{1n} = \rho_2 V_{2n} \tag{7.9}$$

or

$$\frac{\rho_2}{\rho_1} = \frac{V_{1n}}{V_{2n}} \tag{7.10}$$

From Figure 7.10 we see that:

$$V_{1n} = V_t \tan \theta \quad \text{and} \quad V_{2n} = V_t \tan (\theta-\delta) \tag{7.11}$$

Thus, from equations 7.10 and 7.11

$$\frac{\rho_2}{\rho_1} = \frac{V_{1n}}{V_{2n}} = \frac{V_t \tan \theta}{V_t \tan (\theta-\delta)} = \frac{\tan \theta}{\tan (\theta-\delta)} \tag{7.12}$$

From the normal-shock relations that we derived in Unit 6, property ratios across the shock were developed as a function of the approaching (normal) Mach number. Specifically, the density ratio was given in equation 6.26 as

$$\frac{\rho_2}{\rho_1} = \frac{(\gamma+1) \, M_{1n}^2}{(\gamma-1) \, M_{1n}^2 + 2} \tag{6.26}$$

Note that we have added subscripts to the Mach numbers to indicate that these are normal to the shock. Equating (7.12) and (6.26) yields:

$$\frac{\tan \theta}{\tan (\theta-\delta)} = \frac{(\gamma+1) \, M_{1n}^2}{(\gamma-1) \, M_{1n}^2 + 2} \tag{7.13}$$

But

$$M_{1n} = M_1 \sin \theta \tag{7.5}$$

Hence

$$\frac{\tan \theta}{\tan (\theta-\delta)} = \frac{(\gamma+1) \, M_1^2 \sin^2 \theta}{(\gamma-1) \, M_1^2 \sin^2 \theta + 2} \tag{7.14}$$

and we have succeeded in relating the shock angle, deflection angle, and entering Mach number. Unfortunately, equation 7.14 cannot be solved for θ as an explicit function of M, δ, and γ, but we can obtain an explicit solution for

$$\delta = f(M, \theta, \gamma)$$

which is

$$\tan \delta = 2 \cot \theta \left[\frac{M_1^2 \sin^2 \theta - 1}{M_1^2(\gamma + \cos 2\theta) + 2} \right] \tag{7.15}$$

It is interesting to examine equation 7.15 for the extreme values of θ that might accompany any given Mach number.

For $\theta = \theta_{max} = \pi/2$

In this case equation 7.15 yields $\tan \delta = 0$, or $\delta = 0$, which we know to be true for the normal shock.

For $\theta = \theta_{min} = \sin^{-1} \frac{1}{M_1}$

In this case equation 7.15 again yields $\tan \delta = 0$, or $\delta = 0$, which we know to be true for the limiting case of the Mach wave or no shock.

Thus, the relationship developed for the oblique shock includes
as special cases the strongest shock possible (normal shock) and
the weakest shock possible (no shock) as well as all other inter-
mediate strength shocks. Note that for the given deflection angle of $\delta = 0^\circ$ there are two possible shock angles for any given
Mach number. In the next section we shall see that this holds
true for any deflection angle.

7.6 OBLIQUE-SHOCK TABLES AND CHARTS

Equation 7.14 provides a relationship among the shock angle,
deflection angle and entering Mach number. Our motivation to
obtain this relationship was to solve problems in which the shock
angle (θ) is the unknown, but we found that an explicit solution for $\theta = f(M,\delta,\gamma)$ was not possible. The next best thing is
to plot equation 7.14 or 7.15. This can be done in several ways
but it is perhaps most instructive to look at a plot of shock
angle (θ) versus entering Mach number (M_1) for various deflection angles (δ). This is shown in Figure 7.11.

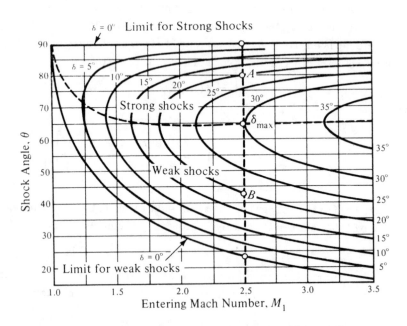

Figure 7.11 Oblique shock relations among θ, M, and δ ($\gamma = 1.4$)
(Adapted with permission from *The Dynamics and
Thermodynamics of Compressible Fluid Flow* by
Ascher H. Shapiro, Copyright 1953, The Ronald Press
Company, New York.)

One can quickly visualize (from Figure 7.11) all possible shocks for any entering Mach number. For example, the dotted vertical line at $M_1 = 2.5$ starts at the top of the plot with the normal shock ($\theta = 90^\circ$, $\delta = 0^\circ$) which is the strongest possible shock. As we move downward the shock angle continually decreases to $\theta_{min} = \mu$ (Mach angle) which means that the shock strength is continually decreasing. Why is this so? What is the "normal Mach number" doing as we move down this line?

It is interesting to note that as the shock angle decreases the deflection angle at first increases from $\delta = 0$ to $\delta = \delta_{max}$ and then the deflection angle decreases back to zero. Thus, for any given Mach number and deflection angle there are two shock situations possible (assuming $\delta < \delta_{max}$). Two such points are labeled A and B. One of these (A) is associated with the higher shock angle and thus has a higher normal Mach number which means that it is a stronger shock with a resulting higher pressure ratio. The other (B) has the lower shock angle and thus is a weaker shock with a lower pressure rise across the shock.

All of the "strong shocks" (above the δ_{max} points) result in subsonic flow after passage through the shock wave. In general, nearly all the region of "weak shocks" (below δ_{max}) result in supersonic flow after the shock although there is a very small region just below δ_{max} where M_2 is still subsonic. Normally, we find the weak shock solution occuring more frequently, although this is entirely dependent upon the boundary conditions that are imposed. This point, along with several applications of oblique shocks, is the subject of the next two sections. For working problems it is suggested that you use the more detailed oblique-shock chart located in the Appendix. One can also use detailed oblique-shock tables such as those by Keenan and Kaye (reference 21).

Example 7.1 Observation of an oblique shock in air reveals that a Mach 2.2 flow at $550^\circ K$ and 2 bars absolute is deflected by 14° . What are the conditions after the shock? Assume that the weak solution prevails.

We enter the chart with $M_1 = 2.2$ and $\delta = 14^\circ$ and find that $\theta = 40^\circ$ and 83° . Knowing that the weak solution exists we select $\theta = 40^\circ$.

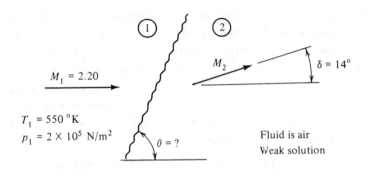

$$M_{1n} = M_1 \sin \theta = 2.2 \sin 40° = \underline{1.414}$$

Enter the normal-shock tables at $M_{1n} = 1.414$ and interpolate:
$M_{2n} = 0.7339$, $T_2/T_1 = 1.2638$ and $p_2/p_1 = 2.1660$

$$T_2 = \frac{T_2}{T_1} T_1 = 1.2638(550) = \underline{695°K}$$

$$p_2 = \frac{p_2}{p_1} p_1 = 2.166(2 \times 10^5) = \underline{4.33 \times 10^5} \ N/m^2$$

$$M_2 = \frac{M_{2n}}{\sin(\theta - \delta)} = \frac{0.7339}{\sin(40 - 14)} = \underline{1.674}$$

7.7 BOUNDARY CONDITION OF FLOW DIRECTION

We have seen that one of the characteristics of an oblique shock is that the flow direction is changed. In fact this is one of only two methods by which a supersonic flow can be turned. (The other method will be discussed in Unit 8.) Consider supersonic flow over a wedge-shaped object as shown in Figure 7.12. For example, this could represent the leading edge of a supersonic airfoil.

In this case the flow is forced to change direction to meet the boundary condition of flow tangency along the wall and this can only be done through the mechanism of an oblique shock. The example in the previous section was just such a situation. (Recall that a flow of $M = 2.2$ was deflected by $14°$.) Now for any given Mach number and deflection angle there are two possible shock angles. Thus, the question naturally arises as to which solution will occur, the "strong" one or the "weak" one. Here is where the surrounding pressure must be considered. Recall that the strong shock occurs at the higher shock angle and results in

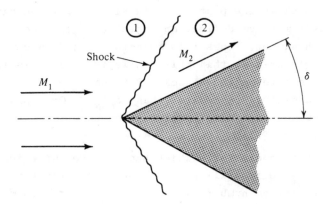

Figure 7.12 Supersonic flow over a wedge

a large pressure change. In order for this solution to occur
a physical situation must exist which creates the necessary pressure differential. It is conceivable that such a case might exist in an internal flow situation. However, for an external flow
situation such as around the airfoil there is no means available
to support the greater pressure difference required by the strong
shock. Thus, in external flow problems (flow around objects) we
always find the weak solution.

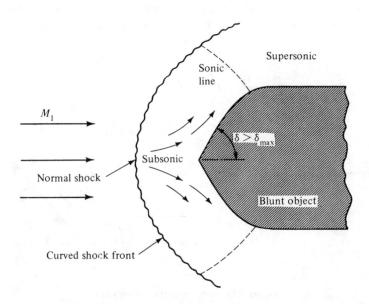

Figure 7.13 Detached shock caused by $\delta > \delta_{max}$

Looking back at the plot shown in Figure 7.11 you may notice that there is a maximum deflection angle (δ_{max}) associated with any given Mach number. Does this mean that the flow cannot turn through an angle greater than this? This is true if we limit ourselves to the simple oblique shock. But what happens if we build a wedge with a half angle greater than δ_{max}? Or suppose we ask the flow to pass over a blunt object? The resulting flow pattern is shown in Figure 7.13.

A "detached shock" forms which has a curved wave front. Behind this wave we find all possible shock solutions associated with the initial Mach number M_1. At the center a normal shock exists with subsonic flow resulting. The subsonic flow has no difficulty adjusting to the large deflection angle required. As the wave front curves around, the shock angle continually decreases with a resultant decrease in shock strength. Eventually we reach a point where supersonic flow exists after the shock front. Although Figures 7.12 and 7.13 illustrate flow over objects, the same patterns result from internal flow along a wall, or "corner flow," as shown in Figure 7.14. The significance of δ_{max} is again seen to be the maximum deflection angle for which the shock can remain "attached" to the corner.

A very practical situation involving a detached shock is caused when a pitot tube is installed in a supersonic tunnel. See Figure 7.15. The tube will reflect the total pressure after the shock front, which at this location is a normal shock. An

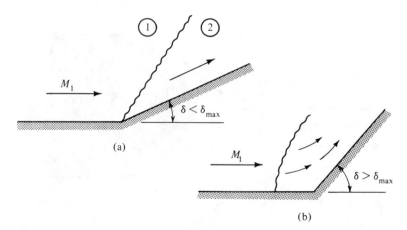

Figure 7.14 Supersonic flow in a corner

additional tap off the side of the tunnel can pick up the static pressure ahead of the shock.

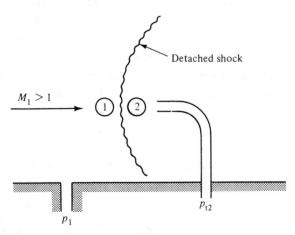

Figure 7.15 Supersonic pitot tube installation

Consider the ratio
$$\frac{p_{t2}}{p_1} = \frac{p_{t2}}{p_{t1}} \frac{p_{t1}}{p_1}$$

p_{t2}/p_{t1} is the total pressure ratio across the shock and is a function of M_1 only (see equation 6.28). p_{t1}/p_1 is also a function of M_1 only (see equation 5.40). Thus, the ratio p_{t2}/p_1 is a function of the initial Mach number and can be found as a parameter in the shock tables.

Example 7.2 A supersonic pitot tube indicates a total pressure of 30 psig and a static pressure of zero gage. Determine the free-stream velocity if the temperature of the air is $450^{\circ}R$.

$$\frac{p_{t2}}{p_1} = \frac{(30+14.7)}{(0+14.7)} = \frac{44.7}{14.7} = 3.041$$

From the shock tables we find M_1 = 1.398

$$a_1 = \left[(1.4)(32.2)(53.3)(450)\right]^{\frac{1}{2}} = 1040 \text{ ft/sec}$$

$$V_1 = M_1 a_1 = 1.398(1040) = \underline{1454} \text{ ft/sec}$$

So far we have discussed oblique shocks that are caused by flow deflections. Another example of this is found in engine inlets of supersonic aircraft. Here, as aircraft and missile

speeds increase, we frequently see two directional changes with
their accompanying shock systems as shown in Figure 7.16. The
losses that occur across the series of shocks shown are less than
those which would occur across a single normal shock at the same
initial Mach number. Further discussion on the design of supersonic diffusers for propulsion systems can be found in Unit 11.
A warning should be given here concerning the application of our
results to inlets with circular cross-sections. These will have
conical "spikes" for flow deflection which cause conical shock
fronts to form. This type of shock has been analyzed and appropriate charts have been developed but we shall not go into them
in this text. For a conical shock analysis see pages 487-495 of
Thompson (reference number 20).

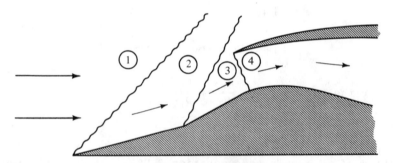

Figure 7.16 Multiple-shock inlet for supersonic aircraft

7.8 BOUNDARY CONDITION OF PRESSURE EQUILIBRIUM

Now let us consider a case where the existing pressure conditions cause an oblique shock to form. Recall our friend, the
converging-diverging nozzle. When it is operating at its second
critical point a normal shock is located at the exit plane. The
pressure rise that occurs across this shock is exactly that which
is required to go from the low pressure that exists within the
nozzle up to the higher receiver pressure which has been imposed
on the system. We again emphasize the fact that the existing
operating pressure ratio is what causes the shock to be located
at this particular position. (If you have forgotten these details,
review Section 6.6.)

We now ask, "What happens when the operating pressure ratio
is between second and third critical?" A normal shock is too
strong to meet the required pressure rise. What is needed is a
compression process that is weaker than a normal shock and our

oblique shock is precisely the mechanism for the job! <u>No</u> <u>matter</u>
<u>what</u> <u>pressure</u> <u>rise</u> <u>is</u> <u>required</u>, the shock can form at an angle
which will produce any desired pressure rise from that of a nor-
mal shock on down to the third critical condition which requires
no pressure change. Figure 7.17 shows a typical oblique shock at
the lip of a two-dimensional nozzle. We have shown only half the
picture as symmetry considerations force the upper half to be the
same. This also permits an alternate viewpoint of thinking of
the central streamline as though it were a solid boundary.

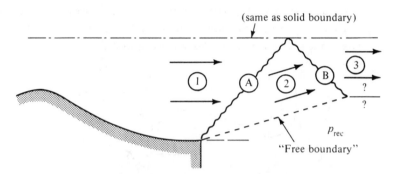

Figure 7.17 Supersonic nozzle operating between second and
third critical

The flow in region 1 is parallel to the centerline and is
at the design conditions for the nozzle; i.e., the flow is supersonic and $p_1 < p_{rec}$. Oblique shock A forms at the appropriate
angle such that the pressure rise that occurs is just sufficient
to meet the boundary condition of $p_2 = p_{rec}$. There is a "free
boundary" between the jet and the surroundings as opposed to a
"physical boundary." Now remember that the flow is also turned
away from the normal and thus will have the direction as indicated in region 2.

This presents a problem since the flow in region 2 cannot
cross the centerline. Something must occur where wave A meets
the centerline and this something must turn the flow parallel to
the centerline. Here it is the boundary condition of flow direction that causes another oblique shock B to form which not only
changes the flow direction but also increases the pressure still
further. Since $p_2 = p_{rec}$, and $p_3 > p_2$, then $p_3 > p_{rec}$ and
pressure equilibrium does not exist between region 3 and the receiver.

Obviously, some type of an expansion is needed which emanates from the point where wave B intersects the free boundary. An "expansion shock" would be just the thing but we know that such an animal cannot exist. Do you recall why not? We shall have to study another phenomenon before we can complete the story of a supersonic nozzle operating between second and third critical. This will be covered in the next unit.

Example 7.3 A converging-diverging nozzle with an area ratio of 5.9 is fed by air from a chamber with a stagnation pressure of 100 psia. Exhaust is to the atmosphere at 14.7 psia. Show that this nozzle is operating between second and third critical and determine the conditions after the first shock.

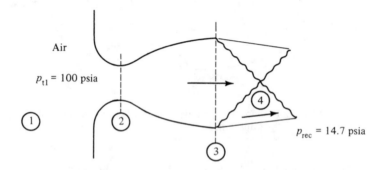

Third critical: $\dfrac{A_3}{A_3^*} = \dfrac{A_3}{A_2}\dfrac{A_2}{A_2^*}\dfrac{A_2^*}{A_3^*} = (5.9)(1)(1) = 5.9$

$M_3 = 3.35$ and $p_3/p_{t3} = 0.01625$

$\dfrac{p_3}{p_{t1}} = \dfrac{p_3}{p_{t3}}\dfrac{p_{t3}}{p_{t1}} = (0.01625)(1) = \underline{0.01625}$

Second critical: Normal shock at $M_3 = 3.35$, $p_4/p_3 = 12.9263$

$\dfrac{p_4}{p_{t1}} = \dfrac{p_4}{p_3}\dfrac{p_3}{p_{t1}} = (12.9263)(0.01625) = \underline{0.2100}$

Since our operating pressure ratio (14.7/100 = 0.147) lies between that of second and third critical an oblique shock must form as shown. Remember, under these conditions the nozzle operates internally as if it were at the third critical point. Thus, the required pressure ratio across the oblique shock is

$$\frac{p_4}{p_3} = \frac{p_{rec}}{p_3} = \frac{14.7}{1.625} = 9.046$$

From the normal-shock tables we see that this pressure ratio requires $M_{3n} = 2.81$, $M_{4n} = 0.4875$

$$\sin \theta = \frac{M_{3n}}{M_3} = \frac{2.81}{3.35} = 0.8388, \quad \theta = \underline{57^\circ}$$

From the oblique-shock chart, $\delta = \underline{34^\circ}$

$$M_4 = \frac{M_{4n}}{\sin(\theta-\delta)} = \frac{0.4875}{\sin(57-34)} = \underline{1.25}$$

Thus, to match the receiver pressure an oblique shock forms at 57°. The flow is deflected 34° and is still supersonic at a Mach number of 1.25 .

7.9 SUMMARY

We have seen how a standing normal shock can be made into a moving normal shock by superposition of a velocity (normal to the shock front) on the entire flow field. Similarly, the superposition of a velocity tangent to the shock front turns a normal shock into an oblique shock. Since velocity superposition does not change the static conditions in a flow field, the normal shock tables may be used to solve oblique-shock problems if we deal with the "normal Mach number." However, to avoid trial and error solutions, oblique-shock tables and charts are available. The following is a significant relation among the variables in an oblique shock:

$$\tan \delta = 2 \cot \theta \left[\frac{M_1^2 \sin^2 \theta - 1}{M_1^2 (\gamma + \cos 2\theta) + 2} \right] \qquad (7.15)$$

Another helpful relation is

$$M_2 = \frac{M_{2n}}{\sin(\theta-\delta)}$$

We summarize the important characteristics of an oblique shock.

1. The flow is always turned "away" from the normal.

2. For given values of M_1 and δ, two values of θ may
    result.
    (a) If a large pressure ratio is available (or required)
         the strong shock at the higher θ will occur and
         M_2 will be subsonic.
    (b) If a small pressure ratio is available (or required)
         the weak shock at the lower θ will occur and M_2
         will be supersonic (except for a small region near
         δ_{max}).
3. A maximum value of δ exists for any given Mach number.
    If δ is physically greater than δ_{max}, then a "de-
    tached" shock will form.

It is important to realize that oblique shocks are caused for two
reasons:
1. To meet a physical boundary condition which causes the
    flow to change direction, or
2. To meet a free boundary condition of pressure equilibrium.

An alternate way of stating this is to say that the flow must be
tangent to any boundary, whether it is a physical wall or a "free
boundary." If it is a free boundary, then pressure equilibrium
must also exist across the flow boundary.

7.10 PROBLEMS

1. A normal shock is traveling into still air (14.7 psia and
    $520^{\circ}R$) at a velocity of 1800 ft/sec.
    (a) Determine the temperature, pressure and velocity that
        exist after passage of the shock wave.
    (b) What is the entropy change experienced by the air?

2. The velocity of a certain atomic blast wave has been deter-
    mined to be approximately 46,000 m/s relative to the ground.
    Assume that it is moving into still air at $300^{\circ}K$ and 1 bar.
    What are the static and stagnation temperatures and pressures
    that exist after the blast wave passes? (Hint: You will
    have to resort to equations as the tables do not cover this
    Mach number range.)

3. Air flows in a duct and a valve is quickly closed. A normal
    shock is observed to propagate back through the duct at a
    speed of 1010 ft/sec. After the air has been brought to rest
    its temperature and pressure are $600^{\circ}R$ and 30 psia, respec-

tively. What were the original temperature, pressure and velocity of the air before the valve was closed?

4. Oxygen at 100°F and 20 psia is flowing at 450 ft/sec in a duct. A valve is quickly shut, causing a normal shock to travel back through the duct.
   (a) Determine the speed of the traveling shock wave.
   (b) What are the temperature and pressure of the oxygen that is brought to rest?

5. A closed tube contains nitrogen at 20°C and a pressure of 1×10^4 N/m^2. A shock wave progresses through the tube at a speed of 380 m/s.
   (a) Calculate the conditions that exist immediately after the shock wave passes a given point. (The fact that this is inside a tube should not bother you as it is merely a normal shock moving into a gas at rest.)
   (b) When the shock wave hits the end wall it is reflected back. What are the temperature and pressure of the gas between the wall and the reflected shock? At what speed is the reflected shock traveling? (This is just like the sudden closing of a valve in a duct.)

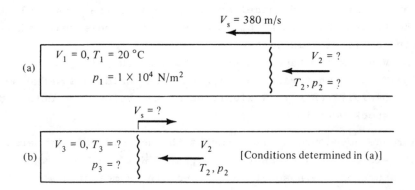

6. An oblique shock forms in air at an angle of $\theta = 30^\circ$. Before passing through the shock the air has a temperature of 60°F, a pressure of 10 psia and is traveling at $M = 2.6$.
   (a) Compute the normal and tangential velocity components before and after the shock.
   (b) Determine the temperature and pressure after the shock.
   (c) What is the deflection angle?

7. Conditions before a shock are $T_1 = 40^\circ C$, $p_1 = 1.2$ bars, and
    $M_1 = 3.0$. An oblique shock is observed at 45° to the
    approaching air flow.
    (a) Determine the Mach number and flow direction after the
        shock.
    (b) What are the temperature and pressure after the shock?
    (c) Is this a weak or strong shock?

8. Air at $800^\circ R$ and 15 psia is flowing at a Mach number of
    $M = 1.8$ and is deflected through a 15° angle. The direc-
    tional change is accompanied by an oblique shock.
    (a) What are the possible shock angles?
    (b) For each shock angle compute the temperature and pressure
        after the shock.

9. The supersonic flow of a gas ($\gamma = 1.4$) approaches a wedge
    with a half angle of 24° ($\delta = 24^\circ$).
    (a) What Mach number will put the shock on the verge of de-
        taching?
    (b) Is this value a minimum or maximum?

10. A simple wedge with a total included angle of 28° is used
    to measure the Mach number of supersonic flows. When in-
    serted into a wind tunnel and aligned with the flow, oblique
    shocks are observed at 50° angles to the free-stream (simi-
    lar to Figure 7.12).
    (a) What is the Mach number in the wind tunnel?
    (b) Through what range of Mach numbers could this wedge be
        useful? (Hint: Would it be of any value if a detached
        shock were to occur?)

11. A round-nosed projectile travels through air at a temperature
    of $-15^\circ C$ and a pressure of 1.8×10^4 N/m^2. The stagnation
    pressure on the nose of the projectile is measured at
    2.1×10^5 N/m^2.
    (a) At what speed (m/s) is the projectile traveling?
    (b) What is the temperature on the projectile's nose?

12. A pitot tube is installed in a wind tunnel in the manner
    shown in Figure 7.15. Tunnel air temperature is $500^\circ R$ and
    the static tap (p_1) indicates a pressure of 14.5 psia.
    (a) Determine the tunnel air velocity if the stagnation probe
        (p_{t2}) indicates 65 psia.

(b) Suppose p_{t2} = 26 psia. What is the tunnel velocity under this condition?

13. A converging-diverging nozzle is designed to produce an exit Mach number of 3.0 when γ = 1.4 . When operating at its second critical point the shock angle is 90° and the deflection angle is zero. Call p_{exit} the pressure at the exit plane of the nozzle just before the shock. As the receiver pressure is lowered both θ and δ change. For the range between second and third critical:
    (a) Plot θ versus p_{rec}/p_{exit} .
    (b) Plot δ versus p_{rec}/p_{exit} .

14. Pictured below is the air inlet to a jet aircraft. The plane is operating at 50,000 ft where the pressure is 243 psfa and the temperature is 392°R. Assume that the flight speed is M_o = 2.5 .
    (a) What are the conditions of the air (temperature, pressure and entropy change) just after it passes through the normal shock?
    (b) If the single 15° wedge is replaced by a double wedge of 7° and 8° (see Figure 7.16), determine the conditions of the air after it enters the diffuser.
    (c) Compare the losses for (a) and (b).

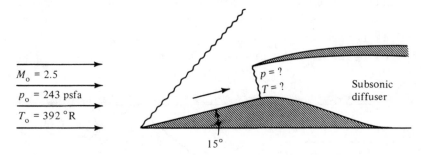

M_o = 2.5
p_o = 243 psfa
T_o = 392 °R

p = ?
T = ?

Subsonic diffuser

15°

15. A converging-diverging nozzle is operating between the second and third critical points as shown in Figure 7.17. M_1 = 2.5 , T_1 = 150°K , p_1 = 0.7 bar , the receiver pressure is 1 bar, and the fluid is nitrogen.
    (a) Compute the Mach number, temperature and flow deflection in region 2.
    (b) Through what angle is the flow deflected as it passes through shock wave B?
    (c) Determine conditions in region 3.

16. For the flow situation shown below, $M_1 = 1.8$, $T_1 = 600°R$, $p_1 = 15$ psia and $\gamma = 1.4$.

   (a) Find conditions in region 2 assuming they are supersonic.

   (b) What must occur along the line marked -----?

   (c) Find conditions in region 3.

   (d) How would the problem change if the flow in region 2 were subsonic?

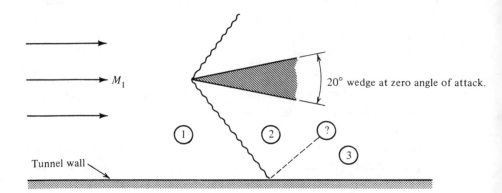

20° wedge at zero angle of attack.

17. Carbon monoxide flows in the duct shown below. The first shock, which turns the flow 15°, is observed to form at a 40° angle. The flow is known to be supersonic in regions 1 and 2 and subsonic in region 3.

   (a) Determine M_3 and β.

   (b) Determine the pressure ratios p_3/p_1 and p_{t3}/p_{t1}.

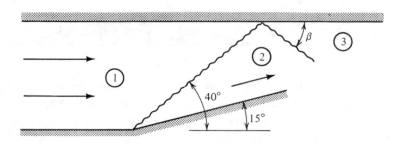

18. A uniform flow of air has a Mach number of 3.3. The bottom of the duct is bent upwards at a 25° angle. At the point where the shock intersects the upper wall the boundary is bent 5° upwards as shown. Assume that the flow is supersonic throughout the system. Compute M_3, p_3/p_1, T_3/T_1 and β.

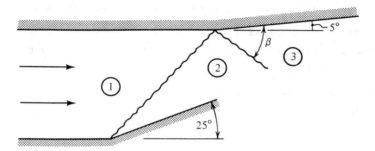

7.11 CHECK TEST

You should be able to complete this test without reference to
material in the unit.

1.

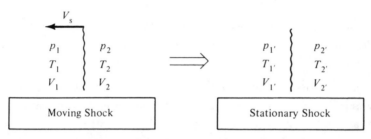

By velocity superposition the moving shock picture shown can
be transformed into the stationary shock problem shown. Cir-
cle the statements below which are true:

(a) $p_1 = p_1'$ $p_1 < p_1'$ $p_1 > p_1'$ $p_1' = p_2'$

(b) $T_{t1}' > T_{t2}'$ $T_{t1}' = T_{t2}'$ $T_{t1}' < T_{t2}'$ $T_{t1} = T_{t1}'$

(c) $\rho_1 > \rho_2$ $\rho_1 = \rho_2$ $\rho_1' < \rho_1$ $\rho_1' > \rho_2'$

(d) $u_2' > u_1'$ $u_2' = u_1'$ $u_2' < u_1'$ $u_2' = u_2$

       ($u \equiv$ internal energy)

2. Fill in the blanks from the choices indicated.

(a) A blast wave will travel through standard air (14.7 psia
   and 60°F) at a speed (less than, equal to, greater than)
   _____ approximately 1118 ft/sec.

(b) If an oblique shock is broken down into components which
   are normal and tangent to the wave front:

   (i)   The normal Mach number (increases, decreases, re-
         mains constant) _____ as the flow passes
         through the wave.

   (ii) The tangential Mach number (increases, decreases,
         remains constant) _____ as the flow
         passes through the wave. (Careful! This deals with
         Mach number, not velocity.)

3. List the conditions which cause an oblique shock to form.

4. Describe the general results of an oblique shock analysis by drawing a plot of shock angle versus inlet Mach number for various deflection angles.

5. Sketch the resulting flow pattern over the nose of the object shown.

$M = 1.5$

15°

6. A normal shock wave travels at 2500 ft/sec into still air at $520°R$ and 14.7 psia. What velocity exists just after the wave passes?

7. Oxygen at 5 psia and $450°R$ is traveling at $M = 2.0$ and leaves a duct as shown. The receiver conditions are 14.1 psia and $600°R$.
    (a) At what angle will the first shocks form? By how much is the flow deflected?
    (b) What are the temperature, pressure and Mach number in region 2?

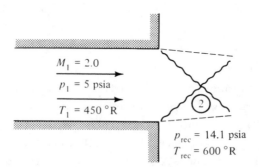

$M_1 = 2.0$

$p_1 = 5$ psia

$T_1 = 450 °R$

2

$p_{rec} = 14.1$ psia

$T_{rec} = 600 °R$

UNIT 8

PRANDTL-MEYER FLOW

8.1 INTRODUCTION

This unit begins with an examination of weak shocks. We will show that for very weak oblique shocks the pressure change is related to the <u>first</u> power of the deflection angle whereas the entropy change is related to the <u>third</u> power of the deflection angle. This will enable us to explain how a smooth turn can be accomplished isentropically — a situation which is called "Prandtl-Meyer Flow." Being reversible, these flows may be expansions or compressions depending on the circumstances.

A detailed analysis of Prandtl-Meyer flow is made for the case of a perfect gas and, as usual, tables are developed to aid in problem solution. Typical flow fields involving Prandtl-Meyer flow will be discussed. In particular, the performance of a converging-diverging nozzle can now be fully explained, as well as supersonic flow around objects.

8.2 OBJECTIVES

After sucessfully completing this unit you should be able to:

1. State how entropy changes and pressure changes vary with deflection angles for weak oblique shocks.

2. Explain how finite turns (with finite pressure ratios) can be
   accomplished isentropically in supersonic flow.

3. Describe and sketch what occurs as fluid flows supersonically
   past a smooth concave corner and a smooth convex corner.

4. Show Prandtl-Meyer flow (both expansions and compressions) on
   a T-s diagram.

5. Develop the differential relation between Mach number (M) and
   flow turning angle (ν) for Prandtl-Meyer flow. (Optional.)

6. Show how tables can be developed for Prandtl-Meyer flow by
   the arbitrary selection of $\nu = 0$ when $M = 1$.

7. Explain the governing boundary conditions and show the results
   when shock waves and P-M waves reflect off: (a) physical
   boundaries, and (b) "free" boundaries.

8. Explain with the aid of diagrams the flow conditions at the
   outlet of a supersonic nozzle created by underexpansion and
   overexpansion.

9. Draw the wave forms created by flow over rounded and/or wedge-
   shaped wings as the angle of attack changes. Be able to solve
   for the flow properties in each region.

10. Demonstrate the ability to solve typical Prandtl-Meyer flow
    problems by use of the appropriate equations and tables.

8.3 ARGUMENT FOR ISENTROPIC TURNING FLOW

A Pressure Change for Normal Shocks

Let us first investigate some special characteristics of any
normal shock. Throughout this section we will assume that the
medium is a perfect gas and this will enable us to develop some
precise relations. We begin by recalling the outcome of the continuity equation.

$$\rho_1 V_1 = \rho_2 V_2 \qquad\qquad (6.2)$$

and from equation 6.7 the result of the momentum analysis can be written as

$$p_1 - p_2 = \rho \frac{V}{g_c} (V_2 - V_1) \qquad (8.1)$$

Choosing $\rho_1 V_1$ for the term ρV in equation 8.1 we have

$$p_1 - p_2 = \frac{\rho_1 V_1}{g_c} (V_2 - V_1) = \frac{\rho_1 V_1^2}{g_c} (V_2/V_1 - 1) \qquad (8.2)$$

We substitute for the density from the perfect gas equation of state (1.13) and solve for the pressure rise across the shock

$$p_2 - p_1 = \left[\frac{p_1}{RT_1}\right] \frac{V_1^2}{g_c} (1 - V_2/V_1) \qquad (8.3)$$

We also know that $V_1^2 = M_1^2 a_1^2 = M_1^2 \gamma g_c R T_1$

Show that equation 8.3 can be written as

$$p_2 - p_1 = \gamma p_1 M_1^2 (1 - V_2/V_1) \qquad (8.4)$$

or in a non-dimensional form

$$\frac{p_2 - p_1}{p_1} = \gamma M_1^2 \left[1 - \frac{V_2}{V_1}\right] \qquad (8.5)$$

We recall further from Unit 6 that the density ratio across the shock can be expressed as

$$\frac{\rho_2}{\rho_1} = \frac{(\gamma+1) M_1^2}{(\gamma-1) M_1^2 + 2} \qquad (6.26)$$

By using continuity (6.2), equation 6.26 can also represent the velocity ratio

$$\frac{V_1}{V_2} = \frac{\rho_2}{\rho_1} = \frac{(\gamma+1) M_1^2}{(\gamma-1) M_1^2 + 2} \qquad (8.6)$$

If we introduce the reciprocal of equation 8.6 into equation 8.5 the result will be

$$\frac{p_2 - p_1}{p_1} = \gamma M_1^2 \left[1 - \frac{(\gamma-1) M_1^2 + 2}{(\gamma+1) M_1^2}\right] \qquad (8.7)$$

Express the right side over a common denominator and show that
this becomes

$$\frac{p_2 - p_1}{p_1} = \frac{2\gamma}{\gamma+1}(M_1^2 - 1) \qquad (8.8)$$

This relation shows that the pressure rise across a normal shock
is directly proportional to the quantity $(M_1^2 - 1)$. We shall
return to this fact later when we apply this to weak shocks at
very small Mach numbers.

B Entropy Change for Normal Shocks

The entropy change for any process with a perfect gas can be
expressed in terms of the temperatures and pressures by

$$s_2 - s_1 = c_p \ln \frac{T_2}{T_1} - R \ln \frac{p_2}{p_1} \qquad (1.53)$$

From the perfect gas equation of state (1.13) we have

$$\frac{T_2}{T_1} = \left[\frac{p_2}{p_2 R}\right]\left[\frac{\rho_1 R}{p_1}\right] = \frac{p_2}{p_1}\frac{\rho_1}{\rho_2} \qquad (8.9)$$

Thus, equation 1.53 becomes

$$s_2 - s_1 = c_p \ln \frac{p_2}{p_1}\frac{\rho_1}{\rho_2} - R \ln \frac{p_2}{p_1} \qquad (8.10)$$

But the specific heat is related to the gas constant by

$$c_p = \frac{\gamma R}{\gamma-1} \qquad (4.15)$$

If you did not develop equation 4.15 when studying Unit 4 it
might be well to do so at this time. Start with equations 1.49
and 1.50.

Introducing (4.15) into equation 8.10 will produce

$$s_2 - s_1 = \frac{\gamma R}{\gamma-1} \ln \frac{p_2}{p_1}\frac{\rho_1}{\rho_2} - R \ln \frac{p_2}{p_1}$$

or in a non-dimensional form

$$\frac{s_2 - s_1}{R} = \frac{\gamma}{\gamma-1} \ln \frac{p_2}{p_1}\frac{\rho_1}{\rho_2} - \ln \frac{p_2}{p_1} \qquad (8.11)$$

Show that this can be written as

$$\frac{s_2 - s_1}{R} = \ln\left[\left(\frac{p_2}{p_1}\right)^{\frac{1}{\gamma-1}}\left(\frac{\rho_2}{\rho_1}\right)^{-\frac{\gamma}{\gamma-1}}\right] \tag{8.12}$$

Now equation 8.8 can be solved for the pressure ratio

$$\frac{p_2}{p_1} = 1 + \frac{2\gamma}{\gamma+1}(M_1^2 - 1) \tag{8.13}$$

If we substitute (8.13) for the pressure ratio and (6.26) for the density ratio, equation 8.12 becomes

$$\frac{s_2 - s_1}{R} = \ln\left\{\left[1 + \frac{2\gamma}{\gamma+1}(M_1^2 - 1)\right]^{\frac{1}{\gamma-1}}\left[\frac{(\gamma+1)\,M_1^2}{(\gamma-1)\,M_1^2 + 2}\right]^{-\frac{\gamma}{\gamma-1}}\right\} \tag{8.14}$$

To aid in simplification let

$$m \equiv M_1^2 - 1 \tag{8.15}$$

and thus also

$$M_1^2 = m + 1 \tag{8.16}$$

Introduce (8.15) and (8.16) into equation 8.14 and show that this becomes

$$\frac{s_2 - s_1}{R} = \ln\left\{\left[1 + \frac{2\gamma m}{\gamma+1}\right]^{\frac{1}{\gamma-1}}\left[1 + m\right]^{-\frac{\gamma}{\gamma-1}}\left[1 + \frac{(\gamma-1)m}{\gamma+1}\right]^{\frac{\gamma}{\gamma-1}}\right\} \tag{8.17}$$

Now each of the bracketed terms in equation 8.17 is of the form $(1 + x)$ and we can take advantage of the expansion

$$\ln(1 + x) = x - \frac{x^2}{2} + \frac{x^3}{3} - \frac{x^4}{4} + \cdots \tag{8.18}$$

Put equation 8.17 into the proper form to expand each bracket according to (8.18). Be careful to retain all terms up to and including the <u>third</u> power. If you have not made any mistakes you will find that all terms involving m and m^2 cancel out and you are left with

$$\frac{s_2 - s_1}{R} = \frac{2\gamma m^3}{3(\gamma+1)^2} + \text{higher order terms in } m \tag{8.19}$$

Or we can say that the entropy rise <u>across</u> <u>a</u> <u>normal</u> <u>shock</u> is proportional to the <u>third</u> <u>power</u> of the quantity $(M_1^2 - 1)$ plus higher order terms.

$$\frac{s_2 - s_1}{R} = \frac{2\gamma(M_1^2 - 1)^3}{3(\gamma+1)^2} + \text{H.O.T.} \qquad (8.20)$$

Note that if we want to consider <u>weak</u> shocks for which $M_1 \rightarrow 1$ or $m \rightarrow 0$ then we could neglect the higher order terms.

C Pressure and Entropy Changes versus Deflection Angles for Weak Oblique Shocks

The developments of Sections A and B were made for normal shocks and thus apply equally to the "normal component" of an oblique shock.

Since $\qquad\qquad\qquad M_{1n} = M_1 \sin \theta \qquad\qquad\qquad (7.5)$

we can rewrite equation 8.8 as

$$\frac{p_2 - p_1}{p_1} = \frac{2\gamma}{\gamma+1} (M_1^2 \sin^2 \theta - 1) \qquad (8.21)$$

and equation 8.20 becomes

$$\frac{s_2 - s_1}{R} = \frac{2\gamma(M_1^2 \sin^2 \theta - 1)^3}{3(\gamma+1)^2} + \text{H.O.T.} \qquad (8.22)$$

We shall now proceed to relate the quantity $(M_1^2 \sin^2 \theta - 1)$ to the deflection angle for the case of weak oblique shocks.

From the analysis that we made in Section 7.5 we have

$$\frac{\tan (\theta-\delta)}{\tan \theta} = \frac{(\gamma-1) M_1^2 \sin^2 \theta + 2}{(\gamma+1) M_1^2 \sin^2 \theta} \qquad (7.14)$$

Break the right hand side up into two fractions and <u>show</u> that equation 7.14 can be rearranged to

$$\frac{(\gamma+1) \tan (\theta-\delta)}{2 \tan \theta} = \frac{(\gamma-1)}{2} + \frac{1}{M_1^2 \sin^2 \theta} \qquad (8.23)$$

With the use of appropriate trigonometric identities one can put equation 8.23 into the following form

$$M_1^2 \sin^2 \theta - 1 = \frac{(\gamma+1)}{2} M_1^2 \frac{\sin \theta \sin \delta}{\cos (\theta-\delta)} \qquad (8.24)$$

Now we shall restrict ourselves to consideration of very weak shocks, in which the deflection angle δ is very small. Thus,

$$\cos (\theta-\delta) \approx \cos \theta \qquad (8.25)$$

$$\sin \delta \approx \delta \qquad (8.26)$$

and equation 8.24 can be written as

$$M_1^2 \sin^2 \theta - 1 = \left[\frac{(\gamma+1)}{2} M_1^2 \frac{\sin \theta}{\cos \theta} \right] \delta$$

or

$$\boxed{M_1^2 \sin^2 \theta - 1 = \left[\frac{(\gamma+1)}{2} M_1^2 \tan \theta \right] \delta} \qquad (8.27)$$

We now substitute equation 8.27 into the important relations previously developed, noting that for very weak oblique shocks the shock angle θ approaches θ_{min} or the Mach angle μ. From equations 8.27 and 8.21 we have

$$\boxed{\frac{p_2 - p_1}{p_1} = \frac{2\gamma}{\gamma+1} \left[\frac{(\gamma+1)}{2} M_1^2 \tan \mu_1 \right] \delta} \qquad (8.28)$$

From equations 8.27 and 8.22 we have (omitting the higher order terms)

$$\boxed{\frac{s_2 - s_1}{R} = \frac{2\gamma}{3(\gamma+1)^2} \left[\frac{(\gamma+1)}{2} M_1^2 \tan \mu_1 \right]^3 \delta^3} \qquad (8.29)$$

Remember, equations 8.27, 8.28, and 8.29 are only valid for very weak oblique shocks (which are associated with very small deflection angles). Let us pause for a moment to interpret these expressions. They really say that for very weak oblique shocks at any arbitrary set of initial conditions

$$\boxed{\Delta p \propto \delta} \qquad (8.30)$$

$$\boxed{\Delta s \propto \delta^3} \qquad (8.31)$$

These are important results which should be remembered.

D Isentropic Turns from Infinitesimal Shocks

We have laid the groundwork to show a remarkable phenomenon. Figure 8.1 shows a finite turn divided into n equal segments of δ each. The total turning angle will be indicated by δ_{Total} or δ_T and thus

$$\delta_T = n\delta \tag{8.32}$$

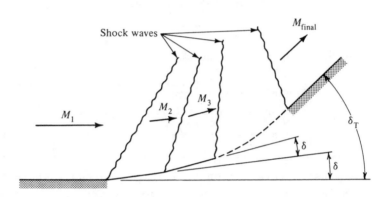

Figure 8.1 Finite turn composed of many small turns

Each segment of the turn causes a shock wave to form with an appropriate change in Mach number, pressure, temperature, entropy, etc. As we increase the number of segments n, δ becomes very small, which means that each shock will become a very weak oblique shock and the results of the previous section are applicable. Thus for each segment we may write

$$\Delta p' \propto \delta \tag{8.33}$$

$$\Delta s' \propto \delta^3 \tag{8.34}$$

where $\Delta p'$ and $\Delta s'$ are the pressure and entropy changes across each segment. Now for the total turn

$$\text{Total } \Delta p = \Sigma \, \Delta p' \propto n\delta \tag{8.35}$$

$$\text{Total } \Delta s = \Sigma \, \Delta s' \propto n\delta^3 \tag{8.36}$$

But from (8.32) we can express $\delta = \delta_T/n$

We now also take the limit as $n \to \infty$

$$\text{Total } \Delta p \propto \lim_{n \to \infty} n \left[\frac{\delta_T}{n} \right] \propto \delta_T \tag{8.37}$$

$$\text{Total } \Delta s \propto \lim_{n \to \infty} n\left[\frac{\delta_T}{n}\right]^3 \to 0 \qquad (8.38)$$

In the limit as $n \to \infty$, we conclude that:

    (a)   the wall makes a smooth turn through angle δ_T .

    (b)   the shock waves approach Mach waves.

    (c)   the Mach number continuously changes.

    (d)   there is a finite pressure change.

    (e)   there is <u>no</u> entropy change!

The final result is shown in Figure 8.2. Note that as the turn progresses, the Mach number is decreasing and thus the Mach waves are at ever increasing angles. (Also, μ_2 is measured from an increasing base line.) Hence, we observe an envelope of Mach lines that forms a short distance from the wall. The Mach waves coalesce to form an oblique shock inclined at the proper angle (θ) corresponding to the initial Mach number and the over-all deflection angle δ_T .

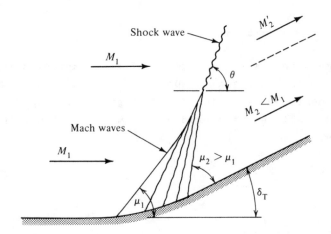

Figure 8.2 Smooth turn—isentropic compression near wall

We return to the flow in the neighborhood of the wall as this is a region of great interest. Here we have an infinite number of infinitesimal compression waves. We have achieved a decrease in Mach number and an increase in pressure <u>without any change in entropy</u>. Since we are dealing with adiabatic flow ($ds_e = 0$), an isentropic process ($ds = 0$) indicates that there are no losses ($ds_i = 0$); i.e., <u>the process is reversible</u>!

   The reverse process (an infinite number of infinitesimal
expansion waves) is shown in Figure 8.3. Here we have a smooth
turn in the other direction from that previously discussed.   In
this case, as the turn progresses, the Mach number increases.
Thus, the Mach angles are decreasing and the Mach waves will
never intersect.

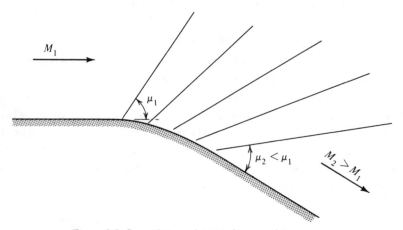

Figure 8.3 Smooth turn—isentropic expansion

   If the corner were sharp, all of the "expansion waves" would
emanate from the corner as illustrated in Figure 8.4.   This is
called a "centered expansion fan."   Figures 8.3 and 8.4 depict
the same overall result providing the wall is turned through the
same angle.

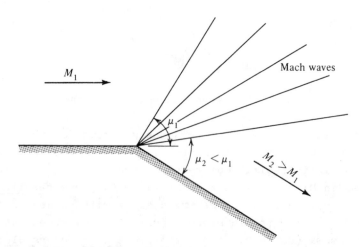

Figure 8.4 Isentropic expansion around sharp corner

All of the above isentropic flows are called "Prandtl-Meyer Flow." At a smooth concave wall (Figure 8.2) we have a Prandtl-Meyer compression. Flows of this type are not too important since boundary layer and other real gas effects interfere with the isentropic region near the wall. At a smooth convex wall (Figure 8.3) or at a sharp convex turn (Figure 8.4) we have Prandtl-Meyer expansions. These expansions are quite prevalent in supersonic flow as the examples given later in this unit will show. Incidentally, you have now discovered the second means by which the flow direction of a supersonic stream may be changed. What was the first?

8.4 ANALYSIS OF PRANDTL-MEYER FLOW

We have already established that the flow is isentropic through a Prandtl-Meyer compression or expansion. If we know the final Mach number, we can use the isentropic equations and tables to compute the final thermodynamic state for any given set of initial conditions. Thus, our objective in this section is to relate the changes in Mach number to the turning angle in Prandtl-Meyer flow.

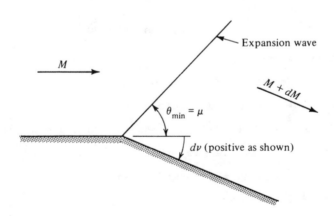

Figure 8.5 Infinitesimal Prandtl-Meyer expansion

Figure 8.5 shows a single Mach wave caused by turning the flow through an infinitesimal angle $d\nu$. It is more convenient to measure ν positive in the direction shown which corresponds to an expansion wave. The pressure difference across the wave front causes a momentum change and hence a velocity change <u>perpendicular</u> to the wave front. There is no mechanism by which the

tangential velocity component can be changed. In this respect
the situation is similar to that of an oblique shock. A detail
of this velocity relationship is shown in Figure 8.6.

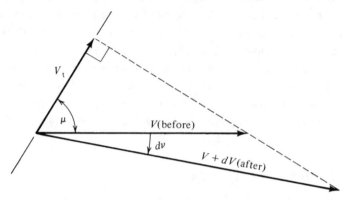

Figure 8.6 Velocities in an infinitesimal Prandtl-Meyer expansion

V represents the magnitude of the velocity before the expansion wave and V + dV is the magnitude after the wave. In both
cases the tangential component of the velocity is V_t . From the
velocity triangles we see that

$$V_t = V \cos \mu \qquad\qquad (8.39)$$

and $$V_t = (V + dV) \cos (\mu + d\nu) \qquad\qquad (8.40)$$

Equating these we obtain

$$V \cos \mu = (V + dV) \cos (\mu + d\nu) \qquad\qquad (8.41)$$

If we expand the cos (μ + dν) this becomes

$$V \cos \mu = (V + dV)(\cos \mu \cos d\nu - \sin \mu \sin d\nu) \qquad (8.42)$$

But dν is a very small angle; thus

$$\cos d\nu \approx 1 \qquad \text{and} \qquad \sin d\nu \approx d\nu$$

and equation 8.42 becomes

$$V \cos \mu = (V + dV)(\cos \mu - d\nu \sin \mu) \qquad\qquad (8.43)$$

By writing each term on the right side we get

$$V \cos \mu = V \cos \mu - V \, d\nu \sin \mu + dV \cos \mu - \overset{\text{H.O.T.}}{dV \, d\nu \sin \mu} \quad (8.44)$$

Canceling like terms and dropping the higher order term yields

$$d\nu = \frac{\cos \mu}{\sin \mu} \frac{dV}{V}$$

or
$$d\nu = \cot \mu \ \frac{dV}{V} \tag{8.45}$$

Now the contangent of μ can easily be obtained in terms of the Mach number. We know that $\sin \mu = 1/M$.

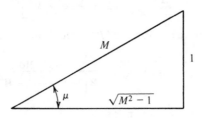

From the triangle we see that
$$\cot \mu = \sqrt{M^2 - 1} \tag{8.46}$$

Substitution of equation 8.46 into 8.45 yields:

$$d\nu = \sqrt{M^2 - 1} \ \frac{dV}{V} \tag{8.47}$$

Recall that our objective is to obtain a relationship between Mach number (M) and the turning angle (dν). Thus, we seek a means of expressing dV/V as a function of Mach number. In order to obtain an explicit expression we shall assume the fluid is a perfect gas. From equations 4.10 and 4.11 we know that

$$V = Ma = M\sqrt{\gamma g_c RT} \tag{8.48}$$

Hence,
$$dV = dM\sqrt{\gamma g_c RT} + \frac{M}{2}\sqrt{\frac{\gamma g_c R}{T}} \ dT \tag{8.49}$$

Show that
$$\frac{dV}{V} = \frac{dM}{M} + \frac{dT}{2T} \tag{8.50}$$

Knowing that
$$T_t = T\left[1 + \frac{(\gamma-1)}{2} M^2\right] \tag{4.18}$$

Then
$$dT_t = dT\left[1 + \frac{(\gamma-1)}{2} M^2\right] + T(\gamma-1)MdM \tag{8.51}$$

But since there is no heat or shaft work transferred to or from the fluid as it passes through the expansion wave, the stagnation enthalpy (h_t) remains constant. For our perfect gas this means that the total temperature remains fixed.

Thus,
$$T_t = \text{constant}$$

or $$dT_t = 0 \qquad (8.52)$$

From equations 8.51 and 8.52 we solve for

$$\frac{dT}{T} = - \frac{(\gamma-1)\ MdM}{\left[1 + \frac{(\gamma-1)}{2}\ M^2\right]} \qquad (8.53)$$

If we insert this result for dT/T into equation 8.50 we have

$$\frac{dV}{V} = \frac{dM}{M} - \frac{(\gamma-1)\ MdM}{2\left[1 + \frac{(\gamma-1)}{2}\ M^2\right]} \qquad (8.54)$$

Show that this can be written as

$$\frac{dV}{V} = \frac{1}{\left[1 + \frac{(\gamma-1)}{2}\ M^2\right]}\ \frac{dM}{M} \qquad (8.55)$$

We can now accomplish our objective by substitution of equation 8.55 into (8.47) with the following result:

$$\boxed{d\nu = \frac{(M^2 - 1)^{\frac{1}{2}}}{\left[1 + \frac{(\gamma-1)}{2}\ M^2\right]}\ \frac{dM}{M}} \qquad (8.56)$$

This is a significant relation for it says that

$$d\nu = f(M,\ \gamma)$$

For a given fluid γ is fixed and equation 8.56 can be integrated to yield

$$\nu + constant = \left[\frac{\gamma+1}{\gamma-1}\right]^{\frac{1}{2}}\ \tan^{-1}\left[\frac{\gamma-1}{\gamma+1}\ (M^2-1)\right]^{\frac{1}{2}} - \tan^{-1}(M^2-1)^{\frac{1}{2}} \qquad (8.57)$$

If we set $\nu = 0$ when $M = 1$, then the constant will be zero and we have

$$\boxed{\nu = \left[\frac{\gamma+1}{\gamma-1}\right]^{\frac{1}{2}}\ \tan^{-1}\left[\frac{\gamma-1}{\gamma+1}\ (M^2-1)\right]^{\frac{1}{2}} - \tan^{-1}(M^2-1)^{\frac{1}{2}}} \qquad (8.58)$$

Establishing the constant as zero, in the above manner, attaches a special significance to the angle ν. This is the angle, measured from the flow direction where $M = 1$, through which the flow has been turned (by an isentropic process) to reach the indicated Mach number. The above expression (8.58) is called the "Prandtl-Meyer function."

8.5 PRANDTL-MEYER TABLES

Equation 8.58 that was developed in the previous section is the basis for solving all problems involving Prandtl-Meyer expansions or compressions. If the Mach number is known, it is relatively easy to solve for the turning angle. However, in a typical problem the turning angle might be prescribed and no explicit solution is available for the Mach number. Fortunately, none is required for the Prandtl-Meyer function can be calculated in advance and tabulated. This function (ν) has been included as a column of the isentropic table. The following examples illustrate how rapidly problems of this type are solved.

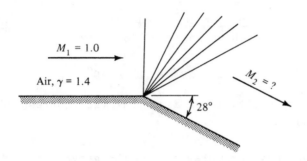

Figure 8.7 Prandtl-Meyer expansion from Mach 1

Example 8.1   The wall in Figure 8.7 turns an angle of 28° with a sharp corner. The fluid, which is initially at $M = 1$, must follow the wall and in so doing executes a Prandtl-Meyer expansion at the corner. Recall that ν represents the angle (measured from the flow direction where $M = 1$) through which the flow has turned.

Since M_1 is unity, then $\nu_2 = 28^{\circ}$

From the isentropic table we see that this Prandtl-Meyer function corresponds to $M_2 \approx 2.06$.

Example 8.2   Now consider flow at a Mach number of 2.06 which expands through a turning angle of 12°. Figure 8.8 shows such a situation and we want to determine the final Mach number M_2.

Now regardless of how the flow with $M_1 = 2.06$ came into existence, we know that <u>it</u> <u>could</u> <u>have</u> <u>been</u> <u>obtained</u> by expanding a flow at $M = 1.0$ around a corner of 28°. This is shown with dotted lines in the figure. It is easy to see that the flow in

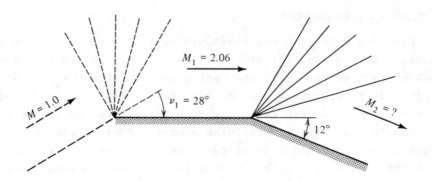

Figure 8.8 Prandtl-Meyer expansion from $M \neq 1$

region 2 <u>could</u> <u>have</u> <u>been</u> <u>obtained</u> by taking a flow at M = 1.0
and turning it through an angle of 28° + 12° or

$$\nu_2 = 28^{\circ} + 12^{\circ} = 40^{\circ}$$

From the isentropic table we find that this corresponds to a flow
at $M_2 \approx 2.54$.

From the above examples we see the general rule for Prandtl-
Meyer flow.

$$\boxed{\nu_2 = \nu_1 + \Delta\nu} \tag{8.59}$$

where $\Delta\nu \equiv$ the turning angle

Note that for an expansion (as shown in Figures 8.7 and 8.8) $\Delta\nu$
is positive and thus both the Prandtl-Meyer function and the Mach
number increase. Once the final Mach number is obtained all properties may be determined easily since it is isentropic flow.

For a turn in the opposite direction $\Delta\nu$ will be negative,
which leads to a Prandtl-Meyer compression. In this case both
the Prandtl-Meyer function and the Mach number will decrease.
An example of this case follows.

Example 8.3 Air at M_1 = 2.40 , T_1 = 325°K , and p_1 = 1.5 bars
approaches a smooth concave turn of 20° as shown in Figure 8.9.
We have previously discussed how the region close to the wall
will be an isentropic compression. We seek the properties in the
flow after the turn.

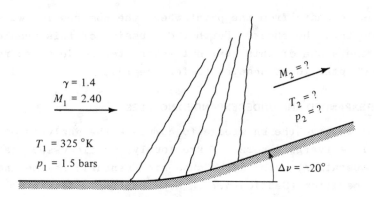

Figure 8.9 Prandtl-Meyer compression

From the tables $\nu_1 = 36.7465^\circ$. Remember, $\Delta\nu$ is negative.

$$\nu_2 = \nu_1 + \Delta\nu = 36.7465^\circ + (-20^\circ) = 16.7465^\circ$$

Again, from the table we see that this corresponds to a Mach number of

$$M_2 = 1.664$$

Since the flow is adiabatic, with no shaft work, and a perfect gas, we know that the stagnation temperature is constant $(T_{t1} = T_{t2})$. In addition there are no losses and thus the stagnation pressure remains constant $(p_{t1} = p_{t2})$. Can you verify these statements with the appropriate equations?

To continue with this example we solve for the temperature and pressure in the usual fashion:

$$p_2 = \frac{p_2}{p_{t2}} \frac{p_{t2}}{p_{t1}} \frac{p_{t1}}{p_1} p_1 = (0.2139)(1) \frac{1}{0.0684} (1.5\times10^5) = \underline{4.69\times10^5} \text{ N/m}^2$$

$$T_2 = \frac{T_2}{T_{t2}} \frac{T_{t2}}{T_{t1}} \frac{T_{t1}}{T_1} T_1 = (0.6436)(1) \frac{1}{0.4647} (325) = \underline{450}^\circ\text{K}$$

As we move away from the wall we know that the Mach waves will coalesce and form an oblique shock. At what angle will the shock be to deflect the flow by 20°? What will the temperature and pressure be after the shock? If you work out this oblique shock problem you should obtain $\theta = 44^\circ$, $M_{1n} = 1.667$, $p_2' = 4.61\times10^5$ N/m^2 , and $T_2' = 466^\circ$K . Since pressure equilibrium does not exist across this free boundary another wave forma-

tion must emanate from the point where the compression waves
coalesce into the shock. Further discussion of this problem is
beyond the scope of this text but interested readers are referred to Chapter 16 of Shapiro (reference 19).

8.6 OVEREXPANDED AND UNDEREXPANDED NOZZLES

Now we have the knowledge to complete the analysis of a
converging-diverging nozzle. Previously, we discussed its isentropic operation, both in the subsonic (venturi) regime and its
design operation (Section 5.7). Non-isentropic operation with
a normal shock standing in the diverging portion was also covered
(Section 6.6). In Section 7.8 we saw that with operating pressure ratios below second critical, oblique shocks come into play
but we were unable to complete the picture.

Figure 8.10 shows an "overexpanded" nozzle; i.e., it is operating someplace between its second and third critical points.
Recall from the summary of Unit 7 that there are two types of
boundary conditions that must be met. One of these concerns flow
direction and the other concerns pressure equilibrium.

1. From symmetry aspects we know that a central streamline
    exists. Any fluid touching this boundary must have a
    velocity that is tangent to the streamline. In this
    respect it is identical with a physical boundary.
2. Once the jet leaves the nozzle there is an outer surface
    that is in contact with the surrounding ambient fluid.
    Since this is a "free" or unrestrained boundary, pres-
    sure equilibrium must exist across this surface.

We can now follow from region to region and, by matching the appropriate boundary condition, determine the flow pattern that
must exist.

Since the nozzle is operating with a pressure ratio between
second and third critical it is obvious that we need a compression process at the exit in order for the flow to end up at the
ambient pressure. However, a normal shock at the exit will produce too strong a compression. What is needed is a shock process
that is weaker than a normal shock and the oblique shock has been
shown to be just this. Thus, at the exit we observe oblique
shock A at the appropriate angle so that $p_2 = p_{amb}$.

We recall that the flow across an oblique shock is always
deflected away from a normal to the shock front and thus the flow

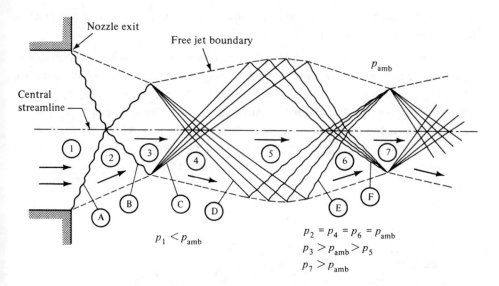

Figure 8.10 Overexpanded nozzle

in region 2 is no longer parallel to the centerline. Wave front
B must deflect the flow back to its original axial direction.
This can easily be accomplished by another oblique shock. (A
Prandtl-Meyer expansion would turn the flow in the wrong direction.) An alternate way of viewing this is that the oblique
shocks from both upper and lower lips of the nozzle "pass through
each other" when they meet at the centerline. If one adopts this
philosophy one should realize that the waves are slightly altered
or "bent" in the process of traveling through one another.

Now since $p_2 = p_{amb}$, passage of the flow through oblique
shock B will make $p_3 > p_{amb}$ and region 3 cannot have a free
surface in contact with the surroundings. Consequently a wave
formation must emanate from the point where wave B meets the
free boundary and the pressure must decrease across this wave.
We now realize that wave form C must be a Prandtl-Meyer expansion so that $p_4 = p_{amb}$.

However, passage of the flow through the expansion fan C
causes it to turn away from the centerline and the flow in region
4 is no longer parallel to the centerline. Thus, as each wave
of the P-M expansion fan meets the centerline a wave form must
emanate to turn the flow parallel to the axis again. If wave
D were a compression, in which direction would the flow turn?
We see that to meet the boundary condition of flow direction wave

D must be another P-M expansion. Thus, the pressure in region
5 is less than ambient.

Can you now reason that to get from 5 to 6 and meet the
boundary condition imposed by the free boundary, E must consist
of P-M compression waves? Similarly F must consist of P-M
compression waves in order to turn the flow from region 6 to
match the direction of the "wall." Now is p_7 equal to, greater
than, or less than p_{amb} ? You should recognize that conditions
in region 7 are similar to those in region 3 and so the cycle
repeats.

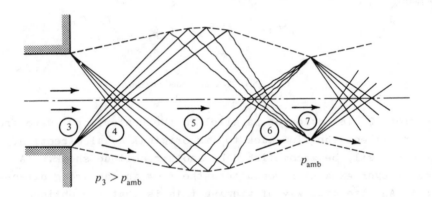

Figure 8.11 Underexpanded nozzle

Now let us examine an "underexpanded" nozzle. This means
that we have an operating pressure ratio below the third critical
or design condition. Figure 8.11 shows such a situation. The
flow leaving this nozzle has a pressure greater than ambient and
the flow is parallel to the axis. Reflect and you will realize
that this condition is exactly the same as region 3 in the
overexpanded nozzle (see Figure 8.10). Thus, the flow patterns
are identical from this point on.

The sketches in Figure 8.10 and 8.11 represent ideal behavior.
The general wave forms described can be seen by special flow visu-
alization techniques such as schlieren photography. Eventually,
the large velocity difference that exists over the free boundary
causes a turbulent shear layer which dissipates the wave patterns.

Example 8.4 Nitrogen issues from a nozzle at a Mach number of
2.5 and a pressure of 10 psia. The ambient pressure is 5 psia.

What is the Mach number and through what angle is the flow turned after passing through the first Prandtl-Meyer expansion fan?

With reference to Figure 8.11 we know that $M_3 = 2.5$, $p_3 = 10$ psia and $p_4 = p_{amb} = 5$ psia.

$$\frac{p_4}{p_{t4}} = \frac{p_4}{p_3} \frac{p_3}{p_{t3}} \frac{p_{t3}}{p_{t4}} = \frac{5}{10} (0.0585)(1) = 0.0293$$

Thus, $M_4 = \underline{2.952}$

$$\Delta\nu = \nu_4 - \nu_3 = 48.8226 - 39.1236 \approx \underline{9.7^{\circ}}$$

WAVE REFLECTIONS

From the above discussions we not only have learned about the details of nozzle jets when operating at off-design conditions but we have also been looking at "wave reflections," although we have not called them such. We could think of the waves as "bouncing" or "reflecting" off the free boundary. Similarly, if the central streamline had been visualized as a solid boundary, we could have thought of the waves as reflecting off that boundary.

In retrospect, we may draw some general conclusions about wave reflections.

1. Reflections from a physical or pseudo-physical boundary (where the boundary condition concerns the flow direction) are of the same "family." That is, shocks reflect as shocks, compression waves reflect as compression waves, and expansion waves reflect as expansion waves.

2. Reflections from a free boundary (where pressure equalization exists) are of the opposite family; i.e., compression waves reflect as expansion waves, and expansion waves reflect as compression waves.

Warning! Care should be taken in viewing waves as reflections. Not only is their character sometimes changed (case 2 above) but the angle of reflection is not quite the same as the angle of incidence. Also, the "strength" of the wave changes somewhat. This can clearly be shown by considering the case of an oblique shock "reflecting" off a solid boundary.

Example 8.5   Air at   Mach = 2.2   passes through an oblique shock
at a 35° angle.   The shock runs into a physical boundary as shown.
Find the angle of "reflection" and compare the strengths of the
two shock waves.

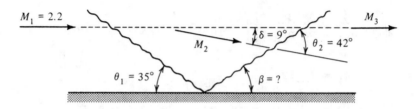

From the shock chart at $M_1 = 2.2$ and $\theta_1 = 35^\circ$ we find
$\delta_1 = 9^\circ$.

$$M_{1n} = 2.2 \sin 35^\circ = \underline{1.262} \text{ , thus } M_{2n} = 0.806$$

$$M_2 = \frac{M_{2n}}{\sin(\theta - \delta)} = \frac{0.806}{\sin(35-9)} = 1.839$$

The reflected shock must turn the flow back parallel to the wall.
Thus, from the chart at $M_2 = 1.839$ and $\delta_2 = 9^\circ$ we find
$\theta_2 = 42^\circ$.

$$\beta = 42^\circ - 9^\circ = \underline{33^\circ}$$

$$M_{2n} = 1.839 \sin 42^\circ = \underline{1.230}$$

Notice that the "angle of incidence" (35°) is not the same as the
"angle of reflection" (33°). Also, the normal Mach number, which
indicates the strength of the wave, has decreased from 1.262 to
1.230 .

8.7 SUPERSONIC AIRFOILS

Airfoils designed for <u>subsonic</u> flight have rounded leading
edges to prevent flow separation. The use of an airfoil of this
type at <u>supersonic</u> speeds would cause a detached shock to form in
front of the leading edge (see Section 7.7). Consequently, all
supersonic airfoil shapes have sharp leading edges. Also, to provide good aerodynamic characteristics, supersonic foils are very
thin. The obvious limiting case of a thin foil with a sharp leading edge is the flat-plate airfoil shown in Figure 8.12. Although
impractical from structural considerations, it provides an interesting study and has characteristics that are typical of all
supersonic airfoils.

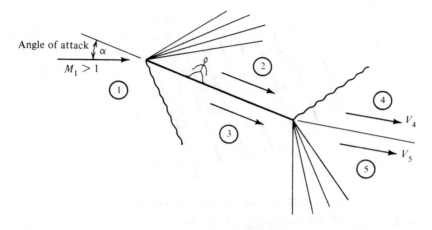

Figure 8.12 Flat plate airfoil

Using the foil as a frame of reference yields a steady flow picture. When operating at an angle of attack (α) the flow must change direction to pass over the foil surface. You should have no trouble recognizing that to pass along the upper surface requires a Prandtl-Meyer expansion through angle α at the leading edge. Thus, the pressure in region 2 is less than atmospheric. To pass along the lower surface necessitates an oblique shock which will be of the weak variety for the required deflection angle α. (Why is it impossible for the strong solution to occur? See Section 7.7.) The pressure in region 3 is greater than atmospheric.

Now consider what happens at the trailing edge. Pressure equilibrium must exist between regions 4 and 5. Thus, a compression must occur off the upper surface and an expansion is necessary on the lower surface. The corresponding wave patterns are indicated in the diagram; an oblique shock from 2 to 4 and a Prandtl-Meyer expansion from 3 to 5. Note that the flows in regions 4 and 5 are not necessarily parallel to that of region 1 nor are the pressures p_4 and p_5 necessarily atmospheric. The boundary conditions that must be met are flow tangency and pressure equilibrium, or

$$V_4 \text{ parallel to } V_5 \quad \text{and} \quad p_4 = p_5$$

The solution at the trailing edge is a trial and error type since neither the final flow direction nor the final pressure is known.

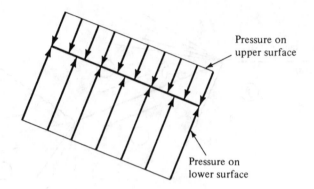

Pressure on
upper surface

Pressure on
lower surface

Figure 8.13 Pressure distribution over flat plate airfoil

A sketch of the pressure distribution is given in Figure
8.13. One can easily see that the center of pressure is at the
middle or mid-chord position. If the angle of attack were
changed the values of the pressures over the upper and lower surfaces would change but the center of pressure would still be at
the mid-chord. Students of aeronautics, who are familiar with
the term "aerodynamic center,"[+] will have no difficulty determining that this important point is also located at the mid-chord.
This is approximately true of all <u>supersonic</u> airfoils providing
they are very thin and operate at small angles of attack.

Example 8.6 Compute the lift per unit span of a flat-plate airfoil with a chord of 2 meters when flying at $M = 1.8$ and an
angle of attack of 5°. Ambient air pressure is 0.4 bars. Use
Figure 8.12 for identification of regions.

The flow over the top is turned 5° by a Prandtl-Meyer expansion.

$$\nu_2 = \nu_1 + \Delta\nu = 20.7251 + 5 = 25.7251^{\circ}$$

Thus, $M_2 = 1.976$ and $p_2/p_{t2} = 0.1327$

$$p_2 = \frac{p_2}{p_{t2}} \frac{p_{t2}}{p_{t1}} \frac{p_{t1}}{p_1} p_1 = (0.1327)(1) \frac{1}{(0.1740)} (0.4) = \underline{0.3051} \text{ bars}$$

[+]The aerodynamic center of an airfoil section is defined as the
point about which the pitching moment is independent of angle of
attack. For subsonic airfoils this is approximately at the 1/4
chord point, or 25% of the chord measured from the leading edge
back towards the trailing edge.

The flow under the bottom is turned 5° by an oblique shock. From the chart at $M = 1.8$ and $\delta = 5^O$ we find $\theta = 38.5^O$.

$$M_{1n} = 1.8 \sin 38.5^O = 1.120 \qquad \text{and} \qquad p_3/p_1 = 1.2968$$

$$p_3 = \frac{p_3}{p_1} p_1 = (1.2968)(0.4) = \underline{0.5187} \text{ bars}$$

The lift force is defined as that component which is perpendicular to the free stream. Thus, the lift force per unit span will be:

$$L = (p_3-p_2)(\text{chord}) \cos \alpha = (0.5187-0.3051)(10^5) \, 2 \cos 5^O$$

$$L = \underline{4.26 \times 10^4} \text{ Newtons/unit span}$$

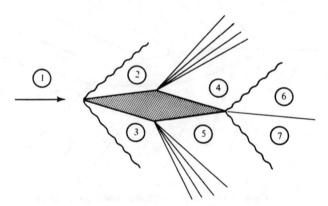

(a) Low angle of attack

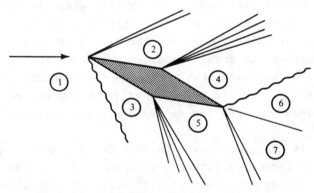

(b) High angle of attack

Figure 8.14 Double wedge airfoil

A more practical design for a supersonic airfoil is shown in Figure 8.14. Here the wave formation depends upon whether or not the angle of attack is less than or greater than the half angle of the wedge at the leading edge. In either case straightforward solutions exist on all surfaces up to the trailing edge. A trial and error solution is required only if one is interested in regions 6 and 7. Fortunately, these regions are only of academic interest as they have no effect on the pressure distribution over the foil. Modifications of the double-wedge airfoil with sections of constant thickness in the center are frequently found in practice.

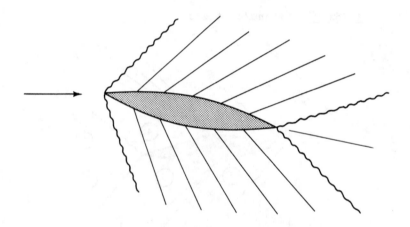

Figure 8.15 Biconvex airfoil at low angle of attack

Another widely used supersonic airfoil shape is the "biconvex" which is shown in Figure 8.15. This is generally constructed of circular or parabolic arcs. The wave formation is quite similar to that on the double-wedge in that the type of waves found at the leading (and trailing) edge is dependent upon the angle of attack. Also, in the case of the biconvex, the expansions are spread out over the entire upper and lower surface.

Example 8.7 It has been suggested that a supersonic airfoil be designed as an isosceles triangle with 10° equal angles and an 8-foot chord. When operating at a 5° angle of attack the air flow appears as shown below. Find the pressures on the various surfaces and the lift and drag forces when flying at $M = 1.5$ through air with a pressure of 8 psia.

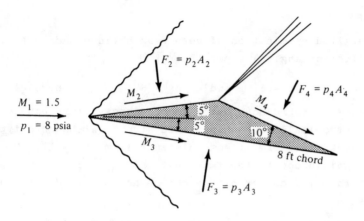

From shock chart at $M_1 = 1.5$ and $\delta = 5^\circ$, $\theta = 48^\circ$

$$M_{1n} = M_1 \sin \theta = 1.5 \sin 48^\circ = 1.115$$

From shock tables, $M_{2n} = 0.900$ and $p_2/p_1 = 1.2838$

$$M_2 = \frac{M_{2n}}{\sin (\theta-\delta)} = \frac{0.900}{\sin (48-5)} = 1.32$$

The Prandtl-Meyer expansion turns the flow by 20°

$$\nu_4 = \nu_2 + 20 = 6.7213 + 20 = 26.7213 \quad \text{and} \quad M_4 = 2.012$$

Note that conditions in region 3 are identical with region 2 . We now find the pressures.

$$p_3 = p_2 = \frac{p_2}{p_1} p_1 = (1.2838)(8) = \underline{10.27} \text{ psia}$$

$$p_4 = \frac{p_4}{p_{t4}} \frac{p_{t4}}{p_{t2}} \frac{p_{t2}}{p_2} p_2 = (0.1254)(1) \frac{1}{(0.3512)} (10.27) = \underline{3.67} \text{ psia}$$

The lift force (perpendicular to the free stream) will be

$$L = F_3 \cos 5^\circ - F_2 \cos 5^\circ - F_4 \cos 15^\circ$$

<u>Show</u> that the lift per unit span will be 3728 lbf.
The drag is that force which is parallel to the free-stream velocity. <u>Show</u> that the drag force per unit span is 999 lbf.

8.8 SUMMARY

A detailed examination of very weak oblique shocks (with small deflection angles) shows that

$$\Delta p \propto \delta \quad \text{and} \quad \Delta s \propto \delta^3 \qquad (8.30) \ \& \ (8.31)$$

This enables us to reason that a smooth concave turn can be negotiated isentropically by a supersonic stream, although a typical oblique shock will form at some distance from the wall. Of even greater significance is the fact that this is a reversible process and turns of a convex nature can be accomplished by isentropic expansions.

The above phenomena are called Prandtl-Meyer flow. An analysis for a perfect gas reveals that the turning angle can be related to the change in Mach number by

$$d\nu = \frac{(M^2-1)^{\frac{1}{2}}}{\left[1 + \frac{(\gamma-1)}{2} M^2\right]} \frac{dM}{M} \qquad (8.56)$$

which when integrated yields the Prandtl-Meyer function.

$$\nu = \left[\frac{\gamma+1}{\gamma-1}\right]^{\frac{1}{2}} \tan^{-1}\left[\frac{\gamma-1}{\gamma+1} (M^2-1)\right]^{\frac{1}{2}} - \tan^{-1}(M^2-1)^{\frac{1}{2}} \qquad (8.58)$$

In establishing equation 8.58 ν was set equal to zero at M = 1.0 which means that ν represents the angle, measured from the direction where M = 1.0 , through which the flow has been turned (isentropically) to reach the indicated Mach number. The above relation has been tabulated which permits easy problem solutions according to the relation

$$\nu_2 = \nu_1 + \Delta\nu \qquad (8.59)$$

in which $\Delta\nu$ is the turning angle. Remember that $\Delta\nu$ will be positive for expansions and negative for compressions.

It must be understood that Prandtl-Meyer expansions and compressions are caused by the same two situations that govern the formation of oblique shocks; i.e., the flow must be tangent to a boundary and pressure equilibrium must exist along the edge of a free boundary. Consideration of these boundary conditions together with any given physical situation should enable you to determine the resulting flow patterns rather quickly.

Waves may sometimes be thought of as "reflecting" off boundaries, in which case it is helpful to remember that

1. Reflections from physical boundaries are of the same "family."
2. Reflections from free boundaries are of the opposite "family."

Remember that all isentropic relations and tables may be
used when dealing with Prandtl-Meyer flow.

8.9 PROBLEMS

1. Air approaches a sharp 15° convex corner (see Figure 8.4)
   with a Mach number of 2.0, temperature of 520°R and pressure
   of 14.7 psia. Determine the Mach number, static and stagnation temperature, static and stagnation pressure of the air
   after it has expanded around the corner.

2. A schleiren photo of the flow around a corner reveals the
   edges of the expansion fan to be indicated by the angles
   shown below. Assume $\gamma = 1.4$.
   (a) Determine the Mach number before and after the corner.
   (b) Through what angle was the flow turned and what is the
   angle of the expansion fan (θ_3)?

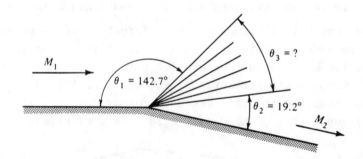

3. A supersonic flow of air has a pressure of 1×10^5 N/m^2 and a
   temperature of 350°K. After expanding through a 35° turn the
   Mach number is 3.5.
   (a) What are the final temperature and pressure?
   (b) Make a sketch similar to that shown for problem 2 and
   determine angles θ_1 , θ_2 and θ_3 .

4. In a problem similar to number 2, θ_1 is unknown but
   $\theta_2 = 15.90^\circ$ and $\theta_3 = 82.25^\circ$. Can you determine the initial Mach number?

5. Nitrogen at 25 psia and $850°R$ is flowing at a Mach number of
    2.54. After expanding around a smooth convex corner the
    velocity of the nitrogen is found to be 4000 ft/sec. Through
    how many degrees did the flow turn?

6. A smooth concave turn similar to that shown in Figure 8.2
    turns the flow through a $30°$ angle. The fluid is oxygen and
    it approaches the turn at $M_1 = 4.0$.
    (a) Compute M_2 , T_2/T_1 and p_2/p_1 via the Prandtl-Meyer
        compression which occurs close to the wall.
    (b) Compute $M_{2'}$, $T_{2'}/T_1$ and $p_{2'}/p_1$ via the oblique shock
        which forms away from the wall. Assume that this flow is
        also deflected by $30°$.
    (c) Draw a T-s diagram showing each process.
    (d) Can these two regions coexist next to one another?

7. A simple flat-plate airfoil has a chord of 8 ft and is flying
    at M = 1.5 and a $10°$ angle of attack. Ambient air pres-
    sure is 10 psia and the temperature is $450°R$.
    (a) Determine the pressures above and below the airfoil.
    (b) Calculate the lift and drag forces per unit span.
    (c) Determine the pressure and flow direction as the air
        leaves the trailing edge (regions 4 and 5 in Figure 8.12).

8. The symmetrical diamond-shaped airfoil shown below is oper-
    ating at a $3°$ angle of attack. The flight speed is M = 1.8
    and the air pressure equals 8.5 psia.
    (a) Compute the pressure on each surface.
    (b) Calculate the lift and drag forces.
    (c) Repeat the problem with a $10°$ angle of attack.

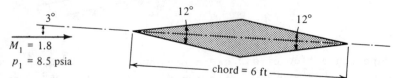

9. A biconvex airfoil (see Figure 8.15) is constructed of circu-
    lar arcs. We shall approximate the curve on the upper sur-
    face by ten straight line segments as shown below.
    (a) Determine the pressure immediately after the oblique
        shock at the leading edge.
    (b) Determine the Mach number and pressure on each segment.
    (c) Compute the contribution to the lift and drag from each
        segment.

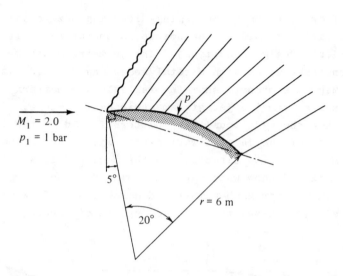

10. Properties of the flow are given at the exit plane of the two-dimensional duct shown. The receiver pressure is 12 psia.

   (a) Determine the Mach number and temperature just past the exit (after the flow has passed through the first wave formation). Assume $\gamma = 1.4$.

   (b) Make a sketch showing the flow direction, wave angles, etc.

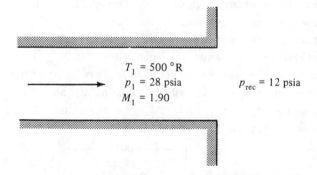

11. Stagnation conditions in a large reservoir are 7 bars and $420°K$. A converging-only nozzle delivers nitrogen from this reservoir into a receiver where the pressure is 1 bar.

   (a) Sketch the first wave formation that will be seen as the nitrogen leaves the nozzle.

   (b) Find the conditions (T, p, V) that exist after the nitrogen has passed through this wave formation.

12. Air flows through a converging-diverging nozzle which has an
     area ratio of 3.5 . The nozzle is operating at its third
     critical (design condition). The jet stream strikes a two-
     dimensional wedge with a total wedge angle of 40° as shown.
     (a) Make a sketch to show the initial wave pattern that re-
         sults from the jet stream striking the wedge.
     (b) Show the additional wave pattern formed by the interac-
         tion of the initial wave system with the free boundary.
         Mark the flow direction in the 'region following each wave
         form and show what happens to the free boundary.
     (c) Compute the Mach number and direction of flow after the
         air jet passes through each system of waves.

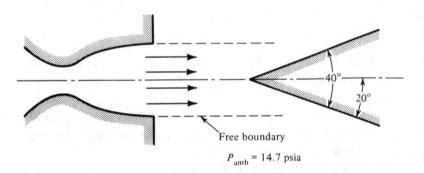

Free boundary

P_{amb} = 14.7 psia

13. Air flows in a two-dimensional channel and exhausts to the
     atmosphere as shown. Note that the oblique shock just
     touches the upper corner.
     (a) Find the deflection angle.
     (b) Determine M_2 and p_2 (in terms of p_{amb}).
     (c) What is the nature of the wave form emanating from the
         upper corner and dividing regions 2 and 3?
     (d) Compute M_3 , p_3 and T_3 (in terms of T_1). Show the
         flow direction in region 3.

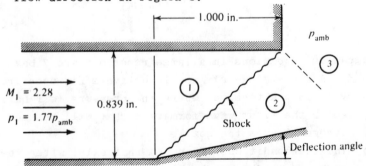

14. A supersonic nozzle produces a flow of nitrogen at $M_1 = 2.0$ and $p_1 = 0.7$ bar . This discharges into an ambient pressure of 1.0 bar producing the flow pattern shown in Figure 8.10.
   (a) Compute the pressures, Mach numbers and flow directions in regions 2, 3 and 4.
   (b) Make a sketch of the exit jet showing all angles to scale (streamlines, shock lines and Mach lines).

15. Consider the expression for the Prandtl-Meyer function which is given in equation 8.58.
   (a) Show that the maximum possible value for ν is

$$\nu_{max} = \frac{\pi}{2}\left[\sqrt{\frac{\gamma+1}{\gamma-1}} - 1\right]$$

   (b) At what Mach number does this occur?
   (c) If $\gamma = 1.4$, what are the maximum turning angles for accelerating flows with initial Mach numbers of 1.0 , 2.0 , 5.0 and 10.0 ?
   (d) If a flow of air at $M = 2.0$, $p = 100$ psia and $T = 600^{\circ}R$ expands through its maximum turning angle, what is the velocity?

16. Flow, initially at a Mach number of unity, expands around a corner through angle ν and reaches Mach number M_2 . Lengths L_1 and L_2 are measured perpendicular to the wall and measure the distance out to the same streamline as shown.
   (a) Derive an equation for the ratio $L_2/L_1 = f(M_2, \gamma)$. (Hints: What fundamental concept must be obeyed? What kind of a process is this?)
   (b) If $M_1 = 1.0$, $M_2 = 1.79$ and $\gamma = 1.67$, compute the ratio L_2/L_1 .

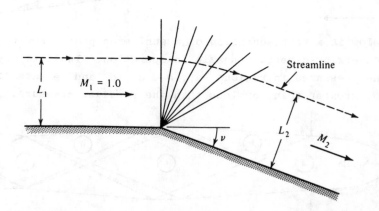

8.10 CHECK TEST

You should be able to complete this test without reference to
material in the unit.

1. For very weak oblique shocks, state how entropy changes and
   pressure changes are related to deflection angles.

2. Explain what the Prandtl-Meyer function represents. (i.e.,
   If someone were to say that $\nu = 36.8°$, what would this mean
   to you?)

3. State the rules for wave reflection.
   (a) Waves reflect off physical boundaries as _____.
   (b) Waves reflect off "free" boundaries as _____.

4. A flow at $M_1 = 1.5$ and $p_1 = 2 \times 10^5$ N/m^2 approaches a sharp
   turn. After negotiating the turn the pressure is 1.5×10^5
   N/m^2 . Determine the deflection angle if the fluid is oxygen.

5. Compute the net force (per square foot of area) acting on the
   flat plate airfoil shown.

$$M_1 = 2.0$$
$$p_1 = 5 \text{ psia} \qquad\qquad 10°$$

6. (a) Sketch the waveforms that you might expect to find over
       the airfoil shown below.
   (b) Identify all wave forms by name.
   (c) State the boundary conditions that must be met as the
       flow comes off the trailing edge of the airfoil.

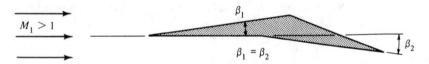

$$M_1 > 1 \qquad\qquad \beta_1 \qquad\qquad \beta_2$$
$$\beta_1 = \beta_2$$

7. Below is a representation of a schlieren photo showing a
   converging-diverging nozzle in operation. Indicate whether
   the pressures in regions a , b , c , d and e are equal
   to, greater than, or less than the receiver pressure.

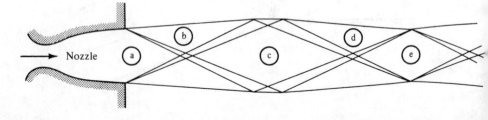

UNIT 9

FANNO
FLOW

9.1 INTRODUCTION

At the start of Unit 5 we mentioned that area changes, friction, and heat transfer are the most important factors which affect the properties in a flow system. Up to this point we have
considered only one of these factors, that of variations in area.
However, we have also discussed the various mechanisms by which
a flow adjusts to meet imposed boundary conditions of either flow
direction or pressure equalization. We now wish to take a look
at the subject of friction losses.

In order to study only the effects of friction, we will analyze flow in a constant-area duct without heat transfer. This
corresponds to many practical flow situations which involve reasonably short ducts. We shall first consider the flow of an arbitrary fluid and will discover that its behavior follows a definite
pattern which.is dependent upon whether the flow is in the subsonic or supersonic regime. Working equations will be developed

for the case of a perfect gas and the introduction of a reference
point allows tables to be constructed. As before, the tables
permit rapid solutions to many problems of this type which are
called "Fanno flow."

9.2 OBJECTIVES

After successfully completing this unit you should be able to:

1. List the assumptions made in the analysis of Fanno flow.

2. Simplify the general equations of continuity, energy, and
   momentum to obtain basic relations valid for any fluid in
   Fanno flow.

3. Sketch a Fanno line in the h-v and the h-s planes. Identify the sonic point and regions of subsonic and supersonic
   flow.

4. Describe the variation of static and stagnation pressure,
   static and stagnation temperature, static density and velocity
   as flow progresses along a Fanno line. Do for both subsonic
   and supersonic flow.

5. Starting with basic principles of continuity, energy, and
   momentum, derive expressions for property ratios such as
   T_2/T_1 , p_2/p_1 , etc., in terms of Mach number (M) and specific
   heat ratio (γ) for Fanno flow with a perfect gas.

6. Describe (include T-s diagram) how Fanno tables are developed with the use of a * reference location.

7. Define friction factor, equivalent diameter, absolute and
   relative roughness, absolute and kinematic viscosity, and
   Reynolds number. Know how to determine each.

8. Compare similarities and differences between Fanno flow and
   normal shocks. Sketch an h-s diagram showing a typical
   Fanno line together with a normal shock for the same mass
   velocity.

9. Explain what is meant by "friction choking."

10. Describe some possible consequences of adding duct in a
    choked Fanno flow situation (for both subsonic and supersonic
    flow).

11. Demonstrate the ability to solve typical Fanno flow problems
    by use of the appropriate tables and equations.

9.3 ANALYSIS FOR A GENERAL FLUID

We shall first consider the general behavior of an arbitrary fluid. In order to isolate the effects of friction we make the following

Assumptions: Steady, one-dimensional flow

Adiabatic	$(\delta q = 0,\ ds_e = 0)$
No shaft work	$(\delta w_s = 0)$
Neglect potential	$(dz = 0)$
Constant area	$(dA = 0)$

We proceed by applying the basic concepts of continuity, energy, and momentum.

CONTINUITY $\qquad\qquad \dot{m} = \rho A V = \text{const.}$ (2.30)

but since the flow area is constant, this reduces to

$$\rho V = \text{constant} \tag{9.1}$$

We assign a new symbol "G" to this constant (the quantity ρV) which is referred to as the "mass velocity," and thus

$$\boxed{\rho V = G = \text{const.}} \tag{9.2}$$

What are the typical units of G ?

ENERGY - We start with

$$h_{t1} + \cancel{q} = h_{t2} + \cancel{w}_s \tag{3.19}$$

For adiabatic and no work this becomes

$$h_{t1} = h_{t2} = h_t = \text{const.} \tag{9.3}$$

If we neglect the potential term, this means that

$$h_t = h + \frac{V^2}{2g_c} = \text{const.} \tag{9.4}$$

Substitute for the velocity from equation 9.2 and show that

$$\boxed{h_t = h + \frac{G^2}{\rho^2 2g_c} = \text{const.}} \tag{9.5}$$

Now for any given flow the constants h_t and G are shown.
Thus, equation 9.5 establishes a unique relationship between h
and ρ . Figure 9.1 is a plot of this equation in the h-v plane
for various values of G (but all for the same h_t). Each curve
is called a Fanno line and represents flow at a particular "mass
velocity." Note carefully that this is constant G and not constant ṁ . Ducts of various sizes could pass the same mass flow
rate but would have different mass velocities.

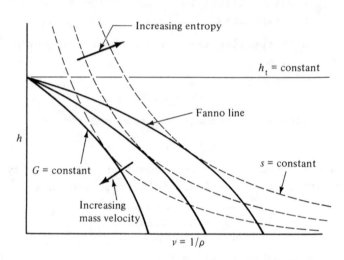

Figure 9.1 Fanno lines in $h-v$ plane

Once the fluid is known, one can also plot lines of constant
entropy on the h-v diagram. Typical curves of s = const. are
shown as dotted lines in the figure. It is much more instructive
to plot these Fanno lines in the familiar h-s plane. Such a
diagram is shown in Figure 9.2. At this point, a significant
fact becomes quite clear. Since we have assumed that there is no
heat transfer (ds_e = 0), the <u>only</u> way that entropy can be generated is through irreversibilities (ds_i). Thus, <u>the flow can only</u>
<u>progress toward increasing values of entropy</u>! Why? Can you lo-
cate the points of maximum entropy for each Fanno line in Figure
9.1?

Let us examine one Fanno line in greater detail. Figure 9.3
shows a given Fanno line together with typical pressure lines.
All points on this line represent states with the same mass flow
rate per unit area (mass velocity) and the same stagnation enthalpy. Due to the irreversible nature of the frictional effects the

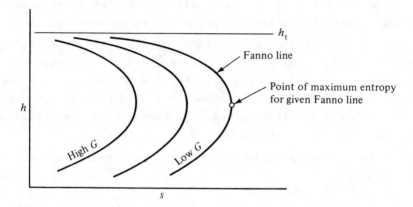

Figure 9.2 Fanno lines in $h-s$ plane

flow can only proceed to the right. Thus, the Fanno line is divided into two distinct parts, an upper and a lower branch, which
are separated by a limiting point of maximum entropy.

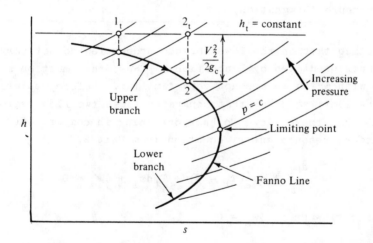

Figure 9.3 Two branches of Fanno line

What does intuition tell us about adiabatic flow in a constant area duct? We normally feel that frictional effects will
show up as an internal generation of heat with a corresponding
reduction in density of the fluid. In order to pass the same flow
rate (with constant area) continuity then forces the velocity to
increase. This increase in kinetic energy must cause a decrease
in enthalpy, since the stagnation enthalpy remains constant. As
can be seen in Figure 9.3, this agrees with flow along the "upper

branch" of the Fanno line. It is also clear that in this case
both the static and stagnation pressure are decreasing.

But what about flow along the "lower branch?" Mark two
points on the lower branch and draw an arrow to indicate the proper movement along the Fanno line. What is happening to the enthalpy? The density (see equation 9.5)? The velocity (see equation 9.2)? From the figure, what is happening to the static pressure? The stagnation pressure?

Fill in the following table with "increase," "decrease," or "remains constant."

Property	Upper Branch	Lower Branch
Enthalpy		
Density		
Velocity		
Pressure (Static)		
Pressure (Stagnation)		

Notice that on the lower branch properties do not vary in
the manner predicted by "intuition." Thus, this must be a flow
regime with which we are not very familiar. Before we investigate the limiting point that separates these two flow regimes let
us note that these flows do have one thing in common. Recall the
stagnation pressure energy equation from Unit 3.

$$\frac{dp_t}{\rho_t} + \cancel{ds}_e(T_t - T) + T_t ds_i + \cancel{\delta w}_s = 0 \qquad (3.25)$$

For Fanno flow $ds_e = \delta w_s = 0$

Thus, any frictional effect must cause a decrease in the total or
stagnation pressure! Figure 9.3 verifies this for flow along
both the upper and lower branches of the Fanno line.

LIMITING POINT

From the energy equation we had developed

$$h_t = h + \frac{V^2}{2g_c} = \text{const.} \qquad (9.4)$$

Differentiating, we obtain

$$dh_t = dh + \frac{VdV}{g_c} = 0 \tag{9.6}$$

From continuity we had found

$$\rho V = G = \text{const.} \tag{9.2}$$

Differentiating this, we obtain

$$\rho dV + Vd\rho = 0 \tag{9.7}$$

which can be solved for

$$dV = - V \frac{d\rho}{\rho} \tag{9.8}$$

Introduce equation 9.8 into (9.6) and <u>show</u> that

$$dh = \frac{V^2 d\rho}{g_c \rho} \tag{9.9}$$

Now recall the property relation

$$Tds = dh - vdp \tag{1.41}$$

which can be written as

$$Tds = dh - \frac{dp}{\rho} \tag{9.10}$$

Substituting for dh from equation 9.9 yields

$$\boxed{Tds = \frac{V^2 d\rho}{g_c \rho} - \frac{dp}{\rho}} \tag{9.11}$$

We hasten to point out that this expression is valid for <u>any</u> fluid and between two differentially separated points <u>anyplace</u> along the Fanno line. Now let's apply equation 9.11 to two adjacent points that surround the limiting point of maximum entropy. At this location s = const ; thus ds = 0 , and (9.11) becomes

$$\frac{V^2 d\rho}{g_c} = dp \qquad \text{at limit point.} \tag{9.12}$$

or $\qquad V^2 = g_c \left(\frac{dp}{d\rho}\right)_{at} = g_c \left(\frac{\partial p}{\partial \rho}\right)_{s \,=\, \text{const}}$ (9.13)
$\qquad\qquad\qquad\qquad$ limit point.

This should be a familiar expression (see equation 4.5) and we recognize that <u>the velocity is sonic at the limiting point</u>. The

upper branch can now be more significantly called the subsonic
branch, and the lower branch is seen to be the supersonic branch.

Now we begin to see a reason for the failure of our intuition to predict behavior on the lower branch of the Fanno line.
From our previous studies in Unit 5 we saw that fluid behavior
in supersonic flow is frequently contrary to our expectations.
This points out the fact that we live most of our lives "subson-
ically" and, in fact, our knowledge of fluid phenomena comes
mainly from experiences with incompressible fluids. It should be
apparent that we cannot use our intuition to guess at what might
be happening, particularly in the supersonic flow regime. We
must learn to get religious and put faith in our carefully derived relations.

MOMENTUM

The above analysis was made using only the continuity and
energy relations. We now proceed to apply momentum concepts to
the control volume shown in Figure 9.4.

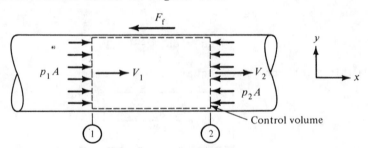

Figure 9.4 Momentum analysis for Fanno flow

The x-component of the momentum equation for steady, one-
dimensional flow is

$$\Sigma F_x = \frac{\dot{m}}{g_c}\left[V_{out_x} - V_{in_x}\right] \qquad (3.46)$$

From Figure 9.4 we see that the force summation is

$$\Sigma F_x = p_1 A - p_2 A - F_f \qquad (9.14)$$

where F_f represents the total wall frictional force on the
fluid between sections 1 and 2. Thus, the momentum equation in
the direction of flow becomes

$$(p_1 - p_2)A - F_f = \frac{\dot{m}}{g_c}(V_2 - V_1) = \frac{\rho A V}{g_c}(V_2 - V_1) \qquad (9.15)$$

Show that equation 9.15 can be written as

$$p_1 - p_2 - \frac{F_f}{A} = \frac{\rho_2 V_2^2}{g_c} - \frac{\rho_1 V_1^2}{g_c} \tag{9.16}$$

or

$$\left[p_1 + \frac{\rho_1 V_1^2}{g_c} \right] - \frac{F_f}{A} = \left[p_2 + \frac{\rho_2 V_2^2}{g_c} \right] \tag{9.17}$$

In this form the equation is not particularly useful except to bring out one significant fact. For the steady, one-dimensional, constant-area flow of any fluid the value of $p + \rho V^2 / g_c$ cannot be constant if frictional forces are present. This fact will be recalled later in this unit when Fanno flow is compared with normal shocks.

Before leaving this section on fluids in general we might say a few words about Fanno flow at low Mach numbers. A glance at Figure 9.3 shows that the upper branch is asymptotically approaching the horizontal line of constant total enthalpy. Thus, the extreme left end of the Fanno line will be nearly horizontal. This indicates that flow at very low Mach numbers will have almost constant velocity. This checks our previous work which indicated that we could treat gases as incompressible fluids if the Mach numbers were very small.

9.4 WORKING EQUATIONS FOR A PERFECT GAS

We have discovered the general trend of property variations that occur in Fanno flow, both in the subsonic and supersonic flow regime. Now we wish to develop some specific working equations for the case of a perfect gas. Recall that these are relations between properties at arbitrary sections of a flow system written in terms of Mach numbers and the specific heat ratio.

ENERGY - We start with the energy equation developed in the previous section since this leads immediately to a temperature ratio.

$$h_{t1} = h_{t2} \tag{9.3}$$

But for a perfect gas enthalpy is a function of temperature only. Therefore,

$$T_{t1} = T_{t2} \tag{9.18}$$

Now for a perfect gas with constant specific heats

$$T_t = T\left[1 + \frac{(\gamma-1)}{2} M^2\right] \qquad (4.18)$$

Hence, the energy equation for Fanno flow can be written as

$$T_1\left[1 + \frac{(\gamma-1)}{2} M_1^2\right] = T_2\left[1 + \frac{(\gamma-1)}{2} M_2^2\right] \qquad (9.19)$$

or

$$\boxed{\frac{T_2}{T_1} = \frac{1 + \frac{(\gamma-1)}{2} M_1^2}{1 + \frac{(\gamma-1)}{2} M_2^2}} \qquad (9.20)$$

CONTINUITY - From the previous section we have

$$\rho V = G = \text{const.} \qquad (9.2)$$

or

$$\rho_1 V_1 = \rho_2 V_2 \qquad (9.21)$$

If we introduce the perfect gas equation of state

$$p = \rho RT \qquad (1.13)$$

the definition of Mach number

$$V = Ma \qquad (4.11)$$

and sonic velocity for a perfect gas

$$a = \sqrt{\gamma g_c RT} \qquad (4.10)$$

equation 9.21 can be solved for

$$\frac{p_2}{p_1} = \frac{M_1}{M_2}\left(\frac{T_2}{T_1}\right)^{\frac{1}{2}} \qquad (9.22)$$

Can you obtain this expression? Now introduce the temperature ratio from (9.20) and you will have the following working relation for static pressures:

$$\boxed{\frac{p_2}{p_1} = \frac{M_1}{M_2}\left[\frac{1 + \frac{(\gamma-1)}{2} M_1^2}{1 + \frac{(\gamma-1)}{2} M_2^2}\right]^{\frac{1}{2}}} \qquad (9.23)$$

The density relation can easily be obtained from equations 9.20, 9.23, and the perfect gas law.

$$\frac{\rho_2}{\rho_1} = \frac{M_1}{M_2}\left[\frac{1 + \frac{(\gamma-1)}{2}M_2^2}{1 + \frac{(\gamma-1)}{2}M_1^2}\right]^{\frac{1}{2}}$$

(9.24)

ENTROPY CHANGE – We start with an expression for entropy change which is valid between any two points.

$$\Delta s_{1-2} = c_p \ln \frac{T_2}{T_1} - R \ln \frac{p_2}{p_1}$$

(1.53)

Equation 4.15 can be used to substitute for c_p and we non-dimensionalize the equation to

$$\frac{s_2 - s_1}{R} = \left[\frac{\gamma}{\gamma-1}\right]\ln \frac{T_2}{T_1} - \ln \frac{p_2}{p_1}$$

(9.25)

If we now utilize the expressions just developed for the temperature ratio (9.20) and the pressure ratio (9.23), the entropy change becomes

$$\frac{s_2 - s_1}{R} = \frac{\gamma}{\gamma-1}\ln \left[\frac{1 + \frac{(\gamma-1)}{2}M_1^2}{1 + \frac{(\gamma-1)}{2}M_2^2}\right] - \ln \frac{M_1}{M_2}\left[\frac{1 + \frac{(\gamma-1)}{2}M_1^2}{1 + \frac{(\gamma-1)}{2}M_2^2}\right]^{\frac{1}{2}}$$

(9.26)

Show that this entropy change between two points in Fanno flow can be written as

$$\frac{s_2 - s_1}{R} = \ln \frac{M_2}{M_1}\left[\frac{1 + \frac{(\gamma-1)}{2}M_1^2}{1 + \frac{(\gamma-1)}{2}M_2^2}\right]^{\frac{\gamma+1}{2(\gamma-1)}}$$

(9.27)

Now recall that in Section 4.5 we integrated the stagnation pressure-energy equation for adiabatic, no-work flow of a perfect gas with the result

$$\frac{p_{t2}}{p_{t1}} = e^{-\Delta s/R}$$

(4.28)

Thus, from equations 4.28 and 9.27 we obtain a simple expression for the stagnation pressure ratio.

$$\frac{p_{t2}}{p_{t1}} = \frac{M_1}{M_2}\left[\frac{1 + \frac{(\gamma-1)}{2}M_2^2}{1 + \frac{(\gamma-1)}{2}M_1^2}\right]^{\frac{\gamma+1}{2(\gamma-1)}}$$

(9.28)

We now have the means to obtain all the properties at a downstream point 2 if we know all the properties at some upstream
point 1 and the Mach number at point 2 . However, in many
situations one does not know both Mach numbers. A typical problem would be to predict the final Mach number, given the initial
conditions and information on duct length, material, etc. Thus,
our next job is to relate the change in Mach number to the friction losses.

MOMENTUM - We turn to the differential form of the momentum equation that was developed in Unit 3.

$$\frac{dp}{\rho} + f\frac{V^2}{2g_c}\frac{dx}{D_e} + \frac{g}{g_c}dz + \frac{dV^2}{2g_c} = 0 \qquad (3.63)$$

Our objective is to get this equation all in terms of Mach number. If we introduce the perfect gas equation of state together
with expressions for Mach number and sonic velocity we obtain

$$\frac{dp}{p}(RT) + f\frac{dx}{D_e}\frac{M^2\gamma g_c RT}{2g_c} + \frac{g}{g_c}dz + \left[\frac{dM^2\gamma g_c RT + M^2\gamma g_c RdT}{2g_c}\right] = 0 \qquad (9.29)$$

or

$$\boxed{\frac{dp}{p} + f\frac{dx}{D_e}\frac{\gamma}{2}M^2 + \frac{g\,dz}{g_c RT} + \frac{\gamma}{2}dM^2 + \frac{\gamma}{2}M^2\frac{dT}{T} = 0} \qquad (9.30)$$

Equation 9.30 is marked since it is a useful form of the
momentum equation that is valid for all steady flow problems involving a perfect gas. We now proceed to apply this to Fanno
flow. From (9.18) and (4.18) we know that

$$T_t = T\left[1 + \frac{(\gamma-1)}{2}M^2\right] = \text{const} \qquad (9.31)$$

Taking the natural logarithm

$$\ln T + \ln\left[1 + \frac{(\gamma-1)}{2}M^2\right] = \ln \text{const} \qquad (9.32)$$

and then differentiating, we obtain

$$\frac{dT}{T} + \frac{d\left[1 + \frac{(\gamma-1)}{2}M^2\right]}{\left[1 + \frac{(\gamma-1)}{2}M^2\right]} = 0 \qquad (9.33)$$

which can be used to substitute for dT/T in (9.30).

The continuity relation (equation 9.2) put in terms of a perfect gas becomes

$$\frac{pM}{\sqrt{T}} = \text{const} \tag{9.34}$$

By logarithmic differentiation (take the natural logarithm and then differentiate) <u>show</u> that

$$\frac{dp}{p} + \frac{dM}{M} - \frac{1}{2}\frac{dT}{T} = 0 \tag{9.35}$$

We can introduce equation 9.33 to eliminate dT/T with the result

$$\frac{dp}{p} = -\frac{dM}{M} - \frac{1}{2}\frac{d\left[1 + \frac{(\gamma-1)}{2}M^2\right]}{\left[1 + \frac{(\gamma-1)}{2}M^2\right]} \tag{9.36}$$

which can be used to substitute for dp/p in (9.30).

Make the indicated substitutions for dp/p and dT/T in the momentum equation, neglect the potential term, and <u>show</u> that equation 9.30 can be put into the following form:

$$f\frac{dx}{D_e} = \frac{d\left[1 + \frac{(\gamma-1)}{2}M^2\right]}{\left[1 + \frac{(\gamma-1)}{2}M^2\right]} - \frac{dM^2}{M^2} + \frac{2}{\gamma}\frac{dM}{M^3} + \frac{1}{\gamma M^2}\frac{d\left[1 + \frac{(\gamma-1)}{2}M^2\right]}{\left[1 + \frac{(\gamma-1)}{2}M^2\right]} \tag{9.37}$$

The last term can be simplified for integration by noting that

$$\frac{1}{\gamma M^2}\frac{d\left[1 + \frac{(\gamma-1)}{2}M^2\right]}{\left[1 + \frac{(\gamma-1)}{2}M^2\right]} = \frac{(\gamma-1)}{2\gamma}\frac{dM^2}{M^2} - \frac{(\gamma-1)}{2\gamma}\frac{d\left[1 + \frac{(\gamma-1)}{2}M^2\right]}{\left[1 + \frac{(\gamma-1)}{2}M^2\right]} \tag{9.38}$$

The momentum equation can now be written as

$$\frac{fdx}{D_e} = \left[\frac{\gamma+1}{2\gamma}\right]\frac{d\left[1 + \frac{(\gamma-1)}{2}M^2\right]}{\left[1 + \frac{(\gamma-1)}{2}M^2\right]} + \frac{2}{\gamma}\frac{dM}{M^3} - \left[\frac{\gamma+1}{2\gamma}\right]\frac{dM^2}{M^2} \tag{9.39}$$

Equation 9.39 is restricted to steady, one-dimensional flow of a perfect gas, with no heat or work transfer, constant area, and negligible potential changes. We can now integrate this equation between two points in the flow and obtain

$$\frac{f(x_2-x_1)}{D_e} = \left[\frac{\gamma+1}{2\gamma}\right]\ln\left[\frac{1 + \frac{(\gamma-1)}{2}M_2^{\,2}}{1 + \frac{(\gamma-1)}{2}M_1^{\,2}}\right] - \frac{1}{\gamma}\left[\frac{1}{M_2^{\,2}} - \frac{1}{M_1^{\,2}}\right] - \left[\frac{\gamma+1}{2\gamma}\right]\ln\left[\frac{M_2^{\,2}}{M_1^{\,2}}\right] \tag{9.40}$$

Note that in performing the integration we have held the friction
factor constant. Some comments will be made on this in a later
section.

9.5   REFERENCE STATE AND FANNO TABLES

The equations developed in the previous section provide the
means of computing the properties at one location in terms of
those given at some other location. The key to problem solution
is predicting the Mach number at the new location through the
use of equation 9.40. The solution of this equation for the unknown M_2 presents a messy task as no explicit relation is possible. Thus, we turn to a technique similar to that used with
isentropic flow in Unit 5.

We introduce <u>another</u> * reference state which is defined in
the same manner as before; i.e., "that thermodynamic state which
would exist if the fluid reached a Mach number of unity <u>by a particular process</u>." In this case, we imagine that we continue <u>by
Fanno flow</u> until the velocity reaches Mach one. Figure 9.5 shows
a physical system together with its T-s diagram for a subsonic
Fanno flow. We know that if we continue along the Fanno line
(remember we always move to the right) we will eventually reach
the limiting point where sonic velocity exists. The dotted lines
show a hypothetical duct of sufficient length to enable the flow
to traverse the remaining portion of the upper branch and reach
the limit point. This is the * reference point <u>for Fanno flow</u>.

The <u>isentropic</u> * reference points have also been included
on the T-s diagram in order to emphasize the fact that the Fanno
* reference is a totally different thermodynamic state. One
other fact should be mentioned. If there is any entropy difference between two points (such as points 1 and 2), then their isentropic * reference conditions are not the same and we have always taken great care to label them separately as 1^* and 2^* .
However, proceeding from either point 1 or point 2 <u>by Fanno
flow</u> will ultimately lead to the same place when Mach one is
reached. Thus, we do not have to talk of 1^* or 2^* but merely
* in the case of Fanno flow. Incidentally, why are all three
* reference points shown on the same horizontal line in Figure
9.5? (You may need to review Section 4.6.)

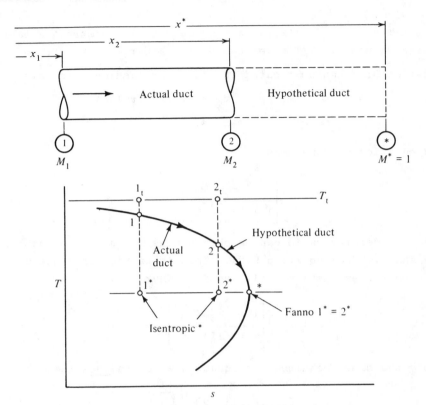

Figure 9.5 The * reference for Fanno flow

We now rewrite the working equations in terms of the Fanno flow * reference condition. Consider first

$$\frac{T_2}{T_1} = \frac{1 + \frac{(\gamma-1)}{2} M_1^2}{1 + \frac{(\gamma-1)}{2} M_2^2} \qquad (9.20)$$

Let point 2 be an arbitrary point in the flow system and let its Fanno * condition be point 1.

Then
$$T_2 \Rightarrow T \qquad M_2 \Rightarrow M \text{ (any value)}$$
$$T_1 \Rightarrow T^* \qquad M_1 \Rightarrow 1$$

and equation 9.20 becomes

$$\frac{T}{T^*} = \frac{\frac{\gamma+1}{2}}{1 + \frac{(\gamma-1)}{2} M^2} = f(M,\gamma) \qquad (9.41)$$

We see that $T/T^* = f(M,\gamma)$ and we can easily construct a table giving values of T/T^* versus M for a particular γ.

Equation 9.23 can be treated in a similar fashion. In this case

$$p_2 \Rightarrow p \qquad\qquad M_2 \Rightarrow M \text{ (any value)}$$
$$p_1 \Rightarrow p^* \qquad\qquad M_1 \Rightarrow 1$$

and equation 9.23 becomes

$$\frac{p}{p^*} = \frac{1}{M}\left[\frac{\frac{\gamma+1}{2}}{1 + \frac{(\gamma-1)}{2} M^2}\right]^{\frac{1}{2}} = f(M,\gamma) \qquad (9.42)$$

The density ratio can be obtained as a function of Mach number and γ from equation 9.24. This is particularly useful since it also represents a velocity ratio. Why?

$$\frac{\rho}{\rho^*} = \frac{V^*}{V} = \frac{1}{M}\left[\frac{1 + \frac{(\gamma-1)}{2} M^2}{\frac{\gamma+1}{2}}\right]^{\frac{1}{2}} = f(M,\gamma) \qquad (9.43)$$

Apply the same techniques to equation 9.28 and <u>show</u> that

$$\frac{p_t}{p_t^*} = \frac{1}{M}\left[\frac{1 + \frac{(\gamma-1)}{2} M^2}{\frac{\gamma+1}{2}}\right]^{\frac{\gamma+1}{2(\gamma-1)}} = f(M,\gamma) \qquad (9.44)$$

We now perform the same type of transformation on equation 9.40; that is

let $\qquad x_2 \Rightarrow x \qquad\qquad M_2 \Rightarrow M \text{ (any value)}$
$\qquad\qquad x_1 \Rightarrow x^* \qquad\qquad M_1 \Rightarrow 1$

with the following result:

$$\frac{f(x-x^*)}{D_e} = \left[\frac{\gamma+1}{2\gamma}\right]\ln\left[\frac{1 + \frac{(\gamma-1)}{2} M^2}{\frac{\gamma+1}{2}}\right] - \frac{1}{\gamma}\left[\frac{1}{M^2} - 1\right] - \left[\frac{\gamma+1}{2\gamma}\right]\ln M^2 \qquad (9.45)$$

But a glance at the physical diagram in Figure 9.5 shows that $(x-x^*)$ will always be a negative quantity; thus, it is more convenient to change all signs in the above equation and simplify it to

$$\frac{f(x^*-x)}{D_e} = \left[\frac{\gamma+1}{2\gamma}\right]\ln\left[\frac{\frac{(\gamma+1)}{2} M^2}{1 + \frac{(\gamma-1)}{2} M^2}\right] + \frac{1}{\gamma}\left[\frac{1}{M^2} - 1\right] = f(M,\gamma) \qquad (9.46)$$

The quantity (x^*-x) represents the amount of duct that would have to be added to cause the flow to reach the Fanno * reference condition. It can alternately be viewed as the maximum duct length which may be added without changing some flow condition. Thus, the expression

$$\frac{f(x^*-x)}{D_e} \qquad \text{is called} \qquad \frac{f\,L_{max}}{D_e}$$

and is listed in the Appendix along with the other Fanno flow parameters T/T^* , p/p^* , V/V^* and p_t/p_t^* . In the next section we shall see how these tables greatly simplify the solution of Fanno flow problems. But first, a word on the determination of friction factors.

Dimensional analysis of the fluid flow problem shows that the friction factor can be expressed as

$$f = f(R, \ \varepsilon/D) \qquad (9.47)$$

where R is the Reynolds number

$$R \equiv \frac{\rho VD}{\mu g_c} \qquad (9.48)$$

and $\varepsilon/D \equiv$ relative roughness

Typical values of ε , the absolute roughness or average height of wall irregularities, are shown in Table 9.1.

TABLE 9.1

Material	ε (in feet)
glass, brass, copper, lead	smooth < 0.00001
steel, wrought iron	0.00015
galvanized iron	0.0005
cast iron	0.00085
riveted steel	0.03

The relationship among f , R , and ε/D is experimentally determined and plotted on a chart similar to Figure 9.6 which is called a Moody diagram. A larger working chart is contained in the Appendix. If the flow rate is known together with the duct size and material, then the Reynolds number and relative roughness can easily be calculated and the value of the friction fac-

tor is taken from the diagram. The curve in the laminar flow
region can be represented by

$$f = \frac{64}{R} \tag{9.49}$$

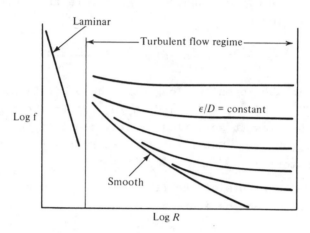

Figure 9.6 Moody diagram for friction factors in ducts
(see Appendix for working chart)

For non-circular cross-sections the "equivalent diameter" as
described in Section 3.8 should be used.

$$D_e \equiv 4A/P \tag{3.61}$$

This equivalent diameter may be used in the determination of relative roughness and Reynolds number, and hence the friction factor.
However, care must be taken to work with the actual average velocity in all computations. Experience has shown that the use of
an equivalent diameter works quite well in the turbulent zone.
In the laminar flow region this concept is not sufficient and
consideration must also be given to the aspect ratio of the duct.

In some problems the flow rate is not known and thus a trial
and error solution results. As long as the duct size is given,
the problem is not too difficult; an excellent approximation to
the friction factor can be made by taking the value corresponding
to where the ε/D curve begins to "level off." This rapidly
converges to the final answer as most engineering problems are
well into the turbulent range.

9.6 APPLICATIONS

The following steps are recommended to develop good problem
solving technique:

1. Sketch the physical situation.

2. Label sections where conditions are known or derived, including the hypothetical * reference point.

3. List all given information with units.

4. Compute the equivalent diameter, relative roughness, and Reynolds number.

5. Find the friction factor from the Moody diagram.

6. Determine the unknown Mach number.

7. Calculate the additional properties desired.

The above procedure may have to be altered depending on what type of information is given and occasionally trial and error solutions are required. You should have no difficulty incorporating these features once the basic straightforward solution has been mastered. In complicated flow systems which involve more than just Fanno flow, a T-s diagram is frequently helpful in solving problems.

For the following examples we are dealing with the steady, one-dimensional flow of air ($\gamma = 1.4$) which can be treated as a perfect gas. Assume $Q = W_s = 0$ and negligible potential changes. Cross-sectional area of the duct remains constant. The physical diagram shown is common to the first three examples.

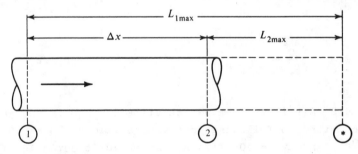

Example 9.1 Given: $M_1 = 1.80$, $p_1 = 40$ psia , and $M_2 = 1.20$
Find: p_2 and $f\Delta x/D$.

Since both Mach numbers are known we can immediately solve for

$$p_2 = \frac{p_2}{p^*} \frac{p^*}{p_1} p_1 = (0.8044) \frac{1}{0.4741}(40) = \underline{67.9} \text{ psia}$$

Check the diagram above to see that

$$\frac{f\Delta x}{D} = \frac{fL_{1max}}{D} - \frac{fL_{2max}}{D} = 0.2419 - 0.0336 = \underline{0.208}$$

Example 9.2   Given: $M_2 = 0.94$, $T_1 = 400^{O}K$, and $T_2 = 350^{O}K$
Find: M_1 and p_2/p_1 .

To determine conditions at section   1   we must establish the
ratio:

$$\frac{T_1}{T^*} = \frac{T_1}{T_2} \frac{T_2}{T^*} = \frac{400}{350} (1.0198) = 1.1655$$

$$\underset{\text{Given}}{\overset{\text{From Fanno tables at } M = 0.94}{\uparrow}}$$

Look up $T/T^* = 1.1655$ in the Fanno table and determine that
$M_1 = 0.385$

Thus, $\frac{p_2}{p_1} = \frac{p_2}{p^*} \frac{p^*}{p_1} = (1.0743) \frac{1}{2.8046} = \underline{0.383}$

Notice that these examples confirm previous statements concerning static pressure changes. In subsonic flow the static
pressure decreases whereas in supersonic flow the static pressure
increases. Compute the stagnation pressure ratio and show that
the friction losses cause p_{t2}/p_{t1} to decrease in each case.

For ex. 9.1 $\frac{p_{t2}}{p_{t1}} =$ $(p_{t2}/p_{t1} = 0.716)$

For ex. 9.2 $\frac{p_{t2}}{p_{t1}} =$ $(p_{t2}/p_{t1} = 0.611)$

Example 9.3 Air flows in a 6-inch-diameter, insulated, galvanized
iron duct with the following initial conditions: $p_1 = 20$ psia,
$T_1 = 70^{O}F$, and $V_1 = 406$ ft/sec. After 70 ft determine the final
Mach number, temperature and pressure.

Since the duct is circular we do not have to compute an equivalent
diameter. From Table 9.1 the absolute roughness ε is 0.0005.

Thus, the relative roughness $\frac{\varepsilon}{D} = \frac{0.0005}{0.5} = \underline{0.001}$

We compute the Reynolds number at section 1 since this is the
only location where information is known.

$$\rho_1 = \frac{p_1}{RT_1} = \frac{(20)(144)}{(53.3)(530)} = 0.102 \text{ lbm/ft}^3$$

$$\mu_1 = 3.8 \times 10^{-7} \text{ lbf-sec/ft}^2 \text{ (from table in Appendix)}$$

Thus, $\quad R_1 = \dfrac{\rho_1 V_1 D_1}{\mu_1 g_c} = \dfrac{(0.102)(406)(0.5)}{(3.8 \times 10^{-7})(32.2)} = \underline{1.69 \times 10^6}$

From the Moody diagram (in the Appendix) at $R = 1.69 \times 10^6$ and $\varepsilon/D = 0.001$ we determine that the friction factor is $f = 0.0198$. To use the Fanno tables (or equations) we need information on Mach numbers.

$$a_1 = \left[\gamma g_c R T_1\right]^{\frac{1}{2}} = \left[(1.4)(32.2)(53.3)(530)\right]^{\frac{1}{2}} = 1128 \text{ ft/sec}$$

$$M_1 = V_1/a_1 = 406/1128 = \underline{0.36}$$

From the Fanno tables at $M_1 = 0.36$ we find $p_1/p^* = 3.0042$, $T_1/T^* = 1.1697$, $fL_{1max}/D = 3.1801$.

The key to completing the problem is in establishing the Mach number at the outlet and this is done through the "friction length"

$$\frac{f \Delta x}{D} = \frac{(0.0198)(70)}{(0.5)} = 2.772$$

Looking at the physical sketch it is apparent (since f and D are constants) that

$$\frac{fL_{2max}}{D} = \frac{fL_{1max}}{D} - \frac{f \Delta x}{D} = 3.1801 - 2.772 = 0.408$$

We enter the Fanno tables with this "friction length" and find that $M_2 = 0.623$, $p_2/p^* = 1.6939$, $T_2/T^* = 1.1136$.

Thus, $\quad p_2 = \dfrac{p_2}{p^*} \dfrac{p^*}{p_1} p_1 = (1.6939) \dfrac{1}{3.0042} (20) = \underline{11.28} \text{ psia}$

and $\quad T_2 = \dfrac{T_2}{T^*} \dfrac{T^*}{T_1} T_1 = (1.1136) \dfrac{1}{1.1697} (530) = \underline{505}^\circ R$

In the above example the friction factor was assumed constant. In fact, this assumption was made when equation 9.39 was integrated to obtain (9.40) and with the introduction of the * reference state this became equation 9.46 which is listed in the Fanno tables. Is this a reasonable assumption? Friction factors

are functions of Reynolds numbers which in turn depend on velocity and density — both of which can change quite rapidly in Fanno
flow. Calculate the velocity at the outlet in example 9.3 and
compare it with that at the inlet. (V_2 = 686 ft/sec   vice
V_1 = 406 ft/sec.)

But don't despair. From continuity we know that the product
of ρV is always a constant and thus the only variable in Reynolds number is the viscosity. Extremely large temperature variations are required to change the viscosity of a gas significantly
and thus variations in the Reynolds number are small. We are
also fortunate in that most engineering problems are well into
the turbulent range where the friction factor is relatively insen-
sitive to Reynolds number. A greater potential error is probably
involved in the estimation of the duct roughness which has a more
significant effect on the friction factor.

Example 9.4 A converging-diverging nozzle with an area ratio of
5.42 connects to an 8-ft-long, constant-area, rectangular duct.
The duct is 8 x 4 inches in cross-section and has a friction factor of f = 0.02. What is the minimum stagnation pressure feeding
the nozzle if the flow is supersonic throughout the entire duct
and it exhausts to 14.7 psia?

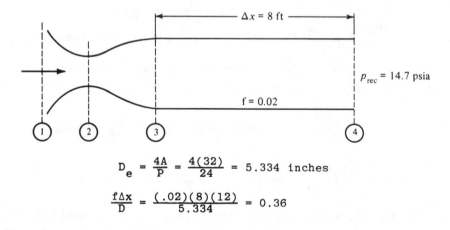

$$D_e = \frac{4A}{P} = \frac{4(32)}{24} = 5.334 \text{ inches}$$

$$\frac{f\Delta x}{D} = \frac{(.02)(8)(12)}{5.334} = 0.36$$

To be supersonic with A_3/A_2 = 5.42 , M_3 = 3.26   and
P_3/P_{t3} = 0.0185 , P_3/p^* = 0.1901 , fL_{3max}/D = 0.5582

$$\frac{fL_{4max}}{D} = \frac{fL_{3max}}{D} - \frac{f\Delta x}{D} = 0.5582 - 0.36 = 0.1982$$

Thus, $M_4 = 1.673$ and $p_4/p^* = 0.5243$

$$p_{t1} = \frac{p_{t1}}{p_{t3}} \frac{p_{t3}}{p_3} \frac{p_3}{p^*} \frac{p^*}{p_4} p_4 = (1) \frac{1}{(0.0185)} (0.1901) \frac{1}{(0.5243)} (14.7)$$

$$= \underline{288} \text{ psia}$$

Any pressure above 288 psia will maintain the flow system as specified but with expansion waves outside the duct. (Recall an underexpanded nozzle.) Can you envision what would happen if the inlet stagnation pressure fell below 288 psia? (Recall the operation of an overexpanded nozzle.)

9.7 CORRELATION WITH SHOCKS

As you have progressed through this unit you may have noticed some similarities between Fanno flow and normal shocks. Let us summarize some pertinent information.

The points just before and after a normal shock represent states with the same mass flow per unit area, the same value of $p + \rho V^2/g_c$, and the same stagnation enthalpy. These facts are the result of applying the basic concepts of continuity, momentum, and energy to any arbitrary fluid. This analysis resulted in equations 6.2, 6.3, and 6.9.

A Fanno line represents states with the same mass flow per unit area and the same stagnation enthalpy. This is confirmed by equations 9.2 and 9.5. To move <u>along</u> a Fanno line requires friction. At the end of Section 9.3 (see equation 9.17) it was pointed out that it is this very friction which causes the value of $p + \rho V^2/g_c$ to change.

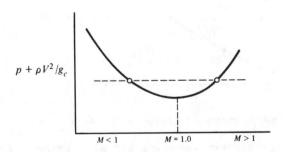

Figure 9.7 Variation of $p + \rho V^2/g_c$ in Fanno flow

The variation of the quantity $p + \rho V^2/g_c$ along a Fanno line is quite interesting. Such a plot is shown in Figure 9.7. You will notice that for every point on the supersonic branch of the Fanno line there is a corresponding point on the subsonic branch with the same value of $p + \rho V^2/g_c$. Thus, these two points satisfy all three conditions for the end points of a normal shock and could be connected by such a shock.

Now we can imagine a supersonic Fanno flow leading into a normal shock. If this is followed by additional duct, then subsonic Fanno flow would occur. Such a situation is shown in Figure 9.8. Note that the shock merely causes the flow to jump from the supersonic branch to the subsonic branch of the <u>same</u> Fanno line.

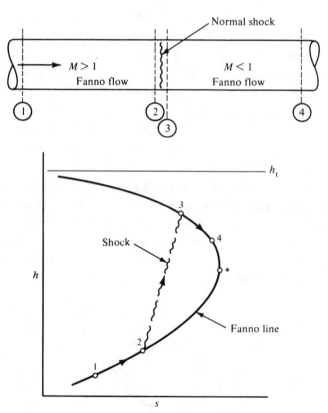

Figure 9.8 Combination of Fanno flow and normal shock

Example 9.5 A large chamber contains air at a temperature of $300°K$ and a pressure of 8 bars absolute. The air enters a converging-diverging nozzle with an area ratio of 2.4. A constant-area duct is attached to the nozzle and a normal shock stands at the exit plane. Receiver pressure is 3 bars. Assume the entire system to be adiabatic and neglect friction in the nozzle. Compute the $f\Delta x/D$ for the duct.

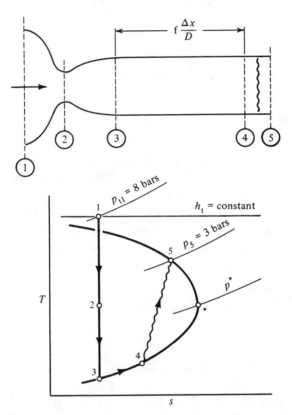

For a shock to occur as specified, the duct flow must be supersonic, which means that the nozzle is operating at its third critical point. The inlet conditions and nozzle area ratio fix conditions at location 3. We can then find p^* at the tip of the Fanno line. Then the ratio p_5/p^* can be computed and the Mach number after the shock is found from the Fanno table. This solution probably would not have occurred to us had we not drawn the T-s diagram and recognized that point 5 is on the same Fanno line as 3 , 4 , and * .

For $A_3/A_2 = 2.4$, $M_3 = 2.4$ and $p_3/p_{t3} = 0.06840$

We proceed immediately to compute p_5/p^*

$$\frac{p_5}{p^*} = \frac{p_5}{p_{t1}} \frac{p_{t1}}{p_{t3}} \frac{p_{t3}}{p_3} \frac{p_3}{p^*} = \frac{(3)}{(8)} \; (1) \; \frac{1}{(0.0684)} \; (0.3111) = 1.7056$$

From the Fanno tables we find $M_5 = 0.619$

and then from the shock tables $M_4 = 1.789$

Returning to the Fanno tables: $fL_{3max}/D = 0.4099$,

$$fL_{4max}/D = 0.2382$$

Thus $\dfrac{f\Delta x}{D} = \dfrac{fL_{3max}}{D} - \dfrac{fL_{4max}}{D} = 0.4099 - 0.2382 = \underline{0.172}$

9.8 FRICTION CHOKING

In Unit 5 we discussed the operation of nozzles that were fed by constant stagnation inlet conditions (see Figures 5.6 and 5.8). We found that as the receiver pressure was lowered the flow through the nozzle increased. When the "operating pressure ratio" reached a certain value the section of minimum area developed a Mach number of unity. The nozzle was then said to be choked. Further reduction in the pressure ratio did not increase the flow rate. This was an example of "area choking."

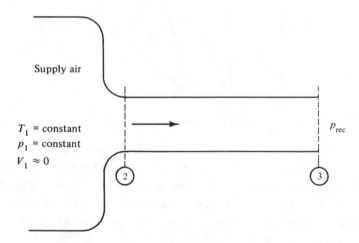

Figure 9.9 Converging nozzle and constant area duct combination

The subsonic Fanno flow situation is quite similar. Figure 9.9 shows a given length of duct fed by a large tank and converging nozzle. If the receiver pressure is below the tank pressure, flow will occur producing a T–s diagram shown as path 1–2–3 in Figure 9.10. Note that we have isentropic flow at the entrance

to the duct and then we move along a Fanno line. As the receiver
pressure is lowered still more, the flow rate and exit Mach number continue to increase while the system moves to Fanno lines
of higher mass velocities (shown as path 1-2'-3'). It is important to recognize that the receiver pressure (or more properly
the operating pressure ratio) is controlling the flow. This is
because in subsonic flow the pressure at the duct exit must equal
that of the receiver.

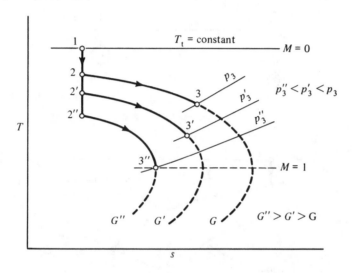

Figure 9.10 $T-s$ diagram for nozzle-duct combination

Eventually, when a certain pressure ratio is reached, the
Mach number at the duct exit will be unity (shown as path 1-2"-
3"). This is called "friction choking" and any further reduction
in receiver pressure would not affect the flow conditions inside
the system. What would occur as the flow leaves the duct and
enters a region of reduced pressure?

Let us consider this last case of choked flow with the exit
pressure equal to the receiver pressure. Now suppose that the
receiver pressure is maintained at this value but more duct is
added to the system. What happens? We know that we cannot move
"around the Fanno line," yet somehow we must reflect the added
friction losses. This is done by moving to a new Fanno line at a
decreased flow rate. The T-s diagram for this is shown as path
1-2'''-3'''-4 in Figure 9.11. Note that pressure equilibrium is
still maintained at the exit but the system is no longer choked
although the flow rate has decreased. If the receiver pressure

were now reduced sufficiently, then sonic velocity could exist at
the exit.

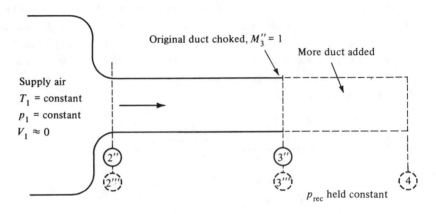

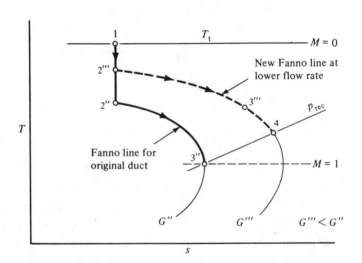

Figure 9.11 Addition of more duct when choked

In summary, when a <u>subsonic</u> Fanno flow has become "friction
choked" and more duct is added to the system, the flow rate must
decrease. Just how much it decreases and whether or not the exit
velocity remains sonic depends upon how much duct is added and
the receiver pressure imposed on the system.

Now suppose we are dealing with <u>supersonic</u> Fanno flow that
is "friction choked." In this case the addition of more duct
causes a normal shock to form inside the duct. The resulting
subsonic flow can accommodate the increased duct length at the

same flow rate. For example, Figure 9.12 shows a Mach 2.18 flow
which has an $fL_{max}/D = 0.356$. If a normal shock were to occur at
this point, the Mach number after the shock would be about 0.550
which corresponds to an $fL_{max}/D = 0.728$. Thus, in this case,
the appearance of the shock permits over twice the duct length to
the choke point. This difference becomes even greater as higher
Mach numbers are reached.

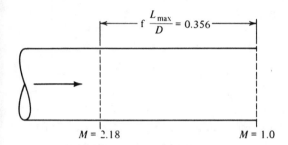

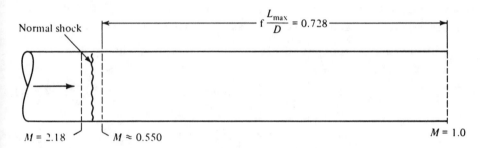

Figure 9.12 Influence of shock on maximum duct length

The shock location is determined by the amount of duct added.
As more duct is added the shock moves upstream and occurs at a
higher Mach number. Eventually, the shock will move into that
portion of the system which precedes the constant-area duct.
(Most likely a converging-diverging nozzle was used to produce
the supersonic flow.) If sufficient friction length is added the
entire system will become subsonic and then the flow rate will
decrease. Whether or not the exit velocity remains sonic will
again depend on the receiver pressure.

9.9 SUMMARY

We have analyzed flow in a constant-area duct with friction
but without heat transfer. The fluid properties change in a pre-

dictable manner dependent upon the flow regime as shown in the
following chart:

Property	Subsonic	Supersonic
Velocity	Inc	Dec
Mach Number	Inc	Dec
Enthalpy [†]	Dec	Inc
Stagnation Enthalpy [†]	Const	Const
Pressure	Dec	Inc
Density	Dec	Inc
Stagnation Pressure	Dec	Dec

[†] Also temperature if fluid is a perfect gas.

The property variations in subsonic Fanno flow follow an intuitive pattern but we note that the supersonic flow behavior is
completely different. The only common occurrence is the decrease
in stagnation pressure which is indicative of the loss.

Perhaps the most significant equations are those that apply
to all fluids.

$$\rho V = G = \text{const.} \qquad\qquad (9.2)$$

$$h_t = h + \frac{G^2}{\rho^2 2g_c} = \text{const.} \qquad\qquad (9.5)$$

Along with these equations you should keep in mind the appearance
of Fanno lines in the h-v and T-s diagrams (see Figures 9.1
and 9.2). Remember that each Fanno line represents points with
the same mass velocity (G) and stagnation enthalpy (h_t), and a
normal shock can connect two points on opposite branches of a
Fanno line which have the same value of $p + \rho V^2/g_c$. "Families"
of Fanno lines could represent:

  (a) different values of G for the same h_t
       (such as those in Figure 9.10) or

  (b) the same G for different values of h_t
       (see homework problem 10.17).

Detailed working equations were developed for perfect gases
and the introduction of a * reference point enabled the construction of Fanno tables which simplify problem solution. The

* condition for Fanno flow has no relation to the one previously
used in isentropic flow (except in general definition).

All Fanno flows proceed toward a limiting point of Mach one.
"Friction choking" of a flow passage is possible in Fanno flow
just as area choking occurs in varying-area isentropic flow. An
h-s (or T-s) diagram is of great help in the analysis of a
complicated flow system. Get into the habit of drawing these
diagrams.

9.10 PROBLEMS

In the problems that follow you may assume that all systems are
completely adiabatic. Also, all ducts are of constant area unless otherwise indicated. You may neglect friction in the
varying-area sections.

1. Conditions at the entrance to a duct are M_1 = 3.0 and
    p_1 = 8x10^4 N/m^2 . After a certain length the flow has
    reached M_2 = 1.5 . Determine p_2 and fΔx/D if γ = 1.4 .

2. A flow of nitrogen is discharged from a duct with M_2 = 0.85 ,
    T_2 = 500°R and p_2 = 28 psia . The temperature at the inlet
    is 560°R. Compute the pressure at the inlet and the mass ve-
    locity (G).

3. Air enters a circular duct with a Mach number of 3.0. The
    friction factor is 0.01.
    (a) How long a duct (measured in diameters) is required to
        reduce the Mach number to 2.0?
    (b) What is the percentage change in temperature, pressure,
        and density?
    (c) Determine the entropy increase of the air.
    (d) Assume the same length of duct as computed in (a) but the
        initial Mach number is 0.5. Compute the percentage
        change in temperature, pressure, density, and the entropy
        increase for this case. Compare the changes in the same
        length duct for subsonic and supersonic flow.

4. Oxygen enters a 6-inch-diameter duct with T_1 = 600°R ,
    p_1 = 50 psia and V_1 = 600 ft/sec . The friction factor is
    f = 0.02 .
    (a) What is the maximum length of duct permitted that will
        not change any of the conditions at the inlet?

(b) Determine T_2 , p_2 and V_2 for the maximum duct length found in (a).

5. Air flows in an 8-cm-diameter (I. D.) pipe that is 4 m long. The air enters with a Mach number of 0.45 and a temperature of $300^O K$.

   (a) What friction factor would cause sonic velocity at the exit?

   (b) If the pipe is made of cast iron estimate the inlet pressure.

6. At one section in a constant-area duct the stagnation pressure is 66.8 psia and the Mach number is 0.80. At another section the pressure is 60 psia and the temperature is $120^O F$.

   (a) Compute the temperature at the first section and the Mach number at the second section if the fluid is air.

   (b) Which way is the air flowing?

   (c) What is the "friction length" ($f\Delta x/D$) of the duct?

7. A 50 x 50 cm duct is 10 m in length. Nitrogen enters at $M_1 = 3.0$ and leaves at $M_2 = 1.7$, with $T_2 = 280^O K$ and $p_2 = 7x10^4 N/m^2$.

   (a) Find the static and stagnation conditions at the entrance.

   (b) What is the friction factor of the duct?

8. A duct of two-foot-by-one-foot cross-section is made of riveted steel and is 500 feet long. Air enters with a velocity of 174 ft/sec , $p_1 = 50$ psia and $T_1 = 100^O F$.

   (a) Determine the temperature, pressure and velocity at the exit.

   (b) Compute the pressure drop assuming the flow to be incompressible. Use the entering conditions and equation 3.29. Note that equation 3.64 can be easily integrated to evaluate

$$\int Tds_i = f \frac{\Delta x}{D_e} \frac{V^2}{2g_c}$$

   (c) How do the results of (a) and (b) compare? Did you expect this?

9. Air enters a duct with a mass flow rate of 35 lbm/sec at $T_1 = 520^O R$ and $p_1 = 20$ psia . The duct is square and has an area of 0.64 ft^2 . The outlet Mach number is unity.

   (a) Compute the temperature and pressure at the outlet.

   (b) Find the length of the duct if it is made of steel.

10. Consider the flow of a perfect gas along a Fanno line. Show that the pressure at the * reference state is given by the relation

$$p^* = \frac{\dot{m}}{A}\left[\frac{2RT_t}{\gamma g_c(\gamma+1)}\right]^{\frac{1}{2}}$$

11. A 10-ft duct 12 inches in diameter contains oxygen flowing at the rate of 80 lbm/sec. Measurements at the inlet give $p_1 = 30$ psia and $T_1 = 800^{\circ}R$. The pressure at the outlet is $p_2 = 23$ psia.
    (a) Calculate M_1 , M_2 , V_2 , T_{t2} and p_{t2} .
    (b) Determine the friction factor and estimate the absolute roughness of the duct material.

12. At the outlet of a 25-cm-diameter duct air is traveling at sonic velocity with a temperature of $16^{\circ}C$ and a pressure of 1 bar. The duct is very smooth and is 15 m long. There are two possible conditions that could exist at the entrance to the duct.
    (a) Find the static and stagnation temperature and pressure for each entrance condition.
    (b) Assuming the surrounding air to be at 1 bar pressure, how much horsepower is necessary to get ambient air into the duct for each case? (You may assume no losses in the work process.)

13. Ambient air at $60^{\circ}F$ and 14.7 psia accelerates isentropically into a 12-inch-diameter duct. After 100 ft the duct transitions into an 8x8 inch square section where the Mach number is 0.50. Neglect all frictional effects except in the constant-area duct where $f = 0.04$.
    (a) Determine the Mach number at the duct entrance.
    (b) What are the temperature and pressure in the square section?
    (c) How much 8x8 inch square duct could be added before the flow chokes? (Assume $f = 0.04$ in this duct also.)

14. Nitrogen with $p_t = 7\times10^5$ N/m^2 and $T_t = 340^{\circ}K$ enters a frictionless converging-diverging nozzle having an area ratio of 4.0 . The nozzle discharges supersonically into a constant-area duct which has a friction length $f\Delta x/D = 0.355$. Determine the temperature and pressure at the exit of the duct.

15. Conditions before a normal shock are M_1 = 2.5 , p_{t1} = 67
    psia and T_{t1} = 700°R. This is followed by a length of Fanno
    flow and a converging nozzle as shown below. The area change
    is such that the system is choked. It is also known that
    p_4 = p_{amb} = 14.7 psia.
    (a) Draw a T-s diagram for the system.
    (b) Find M_2 and M_3 .
    (c) What is $f\Delta x/D$ for the duct?

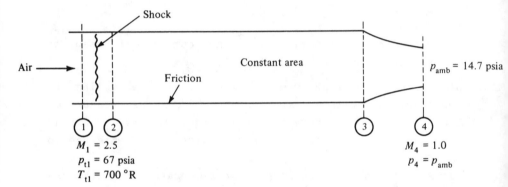

16. A converging-diverging nozzle has an area ratio of 3.0. The
    stagnation conditions of the inlet air are 150 psia and 550°R.
    A constant-area duct with a length of 12 diameters is attached
    to the nozzle outlet. The friction factor in the duct is
    0.025.
    (a) Compute the receiver pressure that would place a shock
        (i) in the nozzle throat.
        (ii) at the nozzle exit.
        (iii) at the duct exit.
    (b) What receiver pressure would cause supersonic flow
        throughout the duct with no shocks within the system (or
        after the duct exit)?
    (c) Make a sketch similar to Figure 6.3 showing the pressure
        distribution for the various operating points of (a) and
        (b).

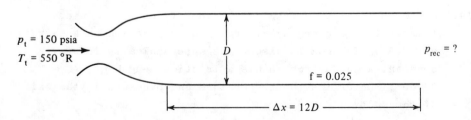

17. For a nozzle-duct system similar to that of problem 16, the nozzle is designed to produce a Mach number of 2.8 with $\gamma = 1.4$. The inlet conditions are p_{t1} = 10 bars and $T_{t1} = 370^{\circ}K$. The duct is 8 diameters in length but the duct friction factor is unknown. The receiver pressure is fixed at 3 bars and a normal shock has formed at the duct exit.

(a) Sketch a T-s diagram for the system.

(b) Determine the friction factor of the duct.

(c) What is the total change in entropy for the system?

18. A large chamber contains air at 65 bars pressure and $400^{\circ}K$. The air passes through a converging-only nozzle and then into a constant-area duct. The friction length of the duct is $f\Delta x/D$ = 1.067 and the Mach number at the duct exit is 0.96.

(a) Draw a T-s diagram for the system.

(b) Determine conditions at the duct entrance.

(c) What is the pressure in the receiver? (Hint: How is this related to the duct exit pressure?)

(d) If the length of the duct is doubled, and the chamber and receiver conditions remain unchanged, what are the new Mach numbers at the entrance and exit of the duct?

19. A constant-area duct is fed by a converging-only nozzle as shown below. The nozzle receives oxygen from a large chamber at p_1 = 100 psia and $T_1 = 1000^{\circ}R$. The duct has a friction length of 5.3 and it is choked at the exit. The receiver pressure is exactly the same as the pressure at the duct exit.

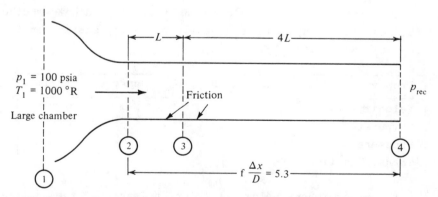

(a) What is the pressure at the end of the duct?

(b) Four-fifths of the duct is removed. (The end of the duct is now at 3.) The chamber pressure, receiver pressure

and friction factor remain unchanged. Now what is the
pressure at the exit of the duct?

(c) Sketch both of the above cases on the same T-s diagram.

20. (a) Plot a Fanno line to scale in the T-s plane for air
       entering a duct with a Mach number of 0.20, a static pres-
       sure of 100 psia and a static temperature of 540^OR. Indi-
       cate the Mach number at various points along the curve.

   (b) On the same diagram plot another Fanno line for a flow
       with the same total enthalpy, the same entering entropy,
       but for double the mass velocity.

21. Which, if any, of the ratios tabulated in the Fanno table
    $(T/T^*$, p/p^* , p_t/p_t^* , etc.) could also be listed in the
    isentropic table with the same numerical values?

22. (a) Introduce the * reference condition into equation 9.27
       and develop an expression for $(s^*-s)/R$.

   (b) Program the expression developed in part (a) and compute
       a table of $(s^*-s)/R$ versus Mach number. Check your val-
       ues with those listed in the Appendix.

9.11 CHECK TEST

You should be able to complete this test without reference to
material in the unit.

1. Sketch a Fanno line in the h-v plane. Include enough additional information as necessary to locate the sonic point and
   then identify the regions of subsonic and supersonic flow.

2. Fill in the blanks in the following table to indicate whether
   the quantities increase, decrease, or remain constant in the
   case of Fanno flow:

Property	Subsonic Regime	Supersonic Regime
Velocity Temperature Pressure Thrust Function		

3. In the system shown below the friction length of the duct is
   $f\Delta x/D = 12.40$ and the Mach number at the exit is 0.8.
   $A_3 = 1.5$ in^2 and $A_4 = 1.0$ in^2 . What is the air pressure
   in the tank if the receiver is at 15 psia?

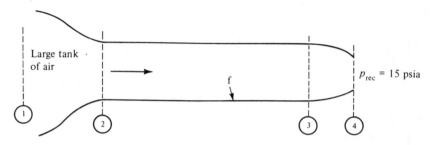

4. Over what range of receiver pressures will normal shocks occur someplace within the system shown below? The area ratio of the nozzle is $A_3/A_2 = 2.403$ and the duct $f\Delta x/D = 0.30$.

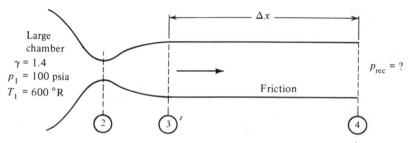

5. There is no friction in the system below except in the constant-area ducts from 3 to 4 and from 6 to 7 . Sketch the T-s diagram for the entire system.

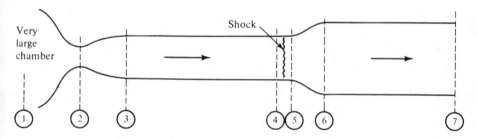

6. Starting with the basic principles of continuity, energy, etc., derive an expression for the property ratio p_2/p_1 in terms of Mach numbers and specific heat ratio for Fanno flow with a perfect gas.

7. Work problem #18 in Section 9.10.

UNIT 10

RAYLEIGH
FLOW

10.1 INTRODUCTION

In this unit we shall consider the consequences of heat
crossing the boundaries of a system. To isolate the effects of
heat transfer from the other major factors we will assume flow in
a constant-area duct without friction. At first this may seem to
be an unrealistic situation, but actually it is a good first approximation to many real problems, as most heat exchangers have
constant-area flow passages. It is also a simple and reasonably
equivalent process for a constant-area combustion chamber. Naturally, in these actual systems frictional effects are present
and what we really are saying is the following:

In systems where high rates of heat transfer occur the entropy
change caused by the heat transfer is much greater than that
caused by friction, or

$$ds_e \gg ds_i \qquad (10.1)$$

Thus,
$$ds \approx ds_e \qquad (10.2)$$

and the frictional effects may be neglected. There are obviously some flows for which this assumption is not reasonable and other methods must be used to obtain more accurate predictions for these systems.

We will first examine the general behavior of an arbitrary fluid and will again find that property variations follow different patterns in the subsonic and supersonic regimes. The flow of a perfect gas will be considered with the now familiar end result of constructing tables. This category of problem is called "Rayleigh flow."

10.2 OBJECTIVES

After successfully completing this unit you should be able to:

1. State the assumptions made in the analysis of Rayleigh flow.

2. Simplify the general equations of continuity, energy, and momentum to obtain basic relations valid for any fluid in Rayleigh flow.

3. Sketch a Rayleigh line in the p-v plane together with lines of constant entropy and constant temperature (for a typical gas). Indicate directions of increasing entropy and temperature.

4. Sketch a Rayleigh line in the h-s plane. Also sketch the corresponding stagnation curves. Identify the sonic point and regions of subsonic and supersonic flow.

5. Describe the variations in fluid properties which occur as flow progresses along a Rayleigh line for the case of heating and also for cooling. Do for both subsonic and supersonic flow.

6. Starting with basic principles of continuity, energy, and momentum, derive expressions for property ratios such as T_2/T_1 , p_2/p_1 , etc., in terms of Mach number (M) and specific heat ratio (γ) for Rayleigh flow with a perfect gas.

7. Describe (include T-s diagram) how Rayleigh tables are developed with the aid of a * reference location.

8. Compare similarities and differences between Rayleigh flow and normal shocks. Sketch an h-s diagram showing a typical Rayleigh line and a normal shock for the same mass velocity.

9. Explain what is meant by "thermal choking."

10. Describe some possible consequences of adding more heat in a choked Rayleigh flow situation (for both subsonic and supersonic flow).

11. Demonstrate the ability to solve typical Rayleigh flow problems by use of the appropriate tables and equations.

10.3 ANALYSIS FOR A GENERAL FLUID

We shall first consider the general behavior of an arbitrary fluid. In order to isolate the effects of heat transfer we make the following

Assumptions: Steady, one-dimensional flow
Negligible friction $(ds_i \approx 0)$
No shaft work $(\delta w_s = 0)$
Neglect potential $(dz = 0)$
Constant area $(dA = 0)$

We proceed by applying the basic concepts of continuity, energy, and momentum.

CONTINUITY $\dot{m} = \rho AV = const$ (2.30)

but since the flow area is constant, this reduces to

$$\rho V = const \qquad (10.3)$$

From our work in Unit 9 we know that this constant is G , the "mass velocity," and thus

$$\boxed{\rho V = G = const} \qquad (10.4)$$

ENERGY - We start with

$$h_{t1} + q = h_{t2} + w_s \qquad (3.19)$$

which for no shaft work becomes

$$\boxed{h_{t1} + q = h_{t2}} \qquad (10.5)$$

Warning! This is the first major flow category for which the total enthalpy has not been constant. By now you have accumulated a store of knowledge — all based on flows for which

h_t = constant. Carefully examine any information that you re-
trieve from your memory bank!

MOMENTUM - We now proceed to apply the momentum equation to the
control volume shown in Figure 10.1.

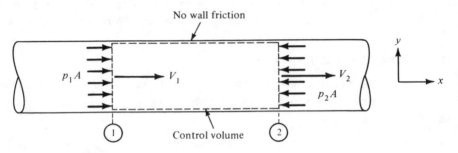

Figure 10.1 Momentum analysis for Rayleigh flow

The x-component of the momentum equation for steady, one-
dimensional flow is

$$\Sigma F_x = \frac{\dot{m}}{g_c}\left[V_{out_x} - V_{in_x}\right] \tag{3.46}$$

From Figure 10.1 we see that this becomes:

$$p_1 A - p_2 A = \frac{\rho A V}{g_c}(V_2 - V_1) \tag{10.6}$$

Canceling the area we have

$$p_1 - p_2 = \frac{\rho V}{g_c}(V_2 - V_1) = \frac{G}{g_c}(V_2 - V_1) \tag{10.7}$$

Show that this can be written as

$$p + \frac{GV}{g_c} = const \tag{10.8}$$

Alternate forms of equation 10.8 are

$$p + \frac{G^2}{g_c \rho} = const \tag{10.9a}$$

$$p + \frac{G^2}{g_c} v = const \tag{10.9b}$$

As an aside we might note that this is the same relation
that holds across a standing normal shock.

Recall
$$p + \rho\frac{V^2}{g_c} = \text{const} \qquad (6.9)$$

In both cases we are led to equivalent results since both analyses deal with constant area and assume negligible friction.

If we multiply equation 6.9 or 10.8 by the constant area we obtain

$$pA + \frac{(\rho AV)V}{g_c} = \text{const} \qquad (10.10)$$

or
$$\boxed{pA + \frac{\dot{m}V}{g_c} = \text{const}} \qquad (10.11)$$

The constant in equation 10.11 is called the "impulse function" or "thrust function" by various authors. We shall see a reason for these names when we study propulsion devices in the next unit. For now let us merely note that the thrust function remains constant for Rayleigh flow and across a normal shock.

We return now to equation 10.9b which will plot as a straight line in the p-v plane (see Figure 10.2). Such a line is called a Rayleigh line and represents flow at a particular mass velocity (G). If the fluid is known, one can also plot lines of constant temperature on the same diagram. Typical isothermals can be easily obtained by assuming the perfect gas equation of state. Some of these pv = const lines are also shown in Figure 10.2.

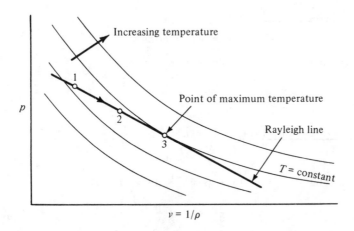

Figure 10.2 Rayleigh line in $p-v$ plane

Does the information depicted by this plot make sense? Normally we would expect the effects of simple heating to increase

the temperature and decrease the density. This appears to be in
agreement with a process from point 1 to point 2 as marked
in Figure 10.2. If we add more heat we move farther along the
Rayleigh line and the temperature increases more. Soon point 3
is reached where the temperature is a maximum. Is this a limiting point of some sort? Have we reached some kind of a "choked"
condition?

To answer these questions we must turn elsewhere. Recall
that the addition of heat causes the entropy of the fluid to increase since

$$ds_e = \delta q/T \tag{3.10}$$

From our basic assumption of negligible friction

$$ds \approx ds_e \tag{10.2}$$

Thus, it appears that the real limiting condition involves entropy (as usual). We can continue to add heat until the fluid
reaches a state of maximum entropy.

It might be that this point of maximum entropy is reached
before the point of maximum temperature, in which case we would
never be able to reach point 3 (of Figure 10.2). We must investigate the shape of constant entropy lines in the p-v diagram. This can easily be done for the case of a perfect gas
which will serve to illustrate the general trend.

For a T = constant line

$$pv = RT = const \tag{10.12}$$

Differentiating $$pdv + vdp = 0 \tag{10.13}$$

and $$\frac{dp}{dv} = -\frac{p}{v} \tag{10.14}$$

For an S = constant line

$$pv^\gamma = const \tag{10.15}$$

Differentiating $$v^\gamma dp + p\gamma v^{\gamma-1} dv = 0 \tag{10.16}$$

and $$\frac{dp}{dv} = -\gamma \frac{p}{v} \tag{10.17}$$

Comparing equations 10.14 and 10.17 and noting that γ is always
greater than 1.0, we see that the isentropic line has the greater
negative slope and thus these lines will plot as shown in Figure

10.3. (Actually, this should come as no great surprise since they were shown this way in Figure 1.2; but did you really believe it then?)

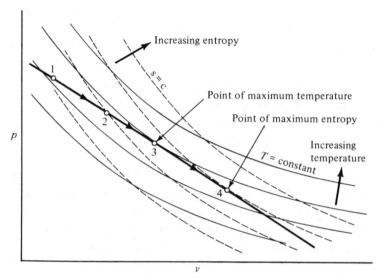

Figure 10.3 Rayleigh line in $p-v$ plane

We now see that not only can we reach the point of maximum temperature, but more heat can be added to take us beyond this point. If desired, we can move (by heating) all the way to the maximum entropy point. It may seem odd that in the region from point 3 to 4 we add heat to the system and its temperature decreases. Let us reflect further on the phenonema occurring.

In a previous discussion we noted that the effects of heat addition are normally thought of as causing the fluid density to decrease. This requires the velocity to increase since $\rho V =$ constant by continuity. This velocity increase automatically boosts the kinetic energy of the fluid by a certain amount. Thus, the chain of events caused by heat addition forces a definite increase in kinetic energy. Some of the heat which is added to the system is converted into this increase in kinetic energy of the fluid, with the heat energy in excess of this amount being available to increase the enthalpy of the fluid.

Noting that kinetic energy is proportional to the square of velocity we realize that as higher velocities are reached the addition of more heat is accompanied by much greater increases in kinetic energy. Eventually we reach a point where <u>all</u> of the

heat energy added is required for the kinetic energy increase.
At this point there is no heat energy left over and the system
is at a point of maximum enthalpy (maximum temperature for a perfect gas). Further addition of heat causes the kinetic energy
to increase by an amount greater than the heat energy being added.
Thus, from this point on, the enthalpy must decrease to provide
the proper energy balance.

Perhaps the foregoing discussion would be more clear if the
Rayleigh lines were plotted in the h-s plane. For any given
fluid this could easily be done and the typical result is shown
in Figure 10.4 along with lines of constant pressure. All points
on this Rayleigh line represent states with the same mass flow
rate per unit area (mass velocity) and the same impulse (or
thrust) function. For heat addition, the entropy must increase,
and the flow moves to the right. Thus, it appears that the Rayleigh line, like the Fanno line, is divided into two distinct
branches which are separated by a limiting point of maximum entropy.

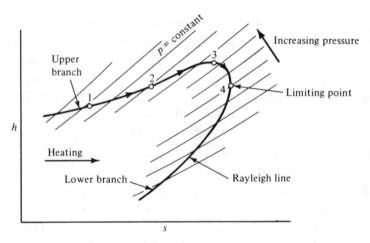

Figure 10.4 Rayleigh line in $h-s$ plane

We have been discussing a "familiar" heating process along
the upper branch. What about the lower branch? Mark two points
along the lower branch and draw an arrow to indicate the proper
movement for a heating process. What is happening to the enthalpy? The static pressure? The density? The velocity? The stagnation pressure? Use the information available in the figures
together with any equations that have been developed and fill in

the following table with "increases," "decreases," or "remains constant."

Property	FOR HEATING	
	Upper Branch	Lower Branch
Enthalpy		
Velocity		
Density		
Pressure (static)		
Pressure (stagnation)		

As was the case for Fanno flow, notice that flow along the lower branch of a Rayleigh line appears to be a regime with which we are not very familiar. The point of maximum entropy is some sort of a limiting point which separates these two flow regimes.

LIMITING POINT

Let's start with the equation of a Rayleigh line in the form

$$p + \frac{G^2}{g_c \rho} = \text{constant} \qquad (10.9a)$$

Differentiating,
$$dp + \frac{G^2}{g_c}\left(-\frac{d\rho}{\rho^2}\right) = 0 \qquad (10.18)$$

Upon introduction of equation 10.4 this becomes

$$\frac{dp}{d\rho} = \frac{G^2}{g_c \rho^2} = \frac{V^2}{g_c} \qquad (10.19)$$

Thus, we have for an <u>arbitrary</u> fluid that

$$V^2 = g_c \frac{dp}{d\rho} \qquad (10.20)$$

which is valid <u>anyplace</u> along the Rayleigh line. Now for a differential movement at the limit point of maximum entropy $ds = 0$, or $s = \text{const}$. Thus, at this point equation 10.20 becomes

$$V^2 = g_c\left(\frac{\partial p}{\partial \rho}\right)_{s=c} \text{(at limit point)} \qquad (10.21)$$

This is immediately recognized as sonic velocity. The upper branch of the Rayleigh line, where property variations appear reasonable, is seen to be a region of subsonic flow and the lower branch is for supersonic flow. Once again we notice that occurrences in supersonic flow are frequently contrary to our expectations.

Another interesting fact can be shown to be true at the limiting point. From equation 10.19 we have

$$dp = \frac{V^2}{g_c} \, d\rho \qquad\qquad (10.22)$$

Differentiating equation 10.4 we can show that

$$d\rho = -\rho\frac{dV}{V} \qquad\qquad (10.23)$$

Combining equations 10.22 and 10.23 we obtain

$$dp = -\rho\frac{V}{g_c} \, dV \qquad\qquad (10.24)$$

This can be introduced into the property relation

$$Tds = dh - \frac{dp}{\rho} \qquad\qquad (1.41)$$

to obtain $$Tds = dh + \frac{VdV}{g_c} \qquad\qquad (10.25)$$

At the limit point where $M = 1$, $ds = 0$, and (10.25) becomes

$$0 = dh + \frac{VdV}{g_c} \quad \text{(at limit point)} \quad (10.26)$$

If we neglect potentials, our definition of stagnation enthalpy is

$$h_t = h + \frac{V^2}{2g_c} \qquad\qquad (3.18)$$

which when differentiated becomes

$$dh_t = dh + \frac{VdV}{g_c} \qquad\qquad (10.27)$$

Therefore, equation 10.26 really tells us that

$$dh_t = 0 \quad \text{(at limit point)} \qquad\qquad (10.28)$$

and thus the limit point is seen to be a point of maximum stagnation enthalpy. This is easily confirmed by looking at equation 10.5. The stagnation enthalpy increases as long as heat can be added. At the point of maximum entropy no more heat can be added and thus h_t must be a maximum at this location.

We have not talked very much of stagnation enthalpy except to note that it is changing. Figure 10.5 shows the Rayleigh line (which represents the locus of static states) together with the corresponding stagnation reference lines. Remember that for a perfect gas this h-s diagram is equivalent to a T-s diagram.

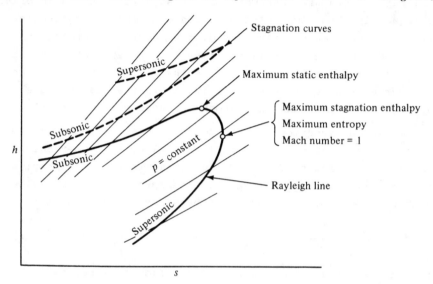

Figure 10.5 Rayleigh line in $h-s$ plane
(including stagnation curves)

Notice that there are two stagnation curves, one for subsonic flow and the other for supersonic flow. You might ask how we know that the supersonic stagnation curve is the top one. We can show this by starting with the differential form of the energy equation.

$$\delta q = \cancel{\delta w_s} + dh_t \qquad (3.20)$$

or

$$\delta q = dh_t \qquad (10.29)$$

Knowing

$$\delta q = Tds_e \qquad (3.10)$$

and

$$ds_e \approx ds \qquad (10.2)$$

thus, we have for Rayleigh flow that

$$dh_t = Tds_e = Tds \qquad (10.30)$$

or

$$\boxed{\frac{dh_t}{ds} = T} \qquad (10.31)$$

Note that equation 10.31 gives the slope of the stagnation curve in terms of the static temperature.

Now draw a constant entropy line on Figure 10.5. This line will cross the subsonic branch of the (static) Rayleigh line at a higher temperature than where it crosses the supersonic branch. Consequently, the slope of the subsonic stagnation reference curve will be greater than that of the supersonic stagnation curve. Since both stagnation curves must come together at the point of maximum entropy this means that the supersonic stagnation curve is a separate curve lying above the subsonic one. In Section 10.7 we shall see another reason why this must be so.

In which direction does a cooling process move along the subsonic branch of the Rayleigh line? Along the supersonic branch? From Figure 10.5 it would appear that the stagnation pressure will increase during a cooling process. This can be substantiated from the stagnation pressure-energy equation.

$$\frac{dp_t}{\rho_t} + ds_e(T_t - T) + T_t ds_i + \delta w_s = 0 \qquad (3.25)$$

With the assumptions made for Rayleigh flow this reduces to

$$\frac{dp_t}{\rho_t} + ds_e(T_t - T) = 0 \qquad (10.32)$$

Now $(T_t - T)$ is always positive. Thus, the sign of dp_t can be seen to depend only on ds_e.

For heating: ds_e + ; thus dp_t - , or p_t decreases.
For cooling: ds_e - ; thus dp_t + , or p_t increases.

In practice, this latter condition is difficult to achieve because the friction which is inevitably present introduces a greater drop in stagnation pressure than the rise created by the cooling process, unless the cooling is done by vaporization of an injected liquid. (See "The Aerothermopressor — A Device for Improving the Performance of a Gas Turbine Power Plant" by Shapiro et al., Transactions of the ASME, April 1956.)

10.4 WORKING EQUATIONS FOR A PERFECT GAS

By this time you should have a good idea of the property changes that are occurring in both subsonic and supersonic Rayleigh flow. Remember that we can progress along a Rayleigh line in either direction depending upon whether the heat is being add-

ed to or removed from the system. We now proceed to develop relations between properties at arbitrary sections. Recall that
we want these working equations to be expressed in terms of Mach
numbers and the specific heat ratio. In order to obtain explicit
relations we assume the fluid to be a perfect gas.

MOMENTUM – We start with the momentum equation developed in the
previous section since this will lead directly to a
pressure ratio.

$$p + \frac{GV}{g_c} = \text{constant} \tag{10.8}$$

or from (10.4) this can be written as

$$p + \frac{\rho V^2}{g_c} = \text{constant} \tag{10.33}$$

Substitute for density from the equation of state

$$\rho = p/RT \tag{10.34}$$

and for the velocity from equations 4.9 and 4.11

$$V^2 = M^2 a^2 = M^2 \gamma g_c RT \tag{10.35}$$

<u>Show</u> that equation 10.33 becomes

$$p(1 + \gamma M^2) = \text{const} \tag{10.36}$$

If we apply this between two arbitrary points we have

$$p_1(1 + \gamma M_1^2) = p_2(1 + \gamma M_2^2) \tag{10.37}$$

which can be solved for

$$\boxed{\frac{p_2}{p_1} = \frac{1 + \gamma M_1^2}{1 + \gamma M_2^2}} \tag{10.38}$$

CONTINUITY – From the previous section we have

$$\rho V = G = \text{const} \tag{10.4}$$

Again, if we introduce the perfect gas equation of state together
with the definition of Mach number and sonic velocity, equation
10.4 can be expressed as

$$\frac{pM}{\sqrt{T}} = \text{const} \tag{10.39}$$

Written between two points this is

$$\frac{p_1 M_1}{\sqrt{T_1}} = \frac{p_2 M_2}{\sqrt{T_2}} \tag{10.40}$$

which can be solved for the temperature ratio

$$\frac{T_2}{T_1} = \frac{p_2{}^2 M_2{}^2}{p_1{}^2 M_1{}^2} \tag{10.41}$$

The introduction of the pressure ratio from (10.38) results in the following working equation for static temperatures:

$$\frac{T_2}{T_1} = \left[\frac{1 + \gamma M_1{}^2}{1 + \gamma M_2{}^2}\right]^2 \frac{M_2{}^2}{M_1{}^2} \tag{10.42}$$

The density relation can easily be obtained from equations 10.38, 10.42, and the perfect gas equation of state.

$$\frac{\rho_2}{\rho_1} = \frac{M_1{}^2}{M_2{}^2}\left[\frac{1 + \gamma M_2{}^2}{1 + \gamma M_1{}^2}\right] \tag{10.43}$$

Does this also represent something else besides the density ratio? (See equation 10.4.)

STAGNATION CONDITIONS

This is the first flow we have examined in which the stagnation enthalpy does not remain constant. Thus, we must seek a stagnation temperature ratio for use with perfect gases. We know that

$$T_t = T\left[1 + \frac{(\gamma-1)}{2} M^2\right] \tag{4.18}$$

If we write this for each location and then divide one equation by the other we will have

$$\frac{T_{t2}}{T_{t1}} = \frac{T_2}{T_1}\left[\frac{1 + \frac{(\gamma-1)}{2} M_2{}^2}{1 + \frac{(\gamma-1)}{2} M_1{}^2}\right] \tag{10.44}$$

Since we already have solved for the static temperature ratio (10.42) this can immediately be written as

$$\frac{T_{t2}}{T_{t1}} = \left[\frac{1 + \gamma M_1^2}{1 + \gamma M_2^2}\right]^2 \frac{M_2^2}{M_1^2}\left[\frac{1 + \frac{(\gamma-1)}{2} M_2^2}{1 + \frac{(\gamma-1)}{2} M_1^2}\right] \qquad (10.45)$$

Similarly, we can obtain an expression for the stagnation pressure ratio since we know that

$$p_t = p\left[1 + \frac{(\gamma-1)}{2} M^2\right]^{\frac{\gamma}{\gamma-1}} \qquad (4.21)$$

which means
$$\frac{p_{t2}}{p_{t1}} = \frac{p_2}{p_1}\left[\frac{1 + \frac{(\gamma-1)}{2} M_2^2}{1 + \frac{(\gamma-1)}{2} M_1^2}\right]^{\frac{\gamma}{\gamma-1}} \qquad (10.46)$$

Substitution for the pressure ratio from equation 10.38 yields

$$\frac{p_{t2}}{p_{t1}} = \left[\frac{1 + \gamma M_1^2}{1 + \gamma M_2^2}\right]\left[\frac{1 + \frac{(\gamma-1)}{2} M_2^2}{1 + \frac{(\gamma-1)}{2} M_1^2}\right]^{\frac{\gamma}{\gamma-1}} \qquad (10.47)$$

Incidently, is this stagnation pressure ratio related to the entropy change in the usual manner?

$$\frac{p_{t2}}{p_{t1}} \overset{?}{=} e^{-\Delta s/R} \qquad (4.28)$$

What assumptions were used to develop equation 4.28? Are these the same assumptions that were made for Rayleigh flow? If not, how would you go about determining the entropy change between two points? Would the method used in Unit 9 for Fanno flow be applicable here? (See equations 9.25 to 9.27.)

In summary, we have developed the means to solve for all properties at one location (2) if we know all the properties at some other location (1) and the Mach number at point (2). Actually, any piece of information about point (2) would suffice. For example, we might be given the pressure at (2). The Mach number at (2) could then be found from equation 10.38 and the solution for the other properties could be carried out in the usual manner.

There are also some types of problems in which nothing is known at the downstream section and our job is to predict the final Mach number given the initial conditions and information on the heat transferred to or from the system. For this we turn to the fundamental relation that involves heat transfer.

ENERGY - From the previous section we have

$$h_{t1} + q = h_{t2} \qquad (10.5)$$

For perfect gases we express enthalpy as

$$h = c_p T \qquad (1.48)$$

which also can be applied to the stagnation conditions

$$h_t = c_p T_t \qquad (10.48)$$

Thus, the energy equation can be written as

$$c_p T_{t1} + q = c_p T_{t2} \qquad (10.49)$$

or

$$\boxed{q = c_p (T_{t2} - T_{t1})} \qquad (10.50)$$

Note carefully that $q = c_p \Delta T_t \neq c_p \Delta T$ $\qquad (10.51)$

In all of the above developments we have not only introduced the perfect gas equation of state but we have made the usual assumption of constant specific heats. In some cases where heat transfer rates are extremely high and large temperature changes result, c_p may vary enough to warrant using an average value of c_p. If, in addition, significant variations in γ occur, it will be necessary to return to the basic equations and derive new working relations by treating γ as a variable.

10.5 REFERENCE STATE AND RAYLEIGH TABLES

The equations developed in the previous section provide the means of predicting properties at one location if sufficient information is known concerning a Rayleigh flow system. Although the relations are straightforward their use is frequently cumbersome and thus we turn to techniques previously used which greatly simplify problem solution.

We introduce still another * reference state defined as
before in that the Mach number of unity must be reached by some
particular process. In this case we imagine that the Rayleigh
flow is continued (i.e., more heat is added) until the velocity
reaches sonic. Figure 10.6 shows a T-s diagram for subsonic
Rayleigh flow with heat addition. A sketch of the physical system is also shown. If we imagine that more heat is added, the
entropy continues to increase and we will eventually reach the
limiting point where sonic velocity exists. The dotted lines
show a hypothetical duct in which the additional heat transfer
takes place. At the end we reach the * reference point for
Rayleigh flow.

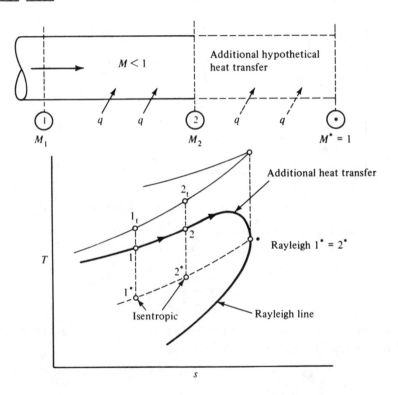

Figure 10.6 The * reference for Rayleigh flow

The isentropic * reference points have also been included
on the T-s diagram in order to emphasize the fact that the
Rayleigh * reference is a completely different thermodynamic
state. Also, we note that proceeding from either point 1 or
point 2 by Rayleigh flow will ultimately lead to the same state

when Mach one is reached. Thus, we do not have to write 1^* or
2^* but simply * in the case of Rayleigh flow. Recall this was
also true for Fanno flow. (You should also realize that the *
reference for Rayleigh flow has nothing to do with the * reference used in Fanno flow.) Notice in Figure 10.6 that the various
* locations are <u>not</u> on a horizontal line as they were for Fanno
flow (see Figure 9.5). Why is this so?

In Figure 10.6 an example of subsonic heating was given.
Consider a case of <u>cooling</u> in the <u>supersonic</u> regime. Figure 10.7
shows such a physical duct. Locate points 1 and 2 on the
accompanying T-s diagram. Also show the hypothetical duct and
the * reference point on the physical system.

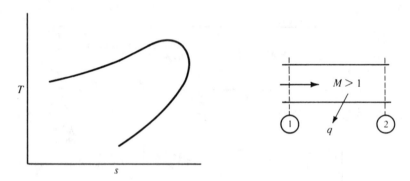

Figure 10.7 Supersonic cooling in Rayleigh flow

We now rewrite the working equations in terms of the Rayleigh
flow * reference condition. Consider first

$$\frac{p_2}{p_1} = \frac{1 + \gamma M_1^2}{1 + \gamma M_2^2} \tag{10.38}$$

Let point 2 be any arbitrary point in the flow system and let
its Rayleigh * condition be point 1 .

Then $p_2 \Rightarrow p$ $M_2 \Rightarrow M$ (any value)

              $p_1 \Rightarrow p^*$ $M_1 \Rightarrow 1$

and equation 10.38 becomes

$$\frac{p}{p^*} = \frac{1 + \gamma}{1 + \gamma M^2} = f(M, \gamma) \tag{10.52}$$

We see that $p/p^* = f(M, \gamma)$ and thus a table can be computed for
p/p^* versus M for a particular γ . By now this scheme is

quite familiar and you should have no difficulty in showing that

$$\frac{T}{T^*} = \frac{M^2(1+\gamma)^2}{(1+\gamma M^2)^2} = f(M,\gamma) \tag{10.53}$$

$$\frac{\rho}{\rho^*} = \frac{1+\gamma M^2}{(1+\gamma)M^2} = f(M,\gamma) \tag{10.54}$$

$$\frac{T_t}{T_t^*} = \frac{2(1+\gamma)M^2}{(1+\gamma M^2)^2}\left[1 + \frac{(\gamma-1)}{2}M^2\right] = f(M,\gamma) \tag{10.55}$$

$$\frac{p_t}{p_t^*} = \frac{1+\gamma}{1+\gamma M^2}\left[\frac{1 + \frac{(\gamma-1)}{2}M^2}{\frac{\gamma+1}{2}}\right]^{\frac{\gamma}{\gamma-1}} = f(M,\gamma) \tag{10.56}$$

Values for the functions represented in equations 10.52 through 10.56 are listed in the Rayleigh tables which are located in the Appendix. Examples of the use of these tables can be found in the next section.

10.6 APPLICATIONS

The procedure for solving Rayleigh flow problems is quite similar to the approach used for Fanno flow except that the tie between the two locations in Rayleigh flow is determined by heat transfer considerations rather than by duct friction. The recommended steps are, therefore, as follows:

1. Sketch the physical situation.
2. Label sections where conditions are known or desired, including the hypothetical * reference point.
3. List all given information with units.
4. Determine the unknown Mach number.
5. Calculate the additional properties desired.

Variations on the above procedure are frequently involved at step 4 depending on what information is known. For example, the amount of heat transferred may be given and a prediction of the downstream Mach number might be desired. On the other hand one of the downstream properties may be known and we could be asked to compute the heat transfer. In flow systems which involve a combination of Rayleigh flow and other phenomena (such as shocks, nozzles, etc.) a T-s diagram is sometimes a great aid to problem solution.

For the following examples we are dealing with the steady, one-dimensional flow of air ($\gamma = 1.4$) which can be treated as a perfect gas. Assume $w_s = 0$, negligible friction, and negligible potential changes. The physical diagram shown is common to the first two examples.

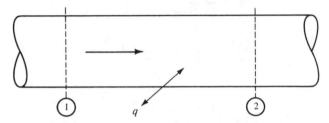

Example 10.1 Given: $M_1 = 1.5$, $p_1 = 10$ psia and $M_2 = 3.0$
            Find: p_2 and the direction of heat transfer.

Since both Mach numbers are known we can immediately solve for

$$p_2 = \frac{p_2}{p^*} \frac{p^*}{p_1} \, p_1 = (0.1765) \, \frac{1}{0.5783} \, (10) = \underline{3.05} \text{ psia}$$

The flow is getting more supersonic, or moving away from the * reference point. A look at Figure 10.5 should confirm that the entropy is decreasing and thus heat is being removed from the system. Alternately, we could compute the ratio T_{t2}/T_{t1} .

$$\frac{T_{t2}}{T_{t1}} = \frac{T_{t2}}{T_t^*} \frac{T_t^*}{T_{t1}} = (0.6540) \, \frac{1}{0.9093} = \underline{0.719}$$

Since this ratio is less than one it indicates a cooling process.

Example 10.2 Given: $M_2 = 0.93$, $T_{t2} = 300^{\circ}C$ and $T_{t1} = 100^{\circ}C$
            Find: M_1 and p_2/p_1

To determine conditions at section 1 we must establish the ratio

$$\frac{T_{t1}}{T_t^*} = \frac{T_{t1}}{T_{t2}} \frac{T_{t2}}{T_t^*} = \frac{(273+100)}{(273+300)} \, (0.9963) = 0.6486$$

Look up T_t/T_t^* = 0.6486 in the Rayleigh table and determine that $M_1 = 0.472$

Thus,
$$\frac{p_2}{p_1} = \frac{p_2}{p^*} \frac{p^*}{p_1} = (1.0856) \, \frac{1}{1.8294} = \underline{0.593}$$

Example 10.3 A constant-area combustion chamber is supplied air at $400°R$ and 10.0 psia. The air stream has a velocity of 402 ft/sec. Determine the exit conditions if 50 Btu/lbm are added in the combustion process and the chamber handles the maximum amount of air possible.

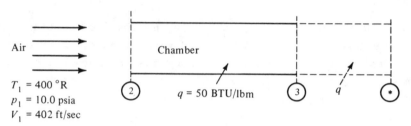

For the chamber to handle the maximum amount of air there will be no "spillover" at the entrance and conditions at 2 will be the same as those of the free stream.

$$T_2 = T_1 = 400°R \; , \; p_2 = p_1 = 10.0 \text{ psia} \; , \; V_2 = V_1 = 402 \text{ ft/sec}$$

$$a_2 = \sqrt{\gamma g_c R T_2} = \left[(1.4)(32.2)(53.3)(400)\right]^{\frac{1}{2}} = 980 \text{ ft/sec}$$

$$M_2 = V_2/a_2 = 402/980 = 0.410$$

$$T_{t2} = \frac{T_{t2}}{T_2} T_2 = \frac{1}{0.9675} (400) = 413°R$$

From the Rayleigh tables at $M_2 = 0.41$ we find

$$T_{t2}/T_t^* = 0.5465 \; , \; T_2/T^* = 0.6345 \; , \; p_2/p^* = 1.9428$$

To determine conditions at the end of the chamber we must work through the heat transfer which fixes the outlet stagnation temperature.

$$\Delta T_t = \frac{q}{c_p} = \frac{50}{0.24} = 208°R$$

Thus, $$T_{t3} = T_{t2} + \Delta T_t = 413 + 208 = 621°R$$

and $$\frac{T_{t3}}{T_t^*} = \frac{T_{t3}}{T_{t2}} \frac{T_{t2}}{T_t^*} = \frac{621}{413} (0.5465) = 0.8217$$

We enter the Rayleigh tables with this T_t/T_t^* and find that

$$M_3 = 0.603 \; , \; T_3/T^* = 0.9196 \; , \; p_3/p^* = 1.5904$$

Thus, $p_3 = \dfrac{p_3}{p^*} \dfrac{p^*}{p_2} p_2 = (1.5904) \dfrac{1}{1.9428} (10.0) = \underline{8.19}$ psia

and $T_3 = \dfrac{T_3}{T^*} \dfrac{T^*}{T_2} T_2 = (0.9196) \dfrac{1}{0.6345} (400) = \underline{580}{}^{\circ}R$

Example 10.4 In the above problem let us ask the question, "How much more heat (fuel) could be added without changing conditions at the entrance to the duct?" We know that as more heat is added we move along the Rayleigh line until the point of maximum entropy is reached. Thus, M_3 will now be 1.0.

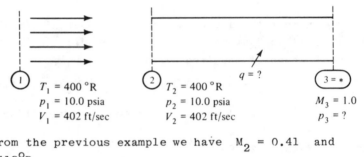

$T_1 = 400\ {}^{\circ}R$ $T_2 = 400\ {}^{\circ}R$ $q = ?$
$p_1 = 10.0$ psia $p_2 = 10.0$ psia $M_3 = 1.0$
$V_1 = 402$ ft/sec $V_2 = 402$ ft/sec $p_3 = ?$

From the previous example we have $M_2 = 0.41$ and $T_{t2} = 413{}^{\circ}R$.

Then $T_{t3} = T_t{}^* = \dfrac{T_t{}^*}{T_{t2}} T_{t2} = \dfrac{1}{0.5465} (413) = \underline{756}{}^{\circ}R$

$p_3 = p^* = \dfrac{p^*}{p_2} p_2 = \dfrac{1}{1.9428} (10.0) = \underline{5.15}$ psia

and $q = c_p \Delta T_t = 0.24(756-413) = \underline{82.3}$ Btu/lbm

or 32.3 Btu/lbm more than the original 50 Btu/lbm.

In these last two examples it has been assumed that the outlet pressure is maintained at the values calculated. Actually, in example 10.4 the receiver pressure could be anyplace below 5.15 psia since sonic velocity exists at the exit.

10.7 CORRELATION WITH SHOCKS

At various places in this unit we have pointed out some similarities between Rayleigh flow and normal shocks. Let us review these points carefully.

The end points before and after a normal shock represent states with the same mass flow per unit area, the same impulse function, and the same stagnation enthalpy.

A Rayleigh line represents states with the same mass flow per unit area and the same impulse function. All points on a Rayleigh line do not have the same stagnation enthalpy because of the heat transfer involved. To move <u>along</u> a Rayleigh line re-quires this heat transfer.

For confirmation of the above compare equations 6.2, 6.3 and 6.9 for the normal shock with equations 10.4, 10.5 and 10.9 for Rayleigh flow. Now check Figure 10.8 and you will notice that for every point on the supersonic branch of the Rayleigh line there is a corresponding point on the subsonic branch with the same stagnation enthalpy. Thus, these two points satisfy all three conditions for the end points of a normal shock and could be connected by such a shock.

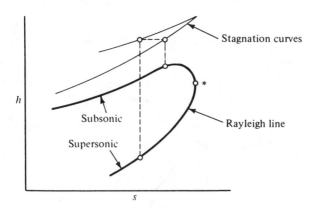

Figure 10.8

We can now picture a supersonic Rayleigh flow followed by a normal shock, with additional heat transfer taking place subson-ically. Such a situation is shown in Figure 10.9. Note that the shock merely jumps the flow from the supersonic branch to the subsonic branch of the <u>same</u> Rayleigh line. This also brings to light another reason why the supersonic stagnation curve must lie above the subsonic stagnation curve. If this were not so a shock would exhibit a decrease in entropy which is not correct.

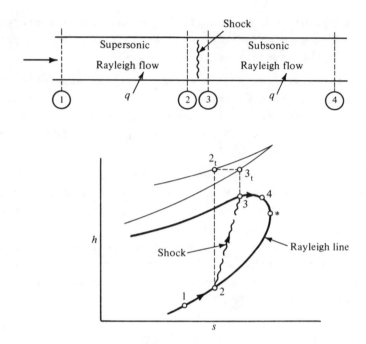

Figure 10.9 Combination of Rayleigh flow and normal shock

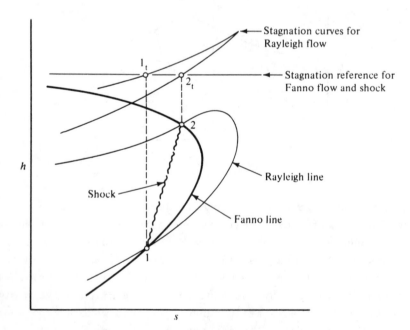

Figure 10.10 Correlation of Fanno flow, Rayleigh flow and normal shock for the same mass velocity

If you recall the information from Section 9.7 which dealt with the correlation of Fanno flow and shocks, it should now be apparent that the end points of a normal shock can represent the intersection of a Fanno line and a Rayleigh line as shown in Figure 10.10. Remember, these Fanno and Rayleigh lines are for the same mass velocity (mass flow per unit area).

Example 10.5 Air enters a constant-area duct with a Mach number of 1.6, a temperature of $200°K$, and a pressure of 0.56 bars. After some heat transfer a normal shock occurs, whereupon the area is reduced as shown. At the exit the Mach number is found to be 1.0 and the pressure is 1.20 bars. Compute the amount and direction of heat transfer.

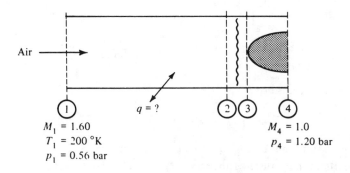

$M_1 = 1.60$
$T_1 = 200°K$
$p_1 = 0.56$ bar

$M_4 = 1.0$
$p_4 = 1.20$ bar

$q = ?$

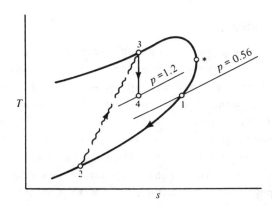

It is not known whether a heating or cooling process is involved. We construct the T-s diagram under the assumption that cooling takes place and will find out if this is correct. The flow from 3 to 4 is isentropic; thus

$$p_{t3} = p_{t4} = \frac{p_{t4}}{p_4} p_4 = \frac{1}{0.5283} (1.20) = 2.2714 \text{ bar}$$

Note that point 3 is on the same Rayleigh line as point 1 and
this permits us to compute M_2 through the use of the Rayleigh
tables. This approach might not have occurred to us had we not
drawn the T-s diagram.

$$\frac{p_{t3}}{p_t^*} = \frac{p_{t3}}{p_1}\frac{p_1}{p_{t1}}\frac{p_{t1}}{p_t^*} = \frac{(2.2714)}{(0.56)}(0.2353)(1.1756) = 1.1220$$

From the Rayleigh tables we find $M_3 = 0.481$
and from the shock tables $M_2 = 2.906$

Now we can compute the stagnation temperatures.

$$T_{t1} = \frac{T_{t1}}{T_1}T_1 = \frac{1}{0.6614}(200) = 302^{\circ}K$$

$$T_{t2} = \frac{T_{t2}}{T_t^*}\frac{T_t^*}{T_{t1}}T_{t1} = (0.6629)\frac{1}{0.8842}(302) = 226^{\circ}K$$

and the heat transfer:

$$q = c_p(T_{t2}-T_{t1}) = 1000(226-302) = \underline{-7.6\times10^4}\ Joules/kg$$

The minus sign indicates a cooling process which is consistent
with the Mach number's increase from 1.60 to 2.906.

10.8 THERMAL CHOKING

We have previously discussed "area choking" (in Section 5.7)
and "friction choking" (in Section 9.8). In Fanno flow, recall
that once sufficient duct was added — or the receiver pressure
was lowered far enough — we reached a Mach number of unity at the
end of the duct. Further reduction of the receiver pressure could
not affect conditions in the flow system. The addition of any
more duct caused the flow to move along a new Fanno line at a
reduced flow rate. You might wish to review Figure 9.11 which
shows this physical situation along with the corresponding T-s
diagram.

Subsonic Rayleigh flow is quite similar. Figure 10.11 shows
a given duct fed by a large tank and converging nozzle. Once
sufficient heat has been added we reach Mach 1 at the end of the
duct. The T-s diagram for this is shown as path 1-2-3. This
is called "thermal choking." It is assumed that the receiver

pressure is at p_3 or below. Reduction of the receiver pressure below p_3 would not affect the flow conditions inside the system. However, the addition of more heat will change these conditions.

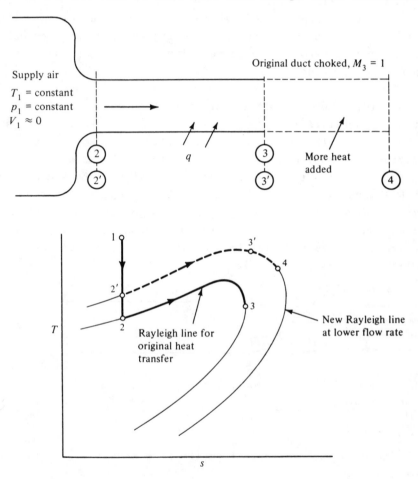

Figure 10.11 Addition of more heat when choked

Now suppose we add more heat to the system. Most likely this would be done by increasing the heat transfer rate through the walls of the original duct. However, it is more convenient to symbolically indicate the additional heat transfer in an extra piece of duct as shown in Figure 10.11. The only way the system can reflect the required additional entropy change is to move to a new Rayleigh line at a <u>decreased</u> flow rate. This is shown as path 1-2'-3'-4 on the T-s diagram. Whether or not the exit

velocity remains sonic depends upon how much extra heat is added
and the receiver pressure imposed on the system.

As a specific example of choked flow we return to the combustion chamber of example 10.4 which had the maximum amount of heat
addition possible, assuming that the free-stream air flow entered
the chamber with no change in velocity. We now consider what
happens as more fuel (heat) is added.

Example 10.6 Let us add sufficient fuel to raise the outlet stagnation temperature to $3000°R$. Assume that the receiver pressure
is very low so that sonic velocity still exists at the exit. The
additional entropy generated by the extra fuel can only be accom-
modated by moving to a new Rayleigh line at a decreased flow rate
which lowers the inlet Mach number. If the chamber is fed by the
same air stream then some "spillage" must occur at the entrance.
This produces a region of external diffusion, as shown, which is
isentropic. We would like to know the Mach number at the inlet
and the pressure at the exit.

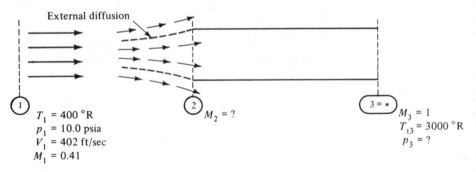

External diffusion

① $T_1 = 400 °R$
   $p_1 = 10.0$ psia
   $V_1 = 402$ ft/sec
   $M_1 = 0.41$

② $M_2 = ?$

③ = * $M_3 = 1$
   $T_{t3} = 3000 °R$
   $p_3 = ?$

Since it is isentropic from the free stream to the inlet, we
know that

$$T_{t2} = T_{t1} = 413°R$$

And since $M_3 = 1$, we know that $T_{t3} = T_t^*$.
Thus, we can determine conditions at 2 by computing

$$\frac{T_{t2}}{T_t^*} = \frac{T_{t2}}{T_{t3}} \frac{T_{t3}}{T_t^*} = \frac{413}{3000} (1) = 0.1377$$

And from the Rayleigh tables $M_2 = 0.176$, $p_2/p^* = 2.3002$.
To find the pressure at the outlet we need to use both the isentropic tables and the Rayleigh tables.

First $\quad p_2 = \dfrac{p_2}{p_{t2}} \dfrac{p_{t2}}{p_{t1}} \dfrac{p_{t1}}{p_1} p_1 = 0.9786 \ (1) \ \dfrac{1}{0.8907} \ (10.0) = \underline{10.99} \ \text{psia}$

Then $\quad p_3 = \dfrac{p_3}{p^*} \dfrac{p^*}{p_2} p_2 = (1) \ \dfrac{1}{2.3002} \ (10.99) = \underline{4.78} \ \text{psia}$

Notice that in order to maintain sonic velocity at the chamber exit the pressure in the receiver must be reduced to at least 4.78 psia.

Suppose that in the previous problem we were unable to lower the receiver pressure to 4.78 psia. Assume that as fuel was added to raise the stagnation temperature to $3000°R$ the pressure in the receiver was maintained at its previous value of 5.15 psia. This would lower the flow rate even further as we move to another Rayleigh line with a lower mass velocity and this time the exit velocity would not be quite sonic. Although both M_2 and M_3 are unknown, two pieces of information are given at the exit. Two simultaneous equations could be written but it is easier to use tables and a trial and error solution.

The important thing to remember is that once a subsonic flow is "thermally choked" the addition of more heat causes the flow rate to decrease. Just how much it decreases and whether or not the exit remains sonic depends on the pressure that exists after the exit.

The parallel between choked Rayleigh and Fanno flow does not quite extend into the supersonic regime. Recall that for choked Fanno flow the addition of more duct introduced a shock in the duct which permitted considerably more friction length to the sonic point. (See Figure 9.12.)

Figure 10.12 shows a Mach 3.53 flow which has a $T_t/T_t^* = 0.6139$. For a given total temperature at this section, the value of T_t/T_t^* is a direct indication of the amount of heat that can be added to the choke point. If a normal shock were to occur at this point, the Mach number after the shock would be 0.450 which also has a $T_t/T_t^* = 0.6139$. Thus, the heat added after the shock is exactly the same as it would be without the shock.

The above situation is not surprising since heat transfer is a function of stagnation temperature and this does not change across a shock. To do any good the shock must occur at some loca-

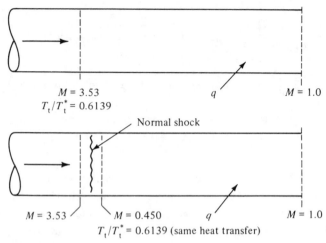

$M = 3.53$
$T_t / T_t^* = 0.6139$
q
$M = 1.0$

Normal shock

$M = 3.53$ $M = 0.450$ q $M = 1.0$
$T_t / T_t^* = 0.6139$ (same heat transfer)

Figure 10.12 Influence of shock on maximum heat transfer

tion <u>preceding</u> the Rayleigh flow. Perhaps this would be in a
converging-diverging nozzle which produces the supersonic flow.
Or if this were a situation similar to example 10.4 (only supersonic) a detached shock would occur in the free stream ahead of
the duct. In either case the resulting subsonic flow could accommodate additional heat transfer.

10.9 SUMMARY

We have analyzed steady, one-dimensional flow in a constant-
area duct with heat transfer but with negligible friction. Fluid
properties can vary in a number of ways depending upon whether
the flow is subsonic or supersonic plus consideration of the direction of heat transfer. However, these variations are easily
predicted and are summarized in the following chart:

Property	Heating		Cooling	
	$M < 1$	$M > 1$	$M < 1$	$M > 1$
Velocity	Inc	Dec	Dec	Inc
Mach Number	Inc	Dec	Dec	Inc
Enthalpy [†]	Inc/Dec	Inc	Inc/Dec	Dec
Stagnation Enthalpy [†]	Inc	Inc	Dec	Dec
Pressure	Dec	Inc	Inc	Dec
Density	Dec	Inc	Inc	Dec
Stagnation Pressure	Dec	Dec	Inc	Inc
Entropy	Inc	Inc	Dec	Dec

[†] Also temperature if fluid is a perfect gas.

As we might expect, the property variations that occur in subsonic Rayleigh flow follow an intuitive pattern but we find that the behavior of a supersonic system is quite different. Notice that even in the absence of friction, heating causes the stagnation pressure to drop. On the other hand, a cooling process predicts an increase in p_t. This is difficult to achieve in practice (except by latent cooling) due to frictional effects that are inevitably present.

Perhaps the most significant equations in this unit are the general ones:

$$\rho V = G \tag{10.4}$$

$$h_{t1} + q = h_{t2} \tag{10.5}$$

$$p + \frac{GV}{g_c} = \text{const} \tag{10.8}$$

An alternate way of expressing the latter equation is to say that the "impulse function" remains constant

$$pA + \frac{\dot{m}V}{g_c} = \text{const} \tag{10.11}$$

Along with these equations you should keep in mind the appearance of Rayleigh lines in the p-v and h-s diagrams (see Figures 10.2 and 10.4) as well as the stagnation reference curves (see Figure 10.5). Remember that each Rayleigh line represents points with the same mass velocity and impulse function, and a normal shock can connect two points on opposite branches of a Rayleigh line which have the same stagnation enthalpy.

Working equations for perfect gases were developed and then simplified with the introduction of a * reference point. This permitted the production of tables which help immeasurably in problem solution. Do not forget that the * condition for Rayleigh flow is not the same as those used for either isentropic or Fanno flow. "Thermal choking" occurs in heat addition problems and the reaction of a choked system to the addition of more heat is quite similar to the way a choked Fanno system reacts to the addition of more duct. Remember, drawing a good T-s diagram helps clarify your thinking on any given problem.

10.10 PROBLEMS

In the problems that follow, you may assume that all ducts are of constant area unless specifically indicated otherwise. In these

constant-area ducts you may neglect friction when heat transfer
is involved and you may neglect heat transfer when friction is
indicated. You may neglect both heat transfer and friction in
sections of varying area.

1. Air enters a constant-area duct with $M_1 = 2.95$ and
    $T_1 = 500°R$. Heat transfer decreases the outlet Mach number
    to $M_2 = 1.60$.
    (a) Compute the exit static and stagnation temperatures.
    (b) Find the amount and direction of heat transfer.

2. At the beginning of a duct the nitrogen pressure is 1.5 bars,
    the stagnation temperature is $280°K$ and the Mach number is
    0.80. After some heat transfer the static pressure is 2.5
    bars. Determine the direction and amount of heat transfer.

3. Air flows at the rate of 39.0 lbm/sec with a Mach number of
    0.30, a pressure of 50 psia, and a temperature of $650°R$. The
    duct has a 0.5 ft^2 cross-sectional area. Find the final
    Mach number, the stagnation temperature ratio T_{t2}/T_{t1} , and
    the density ratio ρ_2/ρ_1 , if heat is added at the rate of
    290 Btu/lbm of air.

4. In a flow of air $p_1 = 1.35 \times 10^5$ N/m^2 , $T_1 = 500°K$ and
    $V_1 = 540$ m/s . Heat transfer occurs in a constant-area duct
    until the ratio $T_{t2}/T_{t1} = 0.639$.
    (a) Compute the final conditions M_2 , p_2 and T_2 .
    (b) What is the entropy change for the air?

5. At some point in a flow system of oxygen $M_1 = 3.0$, $T_{t1} =$
    $800°R$ and $p_1 = 35$ psia. At a section farther along in the
    duct the Mach number has been reduced to $M_2 = 1.5$ by heat
    transfer.
    (a) Find the static and stagnation temperatures and pressures
        at the downstream section.
    (b) Determine the direction and amount of heat transfer that
        took place between these two sections.

6. Show that for a constant-area, frictionless, steady, one-
    dimensional flow of a perfect gas, the maximum amount of heat
    that can be added to the system is given by the expression:

$$\frac{q_{max}}{c_p T_1} = \frac{(M_1^2 - 1)^2}{2M_1^2 (\gamma + 1)}$$

7. Starting with equation 10.53, show that the maximum (static) temperature in Rayleigh flow occurs when the Mach number is $\sqrt{1/\gamma}$.

8. Air enters a 15-cm-diameter duct with a velocity of 120 m/s. The pressure is 1 atmosphere and the temperature is 100°C.
    (a) How much heat must be added to the flow to create the maximum (static) temperature?
    (b) Determine the final temperature and pressure for the conditions of (a).

9. The 12-inch-diameter duct shown below has a friction factor of 0.02 and no heat transfer from section 1 to 2. There is negligible friction from 2 to 3. Sufficient heat is added in the latter portion to just choke the flow at the exit. The fluid is nitrogen.
    (a) Draw a T-s diagram for the system showing the complete Fanno and Rayleigh lines involved.
    (b) Determine the Mach number and stagnation conditions at section 2.
    (c) Determine the static and stagnation conditions at section 3.
    (d) How much heat was added to the flow?

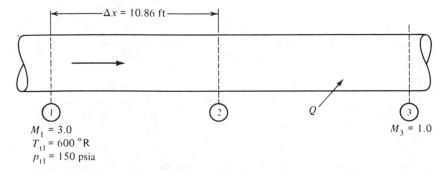

$M_1 = 3.0$
$T_{t1} = 600\,^\circ$R
$p_{t1} = 150$ psia

$M_3 = 1.0$

10. Conditions just prior to a standing normal shock in air are $M_1 = 3.53$ with a temperature of 650°R and a pressure of 12 psia.
    (a) Compute the conditions that exist just after the shock.
    (b) Show that these two points lie on the same Fanno line.
    (c) Show that these two points lie on the same Rayleigh line.

11. Air enters a duct with a Mach number of 2.0, and the temperature and pressure are 170°K and 0.7 bars, respectively. Heat

transfer takes place while the flow proceeds down the duct.
A converging section (A_2/A_3 = 1.45) is attached to the outlet
as shown below and the exit Mach number is 1.0. Assume that
the inlet conditions and the exit Mach number remain fixed.
Find the amount and direction of heat transfer:

(a) If there are no shocks in the system.

(b) If there is a normal shock someplace in the duct.

(c) For part (b) does it make any difference where the shock
     occurs?

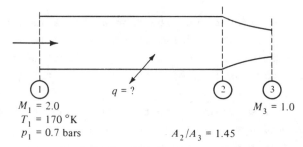

M_1 = 2.0
T_1 = 170 °K
p_1 = 0.7 bars A_2/A_3 = 1.45

M_3 = 1.0

12. In the system shown below, friction exists only from 2   to
     3 and from 5 to 6 . Heat is removed between 7 and 8 .
     The Mach number at section 9 is unity. Draw the T-s dia-
     gram for the system, showing both the static and stagnation
     curves. Are points 4 and 9 on the same horizontal level?

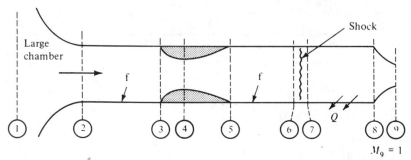

M_9 = 1

13. Oxygen is stored in a large tank where the pressure is 40
     psia and the temperature is 500°R. A DeLaval nozzle with an
     area ratio of 3.5 is attached to the tank and discharges into
     a constant-area duct where heat is transferred. The pressure
     at the duct exit is equal to 15 psia. Determine the amount
     and direction of heat transfer if a normal shock stands where
     the nozzle is attached to the duct.

14. Air enters a converging-diverging nozzle with stagnation
     conditions of 35×10^5 N/m^2 and 450°K . The area ratio of
     the nozzle is 4.0. After passing through the nozzle the flow

enters a duct where heat is added. At the end of the duct
there is a normal shock, after which the static temperature
is found to be $560^\circ K$.

(a) Draw a T-s diagram for the system.

(b) Find the Mach number after the shock.

(c) Determine the amount of heat added in the duct.

15. A converging-only nozzle feeds a constant-area duct in a system similar to that shown in Figure 10.11. Conditions in the
nitrogen supply chamber are $p_1 = 100$ psia and $T_1 = 600^\circ R$.
Sufficient heat is added to choke the flow ($M_3 = 1.0$) and the
Mach number at the duct entrance is $M_2 = 0.50$. The pressure at the exit is equal to that of the receiver.

(a) Compute the receiver pressure.

(b) How much heat is transferred?

(c) Assume that the receiver pressure remains fixed at the
value calculated in part (a) as more heat is added in the
duct. The flow rate must decrease and the flow moves to
a new Rayleigh line as indicated in Figure 10.11. Is the
Mach number at the exit still unity or is it less than
one? (Hint: Assume any lower Mach number at section 2.
From this you can compute a new p^* which should help
answer the question. You can then compute the heat transferred and show this to be greater than the initial value.
A T-s diagram might also help you.)

16. Draw the stagnation curves for both Rayleigh lines that are
shown in Figure 10.11.

17. Recall the expression $p_t A^* = \text{const}$ (see equation 5.35).

(a) Mark the following equations as true or false for the
system shown below.

$$\underline{\hspace{3cm}} \quad p_{t1}A_1^* = p_{t3}A_3^*$$

$$\underline{\hspace{3cm}} \quad p_{t3}A_3^* = p_{t5}A_5^*$$

(b) Draw a T-s diagram for the system below. Include both
static and stagnation curves. Are the flows from 1 to
2 and from 4 to 5 on the same Fanno line?

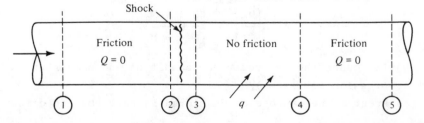

18. Points 1 and 2 represent flows on the same Rayleigh line
    (same mass flow rate, same area, same impulse function) and
    are located such that $s_1 = s_2$ as shown. Now imagine that
    we take the fluid under conditions at 1 and isentropically
    expand to 3 . Further, lets imagine that the fluid at 2
    undergoes an isentropic compression to 4 .
    (a) If 3 and 4 are coincident state points (same T & s)
        prove that A_3 is greater than, equal to, or less than
        A_4 .
    (b) Now suppose that points 3 and 4 are not necessarily
        coincident, but it is known that the Mach number is unity
        at each point. (i.e., $3 \equiv 1_s^*$ and $4 \equiv 2_s^*$.)
        (i) Is V_3 equal to, greater than, or less than V_4 ?
        (ii) Is A_3 equal to, greater than, or less than A_4 ?

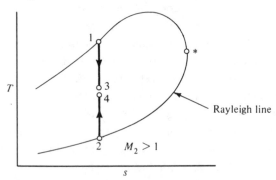

19. (a) Plot a Rayleigh line to scale in the T-s plane for air
        entering a duct with a Mach number of 0.25, a static pres-
        sure of 100 psia and a static temperature of $400^{O}R$. Indi-
        cate the Mach number at various points along the curve.
    (b) Add the stagnation curve to the T-s diagram.

20. (a) By the method of approach used in Section 9.4 (see equa-
        tions 9.25-9.27) show that the entropy change between two
        points in Rayleigh flow can be represented by the follow-
        ing expression if the fluid is a perfect gas:

$$\frac{s_2 - s_1}{R} = \ln\left[\frac{M_2}{M_1}\right]^{\frac{2\gamma}{\gamma-1}} \left[\frac{1 + \gamma\, M_1^{2}}{1 + \gamma\, M_2^{2}}\right]^{\frac{\gamma+1}{\gamma-1}}$$

    (b) Introduce the * reference condition and obtain an ex-
        pression for $(s^*-s)/R$.
    (c) Program the expression developed in part (b) and compute

a table of $(s^*-s)/R$ versus Mach number. Check your values with those listed in the Appendix.

21. Shown below is a portion of a T-s diagram for a system which has steady, one-dimensional flow of a perfect gas with no friction. Heat is added to subsonic flow in the constant-area duct from 1 to 2 . Isentropic, variable-area flow occurs from 2 to 3 . More heat is added in a constant-area duct from 3 to 4 . There are no shocks in the system.
   (a) Complete the diagram of the physical system. (Hint: In order to do this you must prove that A_3 is greater than, equal to, or less than A_2 .)
   (b) Sketch the entire flow system in the p-v plane.
   (c) Complete the T-s diagram for the system.

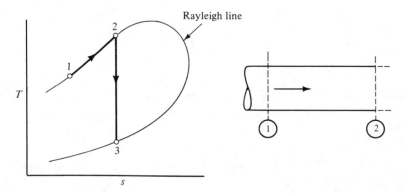
Rayleigh line

10.11 CHECK TEST

You should be able to complete this test without reference to material in the unit.

1. A Rayleigh line represents the locus of points that have the same _____ and _____ .

2. Fill in the blanks in the following table to indicate whether the properties increase, decrease, or remain constant in the case of Rayleigh flow:

Property	Heating		Cooling	
	M<1	M>1	M<1	M>1
Mach Number				
Density				
Entropy				
Stagnation Pressure				

3. Sketch a Rayleigh line in the p-v plane, together with lines
    of constant entropy and constant temperature (for a typical
    perfect gas). Indicate directions of increasing entropy and
    temperature. Show regions of subsonic and supersonic flow.

4. Air flows in the system shown below.
    (a) Find the temperature in the large chamber at location 3.
    (b) Compute the amount and direction of heat transfer.

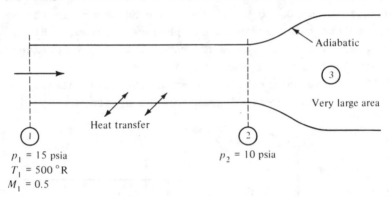

p_1 = 15 psia
T_1 = 500 °R
M_1 = 0.5

p_2 = 10 psia

5. Sketch the T-s diagram for the system shown below. Include
    both the static and stagnation curves in the diagram.

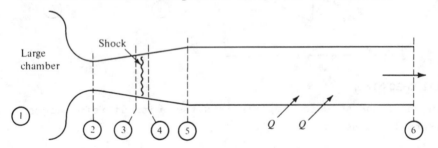

6. Work problem #14 in Section 10.10.

UNIT 11

REACTION PROPULSION SYSTEMS

11.1 INTRODUCTION

All craft which move through a fluid medium must operate by some form of a reaction propulsion system. We will not attempt to discuss all types of these systems but will concentrate on those used for aircraft or missile propulsion and popularly thought of as "jet propulsion devices." Working with these systems permits a natural application of your knowledge in the field of gas dynamics.

These engines can be classified as either air-breathers (such as the turbojet, turbofan, turboprop, ramjet, and pulsejet) or non-air-breathers which are called rockets. Many schemes for rocket propulsion have been proposed but we will only discuss the chemical rocket.

Many of the above air-breathing engines operate on the same basic thermodynamic cycle. Thus, we shall first examine the Brayton cycle to discover its pertinent features. Each of the

propulsion systems will be briefly described and some of their
operating characteristics discussed.

We will then apply momentum principles to an arbitrary propulsive device to develop a general relationship for net propulsive thrust. Other significant performance parameters, such as
power and efficiency criteria, will also be defined and discussed.

The unit will close with an interesting analysis of fixed-
geometry supersonic air inlets.

11.2 OBJECTIVES

After successfully completing this unit you should be able to:

1. Make a schematic of the Brayton cycle and draw h-s diagrams
    for both ideal and real power plants.

2. Analyze both the ideal and real Brayton cycles. Compute all
    work and heat quantities as well as cycle efficiency.

3. State the distinguishing feature of the Brayton cycle that
    makes it ideally suited for turbomachinery. Explain why
    machine efficiencies are so critical in this cycle.

4. Discuss the difference between an open and a closed cycle.

5. Draw a schematic and an h-s diagram (where appropriate)
    and describe the operation of the following propulsion sys-
    tems: turbojet, turbofan, turboprop, ramjet, pulsejet, and
    rocket.

6. Compute all state points in a turbojet or ramjet cycle when
    given appropriate operating parameters, component efficien-
    cies, etc.

7. State the normal operating regimes for various types of pro-
    pulsion systems.

8. Develop the expression for the net propulsive thrust of an
    arbitrary reaction propulsion system.

9. Define or give expressions for effective exhaust velocity,
    input power, propulsive power, thrust power, thermal effi-
    ciency, propulsive efficiency, overall efficiency, specific
    impulse, and specific fuel consumption.

10. Compute the significant performance parameters for a propul-
    sion system when given appropriate velocities, areas, pres-
    sures, etc.

11. Derive an expression for the ideal propulsive efficiency of an air-breathing engine and a rocket engine in terms of the speed ratio ν .

12. Explain why <u>fixed</u>-geometry converging-diverging diffusers are not used for air inlets on supersonic aircraft.

11.3 BRAYTON CYCLE

A. BASIC CLOSED CYCLE

Many small power plants and most air breathing jet propulsion systems operate on a cycle that was developed about 100 years ago by George B. Brayton. Although his first model was a reciprocating engine, this cycle had certain features which destined it to become the basic cycle for all gas turbine plants.

We shall first consider the basic ideal closed cycle in order to develop some of the characteristic operating parameters. A schematic of this cycle is shown in Figure 11.1 and includes a compression process from 1 to 2 with work input designated as w_c , a constant pressure heat addition from 2 to 3 with the heat added denoted by q_a , an expansion process from 3 to 4 with the work output designated as w_t , and a constant pressure heat rejection from 4 to 1 with the heat rejected denoted by q_r .

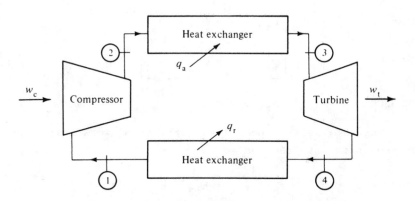

Figure 11.1 Schematic of basic Brayton cycle

For our initial analysis we shall assume no pressure drops in the heat exchangers, no heat loss in the compressor or turbine, and all reversible processes. Our cycle then consists of

two reversible adiabatic processes and
two reversible constant-pressure processes.

An h-s diagram for this cycle is shown in Figure 11.2. Keep in
mind that the working medium for this cycle is in a gaseous form
and thus this h-s diagram is similar to a T-s diagram. In
fact for perfect gases the diagrams are identical.

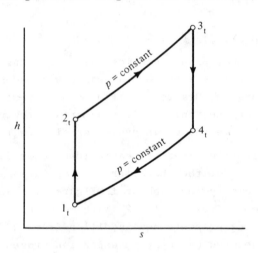

Figure 11.2 *h–s* diagram for ideal Brayton cycle

We shall proceed to make a steady flow analysis of each portion of the cycle.

TURBINE
$$h_{t3} + \cancel{q} = h_{t4} + w_s \tag{11.1}$$

Thus,
$$\boxed{w_t \equiv w_s = h_{t3} - h_{t4}} \tag{11.2}$$

COMPRESSOR
$$h_{t1} + \cancel{q} = h_{t2} + w_s \tag{11.3}$$

Designating w_c as the (positive) quantity of work which the
compressor puts into the system, we have

$$\boxed{w_c \equiv -w_s = h_{t2} - h_{t1}} \tag{11.4}$$

The <u>net</u> work output is

$$w_n \equiv w_t - w_c = (h_{t3} - h_{t4}) - (h_{t2} - h_{t1}) \tag{11.5}$$

HEAT ADDED
$$h_{t2} + q = h_{t3} + \cancel{w_s} \tag{11.6}$$

Thus,

$$q_a \equiv q = h_{t3} - h_{t2}$$ (11.7)

HEAT REJECTED

$$h_{t4} + q = h_{t1} + \cancel{w}_s$$ (11.8)

Denoting q_r as the (positive) quantity of heat which is rejected from the system, we have

$$q_r \equiv -q = h_{t4} - h_{t1}$$ (11.9)

The <u>net</u> heat added is

$$q_n \equiv q_a - q_r = (h_{t3} - h_{t2}) - (h_{t4} - h_{t1})$$ (11.10)

The thermodynamic efficiency of the cycle is defined as

$$\eta_{th} \equiv \frac{\text{Net work output}}{\text{Heat input}} = \frac{w_n}{q_a}$$ (11.11)

For the Brayton cycle this becomes

$$\eta_{th} = \frac{(h_{t3} - h_{t4}) - (h_{t2} - h_{t1})}{(h_{t3} - h_{t2})} = \frac{(h_{t3} - h_{t2}) - (h_{t4} - h_{t1})}{(h_{t3} - h_{t2})}$$

$$\eta_{th} = 1 - \frac{(h_{t4} - h_{t1})}{(h_{t3} - h_{t2})} = 1 - \frac{q_r}{q_a}$$ (11.12)

Notice that the efficiency can be expressed solely in terms of the heat quantities. This latter result can be arrived at much quicker by noting that for any cycle

$$w_n = q_n$$ (1.27)

and the cycle efficiency can be written as

$$\eta_{th} = \frac{w_n}{q_a} = \frac{q_n}{q_a} = \frac{q_a - q_r}{q_a} = 1 - \frac{q_r}{q_a}$$ (11.13)

If the working medium is assumed to be a perfect gas, then additional relationships can be brought into play. For instance, all of the above heat and work quantities can be expressed in terms of temperature differences since

$$\Delta h = c_p \Delta T$$ (1.46)

and similarly $\qquad\qquad\qquad \Delta h_t = c_p \Delta T_t$ $\qquad\qquad\qquad$ (11.14)

Equation 11.12 can thus be written as

$$\eta_{th} = 1 - \frac{c_p(T_{t4} - T_{t1})}{c_p(T_{t3} - T_{t2})} = 1 - \frac{(T_{t4} - T_{t1})}{(T_{t3} - T_{t2})} \qquad (11.15)$$

With a little manipulation this can be put into an extremely simple and significant form. Let us digress for a moment to show
how this can be done.

Looking at Figure 11.2 we notice that the entropy change
calculated between points 2 and 3 will be the same as that calculated between points 1 and 4. Now the entropy change between any
two points, say A and B , can be computed by

$$\Delta s_{A-B} = c_p \ln \frac{T_B}{T_A} - R \, \ln \frac{p_B}{p_A} \qquad (1.53)$$

If we are dealing with a constant-pressure process the last term
is zero and the resulting simple expression is applicable between
2 and 3 as well as between 1 and 4 .

Thus, $\qquad\qquad\qquad\qquad \Delta s_{2-3} = \Delta s_{1-4}$ $\qquad\qquad\qquad$ (11.16)

$$c_p \ln \frac{T_{t3}}{T_{t2}} = c_p \ln \frac{T_{t4}}{T_{t1}} \qquad (11.17)$$

and if c_p is considered constant (which it was to derive equation 1.53)

$$\frac{T_{t3}}{T_{t2}} = \frac{T_{t4}}{T_{t1}} \qquad (11.18)$$

<u>Show</u> that under the condition expressed by (11.18) we can write

$$\frac{T_{t4} - T_{t1}}{T_{t3} - T_{t2}} = \frac{T_{t1}}{T_{t2}} \qquad (11.19)$$

and the cycle efficiency (11.15) can be expressed as

$$\eta_{th} = 1 - \frac{T_{t1}}{T_{t2}} \qquad (11.20)$$

Now since the compression process between 1 and 2 is
isentropic, the temperature ratio can be related to a pressure

ratio. If we designate the <u>pressure</u> ratio of the compression process as r_p

$$r_p \equiv \frac{p_{t2}}{p_{t1}} \tag{11.21}$$

then the <u>ideal</u> Brayton cycle efficiency for a perfect gas becomes (by 1.57)

$$\eta_{th} = 1 - \left(\frac{1}{r_p}\right)^{\frac{\gamma-1}{\gamma}} \tag{11.22}$$

Remember that this relation is only valid for an ideal cycle and when the working medium may be considered a perfect gas. Equation 11.22 is plotted in Figure 11.3 and shows the influence of the compressor pressure ratio on cycle efficiency. Even for real power plants the pressure ratio remains as the most significant basic parameter.

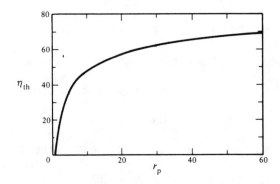

Figure 11.3 Thermodynamic efficiency of ideal Brayton cycle ($\gamma = 1.4$)

Normally in closed cycles all velocities in the flow ducts (stations 1, 2, 3, and 4) are relatively small and may be neglected. Thus, all enthalpies, temperatures, and pressures in the above equations represent static as well as stagnation quantities. However, this is <u>not</u> true for open cycles which are used for propulsion systems. The modifications required for the analysis of various propulsion engines are discussed in Section 11.4.

Example 11.1 Air enters the compressor at 15 psia and 550°R. The pressure ratio is 10. The maximum allowable cycle temperature is 2000°R. Consider an ideal cycle with negligible veloci-

ties and treat the air as a perfect gas with constant specific
heats. Determine the turbine and compressor work and cycle efficiency.

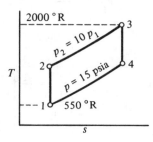

Since velocities are negligible, we shall use static conditions
in all equations.

$$\frac{T_2}{T_1} = \left(\frac{p_2}{p_1}\right)^{\frac{\gamma-1}{\gamma}} = 10^{\frac{1.4-1}{1.4}} = 1.931$$

Thus, $\qquad T_2 = 1.931(550) = 1062°R$

and similarly $T_4 = 2000/1.931 = 1036°R$

$$w_t = c_p(T_3 - T_4) = 0.24(2000 - 1036) = \underline{231} \text{ Btu/lbm}$$

$$w_c = c_p(T_2 - T_1) = 0.24(1062 - 550) = \underline{123} \text{ Btu/lbm}$$

$$w_n = w_t - w_c = 231 - 123 = \underline{108} \text{ Btu/lbm}$$

$$q_a = c_p(T_3 - T_2) = 0.24(2000 - 1062) = \underline{225} \text{ Btu/lbm}$$

$$\eta_{th} = w_n/q_a = 108/225 = \underline{48\%}$$

Notice that even in an ideal cycle the net work is a rather
small proportion of the turbine work. By comparison, in the
Rankine cycle (which is used for steam power plants) over 95% of
the turbine work remains as useful work. This radical difference
is accounted for by the fact that in the Rankine cycle the working medium is compressed as a liquid and in the Brayton cycle the
fluid is always a gas.

This large proportion of "back work" accounts for the basic
characteristics of the Brayton cycle.

(1) Large volumes of gas must be handled in order to obtain
    reasonable capacities. For this reason the cycle is
    particularly suitable for use with turbomachinery.

(2) Machine efficiencies are extremely critical to economical operation. In fact, efficiencies which could be tolerated in other cycles would reduce the net output of a Brayton cycle to zero. (See example at end of this section.)

This latter point highlights the stumbling block which for years prevented exploitation of this cycle, particularly for purposes of aircraft and missile propulsion. Efficient, lightweight, high-pressure ratio compressors were not available until quite recently. Another problem concerns the temperature limitation where the gas enters the turbine. The turbine blading must be able to <u>continuously</u> withstand this temperature while operating under high stress conditions.

B. CYCLE IMPROVEMENTS

The basic cycle performance can be improved by several techniques. If the turbine outlet temperature T_4 is significantly higher than the compressor outlet temperature T_2 , some of the heat that would normally be rejected can be used to furnish part of the heat added. This is called "regeneration" and reduces the heat which must be supplied externally. The net result is a considerable improvement in efficiency. Could a regenerator be used in Example 11.1?

The compression process can be done in stages with "intercooling," or heat removal, between each stage. This reduces the amount of compressor work. Similarly, the expansion can take place in stages with "reheat," or heat addition, between the stages. This increases the amount of turbine work. Unfortunately, this type of staging slightly decreases the cycle efficiency but this can be tolerated in order to increase the net work produced per unit mass of fluid flowing. This parameter is called "specific output" and is an indication of the size of unit required to produce a given amount of power.

The techniques of regeneration and staging with intercooling or reheating are only of use in stationary power plants and thus will not be discussed further. Those interested in more details on these topics may wish to consult a text on gas turbine power plants or Volume II of Zucrow (reference 24).

C. REAL CYCLES

The thermodynamic efficiency of 48% calculated in the previous example is quite high because the cycle was assumed to be ideal. To obtain more meaningful results we must consider the flow losses. We have already touched on the importance of having high machine efficiencies. Relatively speaking, this is not too difficult to accomplish in the turbine where an expansion process takes place but it is quite a task to build an efficient compressor. In addition, pressure drops will be involved in all ducts and heat exchangers (burners, intercoolers, reheaters, regenerators, etc.). An h-s diagram for a real Brayton cycle is given in Figure 11.4 which shows the effects of machine efficiencies and pressure drops. Note that the irreversible effects cause entropy increases in both the compressor and turbine.

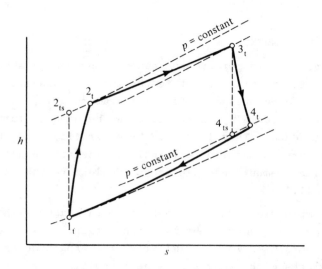

Figure 11.4 *h s* diagram for real Brayton cycle

Turbine efficiency, assuming negligible heat loss, becomes:

$$\eta_t \equiv \frac{\text{Actual work output}}{\text{Ideal work output}} = \frac{h_{t3} - h_{t4}}{h_{t3} - h_{t4s}} \qquad (11.23)$$

For a perfect gas with constant specific heats this can also be represented in terms of temperatures.

$$\eta_t \equiv \frac{c_p(T_{t3} - T_{t4})}{c_p(T_{t3} - T_{t4s})} = \frac{T_{t3} - T_{t4}}{T_{t3} - T_{t4s}} \qquad (11.24)$$

Note that the actual and ideal turbines operate between the same pressures.

The compressor efficiency şimilarly becomes:

$$\eta_c \equiv \frac{\text{Ideal work input}}{\text{Actual work input}} = \frac{h_{t2s} - h_{t1}}{h_{t2} - h_{t1}} \tag{11.25}$$

$$\eta_c = \frac{T_{t2s} - T_{t1}}{T_{t2} - T_{t1}} \tag{11.26}$$

Again, note that the actual and ideal machines operate between the same pressures (see Figure 11.4).

Example 11.2   Assume the same information as given in the previous example except that compressor and turbine efficiencies are both 80%. Neglect any pressure drops in the heat exchangers. Thus, the results will show the effect of low machine efficiencies on the Brayton cycle. We take the ideal values that were calculated in Example 11.1.

$$T_1 = 550^\circ R \qquad T_3 = 2000^\circ R \qquad \eta_t = \eta_c = 0.8$$

$$T_{2s} = 1062^\circ R \qquad T_{4s} = 1036^\circ R$$

$$w_t = (0.8)(0.24)(2000 - 1036) = \underline{185.1} \ \text{Btu/lbm}$$

$$w_c = 0.24(1062 - 550)/0.8 \qquad = \underline{153.6} \ \text{Btu/lbm}$$

$$w_n = 185.1 - 153.6 \qquad = \underline{31.5} \ \text{Btu/lbm}$$

$$T_2 = 550 + 153.6/0.24 \qquad = \underline{1190}^\circ R$$

$$q_a = 0.24(2000 - 1190) \qquad = \underline{194.4} \ \text{Btu/lbm}$$

$$\eta_{th} = w_n/q_a = 31.5/194.4 \qquad = \underline{16.2} \ \%$$

Notice that the introduction of 80% machine efficiencies drastically reduces the net work and cycle efficiency to about 29% and 34% of their respective ideal values. What would the net work and cycle efficiency be if the machine efficiencies were 75%?

D.   OPEN BRAYTON CYCLE FOR PROPULSION SYSTEMS

Most stationary gas turbine power plants operate on the "closed" cycle illustrated in Figure 11.1. Gas turbine engines

used for aircraft and missile propulsion operate on an "open"
cycle; that is, the process of heat rejection (from the turbine
exit to the compressor inlet) does not physically take place
within the engine, but occurs in the atmosphere. Thermodynami-
cally speaking, the open and closed cycles are identical but
there are a number of significant differences in actual hardware
which are enumerated below.

(1) The air enters the system at high velocity and thus
     must be diffused before being allowed to pass into the
     compressor. A significant portion of the compression
     occurs in this diffuser. If flight speeds are superson-
     ic, pressure increases also occur across the shock sys-
     tem at the front of the inlet.

(2) The heat addition is carried out by an internal combus-
     tion process within a burner or combustion chamber.
     Thus, the products of combustion pass through the re-
     mainder of the system.

(3) After passing through the turbine the air leaves the
     system by further expanding through a nozzle. This
     increases the kinetic energy of the exhaust gases which
     aids in producing thrust.

(4) Although the compression and expansion processes gener-
     ally occur in stages (most particularly with axial com-
     pressors) no intercooling is involved. Thrust augmen-
     tation with an "afterburner" could be considered as a
     form of reheat between the last turbine stage and the
     nozzle expansion. The use of regenerators is impracti-
     cal for flight propulsion systems.

The division of the compression process between the diffuser and
compressor and amount of expansion that takes place within the
turbine and the exit nozzle vary greatly depending upon the type
of propulsion system involved. This will be discussed in greater
detail in the next section which describes a number of common
propulsion engines.

11.4 PROPULSION ENGINES

A. TURBOJET

Although the first patent for a jet engine was issued in
1922, the building of practical turbojets did not take place un-

til the next decade. Development work was started in both England
and Germany in 1930 with the British obtaining the first operable
engine in 1937. However, it was not used to power an airplane
until 1941. The thrust of this engine was about 850 lbf. The
Germans managed to achieve the first actual flight of a turbojet
plane in 1939 with an engine of 1100 lbf thrust. (Historical
notes on various engines were obtained from reference 24.)

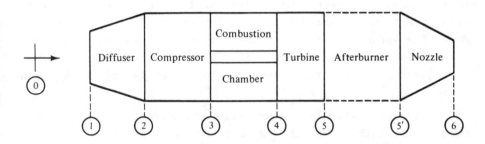

Figure 11.5 Basic parts of a turbojet engine

Figure 11.5 is a schematic showing the basic parts of a turbojet engine and identifying the important section locations.
Air enters the diffuser and is somewhat compressed as its velocity is decreased. The amount of compression that takes place in
the diffuser depends on the flight speed of the vehicle. The
greater the flight speed, the greater the pressure rise within
the diffuser.

After passing through the diffuser the air enters an adiabatic compressor where the remainder of the pressure rise occurs.
The early turbojets used centrifugal compressors as these were
the most efficient type available. Since that time a great deal
more has been learned about aerodynamics and this has enabled the
rapid development of efficient axial flow compressors which are
now widely used in jet engines.

A portion of the air then enters the combustion chamber for
the heat addition by internal combustion which is ideally carried
out at constant pressure. Combustion chambers come in several
configurations; some are annular chambers, but most consist of a
number of small chambers surrounding the central shaft. The remainder of the air is used to cool the chamber and eventually all
excess air is mixed with the products of combustion to cool them
before entering the turbine. This is the most critical tempera-

ture in the entire engine since the turbine blading has reduced
strength at elevated temperatures and operates at high stress
levels. As better materials are developed the maximum allowable
turbine inlet temperature can be raised which will result in more
efficient engines.

The gas is <u>not</u> expanded back to atmospheric pressure within
the turbine. It is only expanded enough to produce sufficient
shaft work to run the compressor plus the engine auxiliaries.
This expansion is essentially adiabatic. In most jet engines the
gases are then exhausted to the atmosphere through a nozzle.
Here, the expansion permits conversion of enthalpy into kinetic
energy and the resulting high velocities produce thrust. Normally, converging nozzles are used and they operate in a choked condition.

Many jet engines used for military aircraft have a section
between the turbine and the exhaust nozzle which includes an
"afterburner." Since the gases contain a large amount of excess
air, additional fuel can be added in this section. The temperature can be raised quite high since the surrounding material operates at a low stress level. The use of an afterburner enables
much greater exhaust velocities to be obtained from the nozzle
with high resultant thrusts. However, this increase in thrust
is obtained at the expense of an extremely high rate of fuel consumption.

An h-s diagram for a turbojet is shown in Figure 11.6
which, for the sake of simplicity, indicates all processes as
ideal. The station numbers refer to those marked in the schematic of Figure 11.5. The diagram represents <u>static</u> values. The
free stream exists at state 0 and has a high velocity (relative
to the engine). These same conditions may or may not exist at
the actual inlet to the engine. An external diffusion with spillage or an external shock system would cause the thermodynamic
state at 1 to differ from that of the free stream. Notice that
point 1 does not even appear on the h-s diagram. This is
because the performance of an air inlet is usually given with respect to the free stream conditions, enabling one to immediately
compute properties at section 2.

Operation both with and without an afterburner is shown on
Figure 11.6, the process from 5 to 5' indicating the use of
an afterburner with 5' to 6' representing the subsequent flow

through the exhaust nozzle. In this case a nozzle with a variable exit area is required to accommodate the flow when in the
afterburning mode. Since the converging nozzle is usually choked
we have indicated point 6 (and 6') at a pressure greater than
atmospheric. High velocities exist at the inlet and outlet (0,
1 and 6 or 6') and relatively low velocities exist at all other
sections. Thus, points 2 through 5 (and 5') also represent
approximate stagnation values. (These internal velocities may
not always be negligible especially in the afterburner region.)

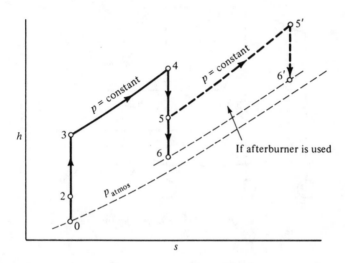

Figure 11.6 $h-s$ diagram for ideal turbojet
(for schematic see Figure 11.5)

A detailed analysis of a turbojet is identical with that of
the primary air passing through a turbofan engine. An example
problem for this case is worked out in the next sub-section which
discusses the turbofan.

A turbojet engine has a high fuel consumption because it
creates thrust by accelerating a relatively small amount of air
through a large velocity differential. In a later section we
shall see that this creates a low propulsion efficiency unless
the flight velocity is very high. Thus, the profitable application of the turbojet is in the speed range from $M_o = 1.0$ up to
about $M_o = 2.5$ or 3.0. At flight speeds above approximately
$M_o = 3.0$ the ramjet appears to be more desirable. In the subsonic speed range other variations of the turbojet are more economical and these will be discussed next.

B.   TURBOFAN

This is a comparatively recent modification to the basic
turbojet engine. The concept is to move a great deal more air
through a smaller velocity differential, thus increasing the propulsion efficiency at low flight speeds. This is accomplished
by adding a large shrouded fan to the engine. Figure 11.7 shows
a schematic of a turbofan engine.

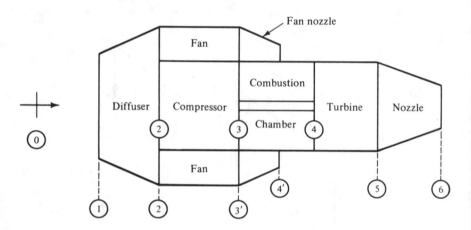

Figure 11.7 Basic parts of a turbofan engine

The flow through the central portion, or basic gas generator,
(0 - 1 - 2 - 3 - 4 - 5 - 6) is identical to that previously discussed for the pure jet (without an afterburner). Additional air,
often called secondary or "by-pass air," is drawn in through a
diffuser and passed to the fan section where it is compressed
through a relatively low pressure ratio. It is then exhausted
through a nozzle to the atmosphere. Many variations of this configuration are found. Some fans are located near the rear with
their own inlet and diffuser. In some models the by-pass air
from the fan is mixed with the main air from the turbine and the
total air flow exits through a common nozzle.

The "by-pass ratio" is defined as

$$\beta \equiv \frac{\dot{m}_a{}'}{\dot{m}_a} \tag{11.27}$$

where $\dot{m}_a \equiv$ mass flow rate of primary air (through compressor).

$\dot{m}_a' \equiv$ mass flow rate of secondary air (through fan).

An h-s diagram for the primary air is shown in Figure 11.8 and for the secondary air in Figure 11.9. In these diagrams both the actual and ideal processes are shown so that a more accurate picture of the losses can be obtained. These diagrams are for the configuration shown in Figure 11.7 in which a common diffuser is used for all entering air and separate nozzles are used for the fan and turbine exhaust.

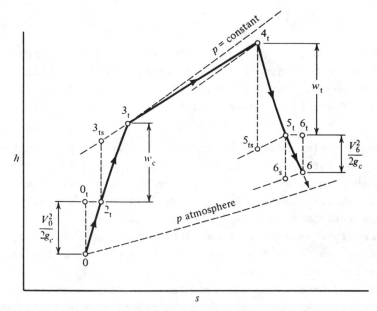

Figure 11.8 *h–s* diagram for primary air of turbofan
(for schematic see Figure 11.7)

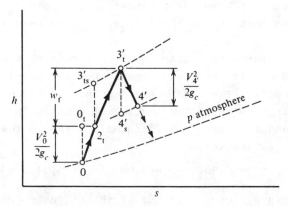

Figure 11.9 *h–s* diagram for secondary air of turbofan
(for schematic see Figure 11.7)

The analysis of a fanjet is identical to that of a pure jet with the exception of sizing the turbine. In the fanjet the turbine must produce enough work to run both the compressor and fan.

Turbine work   = Compressor work + Fan work

$$\dot{m}_a(h_{t4} - h_{t5}) = \dot{m}_a(h_{t3} - h_{t2}) + \dot{m}_a'(h_{t3'} - h_{t2}) \qquad (11.28)$$

If we divide by $\dot{m}_a$ and introduce the by-pass ratio β (see 11.27) this becomes

$$(h_{t4} - h_{t5}) = (h_{t3} - h_{t2}) + \beta(h_{t3'} - h_{t2}) \qquad (11.29)$$

Note that the mass of the fuel has been neglected in computing the turbine work. This is not unrealistic since air bled from the compressor for cabin pressurization and air-conditioning plus operation of auxiliary power amounts to approximately the mass of fuel that is added in the burner.

The following example will serve to illustrate the method of analysis for turbojet and turbofan engines. Some simplication is made in that the working medium is treated as a perfect gas with constant specific heats. These assumptions would actually yield fairly satisfactory results if two values of c_p (and γ) were used — one for the cold section (diffuser, compressor, fan, and fan nozzle) and another one for the hot section (turbine and turbine nozzle). For the sake of simplicity we shall use only one value of c_p (and γ) in the example which follows. If more accurate results were desired, we could resort to gas tables which give precise enthalpy versus temperature relations, not only for the entering air but also for the particular products of combustion which pass through the turbine, etc. (See reference 21.)

Example 11.3 A turbofan engine is operating at Mach 0.9 at an altitude of 33,000 ft where the temperature and pressure are $400^{\circ}R$ and 546 psfa. The engine has a by-pass ratio of 3.0 and the primary air flow is 50 lbm/sec. Exit nozzles for both the main and by-pass flow are converging only. The following component efficiencies are known:

$$\eta_c = 0.88 \; , \; \eta_f = 0.90 \; , \; \eta_b = 0.96 \; , \; \eta_t = 0.94 \; , \; \eta_n = 0.95$$

The total-pressure recovery factor of the diffuser (related to the free stream) is $\eta_r = 0.98$, compressor total pressure ratio is 15, fan total pressure ratio is 2.5, maximum allowable turbine inlet temp is $2500^\circ R$, total pressure loss in the combustor is 3%, and heating value of the fuel is 18,900 Btu/lbm. Assume the working medium to be air and treat it as a perfect gas with constant specific heats. Compute the properties at each section (see Figure 11.7 for section numbers).

<u>Diffuser</u>: $M_o = 0.9$, $T_o = 400^\circ R$, $p_o = 546$ psfa

$$a_o = \sqrt{(1.4)(32.2)(53.3)(400)} = 980 \text{ ft/sec}$$

$$V_o = M_o a_o = (0.9)(980) = 882 \text{ ft/sec}$$

$$p_{to} = \frac{p_{to}}{p_o} p_o = \frac{1}{0.5913} (546) = 923 \text{ psfa}$$

$$T_{to} = \frac{T_{to}}{T_o} T_o = \frac{1}{0.8606} (400) = 465^\circ R = T_{t2}$$

It is common practice to base the performance of an air inlet on the free-stream conditions.

$$P_{t2} = \eta_r \, p_{to} = (0.98)(923) = 905 \text{ psfa}$$

<u>Compressor</u>:

$$p_{t3} = 15 \, p_{t2} = 15(905) = 13,575 \text{ psfa}$$

$$\frac{T_{t3s}}{T_{t2}} = \left(\frac{p_{t3}}{p_{t2}}\right)^{\frac{\gamma-1}{\gamma}} = 15^{.286} = 2.170$$

$$T_{t3s} = 2.17(465) = 1009^\circ R$$

$$\eta_c = \frac{h_{t3s} - h_{t2}}{h_{t3} - h_{t2}} = \frac{T_{t3s} - T_{t2}}{T_{t3} - T_{t2}}$$

Thus, $$T_{t3} - T_{t2} = \frac{(1009 - 465)}{0.88} = 618^\circ R$$

and $$T_{t3} = T_{t2} + 618 = 465 + 618 = 1083^\circ R$$

<u>Fan</u>: $$p_{t3'} = 2.5 \, p_{t2} = (2.5)(905) = 2263 \text{ psfa}$$

$$\frac{T_{t3s'}}{T_{t2}} = \left(\frac{p_{t3'}}{p_{t2}}\right)^{\frac{\gamma-1}{\gamma}} = 2.5^{.286} = 1.300$$

$$T_{t3s'} = 1.3(465) = 604^{O}R$$

$$T_{t3'} - T_{t2} = \frac{T_{t3s'} - T_{t2}}{\eta_f} = \frac{(604 - 465)}{0.90} = 154.4^{O}R$$

and $\quad T_{t3'} = T_{t2} + 154.4 = 465 + 154.4 = 619^{O}R$

Burner: $\quad p_{t4} = 0.97 \, p_{t3} = 0.97(13,575) = 13,168 \text{ psfa}$

$$T_{t4} = 2500^{O}R \text{ (max allowable)}$$

An energy analysis of the burner reveals:

$$(\dot{m}_f + \dot{m}_a) \, h_{t3} + \eta_b \, (HV) \, \dot{m}_f = (\dot{m}_f + \dot{m}_a) \, h_{t4} \qquad (11.30)$$

where $\quad HV \equiv$ heating value of the fuel

$\qquad\quad \eta_b \equiv$ combustion efficiency

Let $\qquad f \equiv \dot{m}_f / \dot{m}_a \quad$ denote the fuel-air ratio

Then $\qquad \eta_b \, (HV) \, f = (1 + f) \, c_p (T_{t4} - T_{t3}) \qquad (11.31)$

or $\quad f = \dfrac{1}{\dfrac{\eta_b(HV)}{c_p(T_{t4} - T_{t3})} - 1} = \dfrac{1}{\dfrac{0.96(18,900)}{0.24(2500 - 1083)} - 1} = 0.0191$

Turbine:

If we neglect the mass of fuel added, we have from equation 11.29 (for constant specific heats)

$$(T_{t4} - T_{t5}) = (T_{t3} - T_{t2}) + \beta \, (T_{t3'} - T_{t2})$$

$$T_{t4} - T_{t5} = (1083 - 465) + (3)(619 - 465) = 1080^{O}R$$

and $\quad T_{t5} = T_{t4} - 1080 = 2500 - 1080 = 1420^{O}R$

$$\eta_t = \frac{h_{t4} - h_{t5}}{h_{t4} - h_{t5s}} = \frac{T_{t4} - T_{t5}}{T_{t4} - T_{t5s}}$$

$$T_{t4} - T_{t5s} = \frac{1080}{0.94} = 1149^{O}R$$

and $\quad T_{t5s} = T_{t4} - 1149 = 2500 - 1149 = 1351^{O}R$

$$\frac{p_{t4}}{p_{t5}} = \left(\frac{T_{t4}}{T_{t5s}}\right)^{\frac{\gamma}{\gamma-1}} = \left(\frac{2500}{1351}\right)^{3.5} = 8.62$$

$$p_{t5} = \frac{p_{t4}}{8.62} = \frac{13,168}{8.62} = 1528 \text{ psfa}$$

Turbine Nozzle:

Operating pressure ratio for the nozzle will be

$$\frac{p_o}{p_{t5}} = \frac{546}{1528} = 0.357 < 0.528$$

which means that the nozzle is choked and has sonic velocity at the exit.

$$T_{t6} = T_{t5} = 1420^{\circ}R \qquad M_6 = 1 \qquad \text{and thus} \qquad \frac{T_6}{T_{t6}} = 0.8333$$

$$T_6 = 0.8333(1420) = 1183^{\circ}R$$

$$V_6 = a_6 = \sqrt{(1.4)(32.2)(53.3)(1183)} = 1686 \text{ ft/sec}$$

$$\eta_n = \frac{h_{t5} - h_6}{h_{t5} - h_{6s}} = \frac{T_{t5} - T_6}{T_{t5} - T_{6s}}$$

Thus, $\qquad T_{t5} - T_{6s} = \dfrac{1420 - 1183}{0.95} = \dfrac{237}{0.95} = 249^{\circ}R$

and $\qquad T_{6s} = T_{t5} - 249 = 1420 - 249 = 1171^{\circ}R$

$$\frac{p_{t5}}{p_{6s}} = \left(\frac{T_{t5}}{T_{6s}}\right)^{\frac{\gamma}{\gamma-1}} = \left(\frac{1420}{1171}\right)^{3.5} = 1.964$$

$$p_6 = p_{6s} = \frac{p_{t5}}{1.964} = \frac{1528}{1.964} = 778 \text{ psfa}$$

Fan Nozzle:

$$\frac{p_o}{p_{t3'}} = \frac{546}{2263} = 0.241 < 0.528 \text{ (nozzle is choked)}$$

$$T_{t4'} = T_{t3'} = 619^{\circ}R$$

$$M_{4'} = 1, \quad T_{4'} = 0.8333(619) = 516^{\circ}R$$

$$V_{4'} = a_{4'} = \sqrt{(1.4)(32.2)(53.3)(516)} = 1113 \text{ ft/sec}$$

$$T_{t3'} - T_{4s'} = \frac{T_{t3'} - T_{4'}}{\eta_n} = \frac{619 - 516}{0.95} = 108^{\circ}R$$

$$T_{4s'} = 619 - 108 = 511^{\circ}R$$

$$\frac{P_{t3'}}{P_{4s'}} = \left(\frac{T_{t3'}}{T_{4s'}}\right)^{\frac{\gamma}{\gamma-1}} = \left(\frac{619}{511}\right)^{3.5} = 1.956$$

$$p_{4'} = p_{4s'} = \frac{2263}{1.956} = 1157 \text{ psfa}$$

In a later section we shall continue with this example to determine the thrust and other performance parameters of the engine.

C. TURBOPROP

Figure 11.10 is a schematic of a turboprop engine. It is quite similar to the turbofan engine except for the following:

1. As much power as possible is developed in the turbine and thus more power is available to operate the propeller. In essence, the engine is operating as a stationary power plant — but on an open cycle.

2. The propeller operates through reduction gears at a relatively low r.p.m. (compared with a fan).

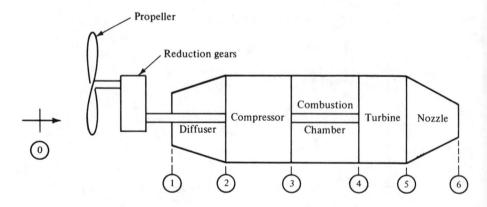

Figure 11.10 Basic parts of a turboprop engine

As a result of extracting so much power from the turbine very little expansion can take place in the nozzle and consequently the exit velocity is relatively low. Thus, little thrust (about 10 - 20% of the total) is obtained from the jet.

On the other hand, the propeller accelerates very large quantities of air (compared to the turbofan and turbojet) through a very small velocity differential. This makes an extremely efficient propulsion device for the lower subsonic flight regime.

Another operating characteristic of a propeller-driven aircraft
is that of high thrust and power available for take-off. The
turboprop engine is both considerably smaller in diameter and
lighter in weight than a reciprocating engine of comparable power
output.

D. RAMJET

The ramjet cycle is basically the same as that of the turbojet. Air enters the diffuser and most of its kinetic energy is
converted into a pressure rise. If the flight speed is supersonic, part of this compression actually occurs across a shock system which precedes the inlet (see Figure 7.16). When flight
speeds are high, sufficient compression can be attained at the
inlet and in the diffusing section and thus a compressor is not
needed. Once the compressor is eliminated, the turbine is no
longer required and it can also be omitted. The result is a ramjet engine which is pictured in Figure 11.11.

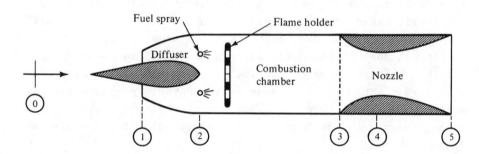

Figure 11.11 Basic parts of a ramjet engine

The combustion region in a ram-jet is generally a large single chamber, similar to an afterburner. Since the cross-
sectional area is relatively small, velocities are much higher
in the combustion zone than are experienced in a turbojet. Thus,
"flame holders" (similar to those used in afterburners) must be
introduced to stabilize the flame and prevent blow-outs. Experi-
mental work is presently being carried out with solid-fuel ramjets.

Although a ramjet engine can operate at speeds as low as
$M_o = 0.2$ the fuel consumption is horrendous at these low velocities. The operation of a ramjet does not become competitive with
that of a turbojet until speeds of about $M_o = 2.5$ or above are

reached. Another disadvantage of a ramjet is that it cannot operate at zero flight speed and thus requires some auxiliary means
of starting; it may be dropped from a plane or launched by rocket
assist. Development work is currently underway on combination
turbojet and ramjet engines for high-speed piloted craft. This
would solve the launch problem as well as the inefficient operation at low speeds.

The ramjet was invented in 1913 by a Frenchman named Lorin.
Various other patents were obtained in England and Germany in the
1920's. The first plane to be powered by a ramjet was designed
in France by Leduc in 1938 but its construction was delayed by
the war and it did not fly until 1949. Ramjets are very simple
and lightweight, and thus are ideally suited as expendable engines
for high-speed target drones or guided missiles.

Example 11.4 A ramjet has a flight speed of $M_o = 1.8$ at an
altitude of 13,000 m where the temperature is $218°K$ and the pressure is 1.7×10^4 N/m^2 . Assume a two-dimensional inlet with a
deflection angle of $10°$. Neglect frictional losses in the diffuser and combustion chamber. The inlet area is $A_1 = 0.2$ m^2 ,
sufficient fuel is added to increase the total temperature to
$2225°K$. Heating value of the fuel is 4.42×10^7 J/kg with
$\eta_b = 0.98$. The nozzle expands to atmospheric pressure for maximum thrust with $\eta_n = 0.96$. The velocity entering the combustion chamber is to be kept as large as possible but not greater
than $M_2 = 0.25$.

Assume the fluid to be air and treat as a perfect gas with
$\gamma = 1.4$. Compute significant properties at each section, mass
flow rate, fuel-air ratio, and diffuser total-pressure recovery
factor.

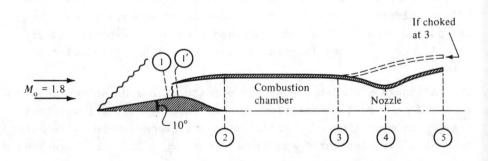

<u>Oblique Shock</u> — for $M_o = 1.8$ and $\delta = 10^o$, $\theta = 44^o$

$$M_{on} = M_o \sin \theta = 1.8 \sin 44^o = 1.250$$

$$M_{1n} = 0.8126 \ , \ p_1/p_o = 1.6562 \ , \ T_1/T_o = 1.1594$$

$$M_1 = \frac{M_{1n}}{\sin(\theta-\delta)} = \frac{0.8126}{\sin(44-10)} = \underline{1.453}$$

<u>Normal Shock</u> — for $M_1 = 1.453$

$$M_{1'} = 0.7184 \ , \ p_{1'}/p_1 = 2.2964 \ , \ T_{1'}/T_1 = 1.2892$$

$$p_{1'} = \frac{p_{1'}}{p_1} \frac{p_1}{p_o} p_o = (2.2964)(1.6562)p_o = \underline{3.803} \ p_o$$

$$T_{t2} = T_{to} = T_o \frac{T_{to}}{T_o} = (218) \frac{1}{0.6068} = \underline{359.3}^o K$$

<u>Rayleigh Flow</u> — If $M_2 = 0.25$

$$T_t^* = T_{t2} \frac{T_t^*}{T_{t2}} = (359.3) \frac{1}{0.2568} = 1399^o K$$

Thus, adding fuel to make $T_{t3} = 2225^o K$ means that the flow will be choked ($M_3 = 1.0$) and $M_2 < 0.25$. We proceed to find M_2 .

$$\frac{T_{t2}}{T_t^*} = \frac{T_{t2}}{T_{t3}} \frac{T_{t3}}{T_t^*} = \frac{359.3}{2225} (1) = 0.1615$$

$$M_2 = \underline{0.192}$$

<u>Diffuser</u>

$$p_2 = \frac{p_2}{p_{t2}} \frac{p_{t2}}{p_{t1'}} \frac{p_{t1'}}{p_{1'}} p_{1'} = (0.9746)(1) \frac{1}{0.7091} (3.803 \ p_o) = 5.227 \ p_o$$

$$T_2 = \frac{T_2}{T_{t2}} T_{t2} = (0.9927)(359.3) = 356.7^o K$$

<u>Combustion Chamber</u>

$$p_3 = p^* = \frac{p^*}{p_2} p_2 = \frac{1}{2.2822} (5.227 p_o) = 2.29 p_o$$

$$T_3 = T_{t3} \frac{T_3}{T_{t3}} = 2225 (0.8333) = 1854^o K$$

Nozzle - Since $M_3 = 1.0$, the nozzle diverges immediately.

$$T_{5s} = T_3 \left(\frac{p_3}{p_{5s}}\right)^{\frac{1-\gamma}{\gamma}} = 1854\left(\frac{2.29p_o}{p_o}\right)^{\frac{1-1.4}{1.4}} = 1463^{\circ}K$$

$$T_5 = T_3 - \eta_n(T_3 - T_{5s}) = 1854 - 0.96(1854 - 1463) = 1479^{\circ}K$$

$$\frac{T_5}{T_{t5}} = \frac{1479}{2225} = 0.6647 \quad\text{and}\quad M_5 = \underline{1.588}$$

Flow Rate

$$p_1 = \frac{p_1}{p_o} p_o = (1.6562)(1.7\text{x}10^4) = 2.816\text{x}10^4 \text{ N/m}^2$$

$$T_1 = \frac{T_1}{T_o} T_o = (1.1594)(218) = 253^{\circ}K$$

$$\rho_1 = \frac{p_1}{RT_1} = \frac{2.816\text{x}10^4}{(287)(253)} = 0.388 \text{ kg/m}^3$$

$$V_1 = M_1 a_1 = (1.453)\left[(1.4)(1)(287)(253)\right]^{\frac{1}{2}} = 463 \text{ m/s}$$

$$\dot{m} = \rho_1 A_1 V_1 = (0.388)(0.2)(463) = \underline{35.9} \text{ kg/s}$$

Fuel-Air Ratio

$$f = \frac{1}{\dfrac{\eta_b(HV)}{c_p(T_{t3} - T_{t2})} - 1} = \frac{1}{\dfrac{(0.98)(4.42\text{x}10^7)}{1000(2225-359.3)} - 1} = 0.0450$$

Total-Pressure Recovery Factor

$$\eta_r = \frac{p_{t2}}{p_{to}} = \frac{p_{t2}}{p_2}\frac{p_2}{p_o}\frac{p_o}{p_{to}} = \frac{1}{0.9746}\frac{(5.227p_o)}{p_o}(0.17404) = \underline{0.933}$$

In a later section we shall continue with this example to determine the thrust and other performance parameters.

E. PULSEJET

The turbojet, turbofan, turboprop, and the ramjet all operate on variations of the Brayton cycle. The pulsejet is a totally different animal and is shown in Figure 11.12.

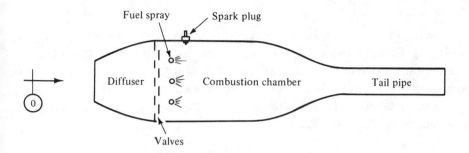

Figure 11.12 Basic parts of a pulsejet engine

A key feature in the design of the pulsejet is a bank of spring-loaded check valves that forms the wall between the diffuser and the combustion chamber. These valves are normally closed, but if a predetermined pressure differential exits they will open to permit high-pressure air from the diffusing section to pass into the combustion chamber. They never permit flow from the chamber back into the diffuser. A spark plug initiates combustion which occurs at something approaching a constant volume process. The resultant high temperature and pressure cause the gases to flow out the tail pipe at high velocity. The inertia of the exhaust gases creates a slight vacuum in the combustion chamber. This vacuum combined with the ram pressure developed in the diffuser causes sufficient pressure differential to open the check valves. A new charge of air enters the chamber and the cycle repeats.

The frequency of the above cycle depends on the size of the engine and the dynamic characteristics of the valves must be carefully matched to this frequency. Small engines operate as high as 300 to 400 cycles per second and large engines have been built which operate as low as 40 cycles per second.

The idea of a pulsejet first originated in France in 1906 but the modern configuration was not developed until the early 1930's in Germany. Perhaps the most famous pulsejet was the V-1 engine which powered the German "buzz-bombs" of World War II.

The speed range of pulsejets is limited to the subsonic regime since the large frontal area required (because the air is admitted intermittently) causes high drag. Its extreme noise and vibration levels render it useless for piloted craft. However, its ability to develop thrust at zero speed gives it a distinct advantage over the ramjet.

F. ROCKET

All of the propulsion systems discussed so far belong to the
category of air-breathing engines. As such their application is
limited to altitudes of about 100,000 feet or less. On the other
hand, rockets carry oxidizer on board as well as fuel, and thus
can function outside the atmosphere. Schematics of rocket engines
are shown in Figure 11.13.

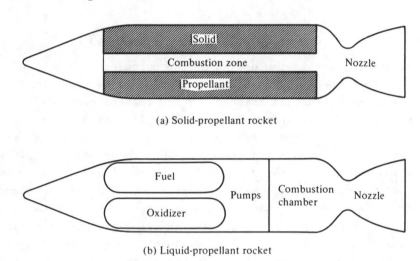

(a) Solid-propellant rocket

(b) Liquid-propellant rocket

Figure 11.13 Basic parts of a rocket engine

Chemical rocket propellants are either solid or liquid. In
a liquid system the fuel and oxidizer are separately stored and
are sprayed under high pressure (300 to 800 psia) into the combustion chamber where burning takes place. When solid propellants
are used both fuel and oxidizer are contained in the propellant
grain and the burning takes place on the surface of the propellant. Thus, the "combustion chamber" continually increases in
volume. Some solid propellants are "internal-burning" as shown
in Figure 11.13(a) whereas others are "end-burning" (like a ciga-
rette). Solid propellants develop chamber pressures of from 500
to 3000 psia.

The combustion products are exhausted through a converging-
diverging nozzle with exit velocities ranging from 5,000 to
10,000 ft/sec. The extremely high temperatures reached during
the combustion process plus the high rate of fuel consumption
limit the use of a rocket engine to short times (on the order of
seconds or minutes).

The invention of the rocket is generally attributed to the Chinese around the year 1200, although there is some evidence that rockets may have been used by the Greeks as much as 500 years previous to that time. The father of modern rockets is generally considered to be an American named Robert Goddard. His experiments started in 1915 and extended well into the 1930's.

Some of the first successful American rockets were the JATO (Jet-Assisted Take-Off) units used during the war (solid in 1941 and liquid in 1942). Also famous was the V-2 rocket developed by Wernher von Braun in Germany. This first flew in 1942 and had a liquid propulsion system that developed 56,000 pounds of thrust. The first rocket-propelled aircraft was the German ME-163.

11.5 THRUST, POWER, AND EFFICIENCY

In this section we shall examine reaction propulsion systems and obtain a general expression for their net propulsive thrust. We shall then continue to develop some significant performance parameters such as power and efficiency and note the form that these relations take for individual propulsive devices.

A. THRUST CONSIDERATIONS

Consider an airplane or missile which is traveling to the left at a constant velocity V_0 as shown in Figure 11.14. The thrust force is the result of interaction between the fluid and the propulsive device. The fluid pushes on the propulsive device and provides thrust to the left, or in the direction of motion, whereas the propulsive device pushes on the fluid opposite to the direction of flight.

Figure 11.14 Direction of flight and net propulsive force

Analysis of Fluid

We start by analyzing the fluid as it passes through the propulsive device. We define a control volume which surrounds all of the fluid inside the propulsion system. See Figure 11.15. Velocities are shown relative to the device, which is used as a frame of reference in order to make a steady-flow picture.

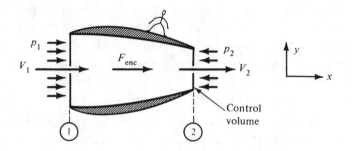

Figure 11.15 Forces on the fluid inside the propulsion system

The x-component of the momentum equation for steady flow is (from 3.42)

$$\Sigma F_x = \int_{cs} \frac{\rho V_x}{g_c} (\vec{V} \cdot \hat{n}) dA \qquad (11.32)$$

and for one-dimensional flow this becomes

$$\Sigma F_x = \frac{\dot{m}_2 V_{2x}}{g_c} - \frac{\dot{m}_1 V_{1x}}{g_c} \qquad (11.33)$$

We define an "enclosure" force as the vector sum of the friction forces and the pressure forces of the wall on the fluid within the control volume. We shall designate F_{enc} as the x-component of this enclosure force.

$$F_{enc} \equiv \text{x-component of the enclosure force}$$
$$\text{on the fluid inside the control volume}$$

Then $$\Sigma F_x = F_{enc} + p_1 A_1 - p_2 A_2 \qquad (11.34)$$

and $$p_1 A_1 - p_2 A_2 + F_{enc} = \frac{\dot{m}_2 V_2}{g_c} - \frac{\dot{m}_1 V_1}{g_c} \qquad (11.35)$$

or $$F_{enc} = \left(p_2 A_2 + \frac{\dot{m}_2 V_2}{g_c} \right) - \left(p_1 A_1 + \frac{\dot{m}_1 V_1}{g_c} \right) \qquad (11.36)$$

Notice that the enclosure force, which is an extremely complicated summation of internal pressure and friction forces, can be easily expressed in terms of known quantities at the inlet and exit. This shows the great power of the momentum equation.

You may recall from Unit 10 (see equation 10.11) that the combination of variables found in equation 11.36 is called the thrust function. Perhaps now you can see a reason for this name.

Analysis of Enclosure:

We now analyze the forces on the enclosure or the propulsive device. If the enclosure is pushing on the fluid with a force of magnitude F_{enc} to the right, then the fluid must be pushing on the enclosure with a force of equal magnitude to the left. This is the internal reaction of the fluid and is shown in Figure 11.16 as F_{int} .

$F_{int} \equiv$ Positive thrust on enclosure from internal forces

$$|F_{int}| = |F_{enc}| \qquad (11.37)$$

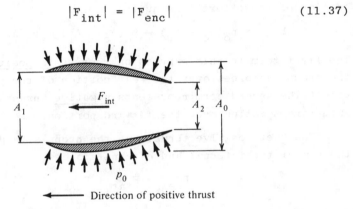

Figure 11.16 Forces on the propulsion device

In Figure 11.16 we have indicated the external forces as being ambient pressure over the entire enclosure. At first you might say this is incorrect since the pressure is not constant over the external surface. Furthermore, we have not shown any friction forces over the external surface. The answer is that these differences are accounted for when the drag forces are computed, since the drag force includes an integration of the shear stresses along the surface and also a pressure drag term which is normally put in the following form:

$$\text{Pressure Drag} = \int_{1}^{2} (p - p_0)dA_x \qquad (11.38)$$

In equation 11.38 the integration is carried out over the entire external surface of the device and dA_x represents the projection of the increment of area on a plane perpendicular to the x-axis .

Recognizing how the friction and pressure drag calculations are made we can agree that the constant-pressure representation over the external surfaces as shown in Figure 11.16 is proper for computing the net positive thrust. We define F_{ext} as the positive thrust which arises from the external forces pushing on the enclosure.

$F_{ext} \equiv$ Positive thrust on enclosure from external forces

Since this has been represented as a constant pressure, the integration of these forces is quite simple.

$$F_{ext} = p_o(A_o - A_2) - p_o(A_o - A_1) = p_o(A_1 - A_2) \quad (11.39)$$

The first term in this expression represents <u>positive</u> thrust from the pressure forces over the rear portion of the propulsive device. The second term represents <u>negative</u> thrust from the pressure forces acting over the forward portion.

The "net positive thrust" on the propulsive device will be the sum of the internal and external forces.

$$F'_{net} = F_{int} + F_{ext} \quad (11.40)$$

<u>Show</u> that the net positive thrust can be expressed as:

$$F'_{net} = \left(p_2 A_2 + \frac{\dot{m}_2 V_2}{g_c} \right) - \left(p_1 A_1 + \frac{\dot{m}_1 V_1}{g_c} \right) + p_o(A_1 - A_2) \quad (11.41)$$

or

$$\boxed{F'_{net} = \frac{\dot{m}_2 V_2}{g_c} - \frac{\dot{m}_1 V_1}{g_c} + A_2(p_2 - p_o) - A_1(p_1 - p_o)} \quad (11.42)$$

Notice that equation 11.42 has been left in a general form and as such can apply to all cases; i.e., $\dot{m}_2$ can be different from $\dot{m}_1$ if it is desired to account for the fuel added, p_2 may be different than p_o for the case of sonic or supersonic exhausts, and p_1 may not be the same as p_o. If $p_1 \neq p_o$, then $V_1 \neq V_o$. An example of this is shown for subsonic flight in Figure 11.17. Here the flow system is choked and an external diffusion with flow "spill-over" occurs. The fluid which actually enters the engine is said to be contained within the "pre-entry streamtube."

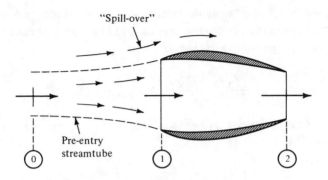

Figure 11.17 External diffusion prior to inlet

It is customary in the field of propulsion to work with the free-stream conditions (p_o and V_o) that exist far ahead of the actual inlet. Thus, by applying equation 11.42 between sections 0 and 2 (vice between 1 and 2) we obtain a simpler expression which is much more convenient to use. We <u>call</u> <u>this</u> <u>the</u> "<u>net</u> <u>propulsive</u> <u>thrust</u>."

$$F_{net} = \frac{\dot{m}_2 V_2}{g_c} - \frac{\dot{m}_o V_o}{g_c} + A_2(p_2 - p_o) \qquad (11.43)$$

It should be clearly noted that equations 11.42 and 11.43 are <u>not</u> equal since the last one, in effect, considers the region from zero to one as a part of the propulsive device. Thus, this equation includes the "pre-entry thrust" or propulsive force that the internal fluid exerts on the boundary of the pre-entry streamtube. This error will be compensated for when the drag is computed since the pressure drag must now be integrated from o to 2 as follows:

$$\text{Pressure Drag} = \int_0^1 (p - p_o)dA_x + \int_1^2 (p - p_o)dA_x \qquad (11.44)$$

The integral from o to 1 is called the "pre-entry drag" or "additive drag" and this exactly balances the pre-entry thrust.

Effective Exhaust Velocity:

In most propulsion systems the exit pressure (p_2) is greater than ambient (p_o) and thus the pressure term in equation 11.43 represents positive thrust. If we omit this pressure thrust term

we would need a higher exhaust velocity to produce the same net
thrust. This fictitious velocity is called the "effective exhaust
velocity" and is given the symbol V_j .

$$\frac{\dot{m}_2 V_j}{g_c} \equiv \frac{\dot{m}_2 V_2}{g_c} + A_2(p_2 - p_o) \tag{11.45}$$

Introducing this concept permits writing the thrust equation in a
simpler form.

$$F_{net} = \frac{\dot{m}_2 V_j}{g_c} - \frac{\dot{m}_o V_o}{g_c} \tag{11.46}$$

As shown here it is directly applicable to air-breathing engines.
This can be further simplified if we assume that the flow rates
are equal.

$$\boxed{F_{net} = \frac{\dot{m}}{g_c} (V_j - V_o)} \tag{11.47}$$

This form of the thrust equation reveals an interesting characteristic of all air-breathing propulsion systems. As their
flight speed approaches the effective exhaust velocity the thrust
goes to zero. Even long before reaching this point the thrust
drops below the drag force (which is rapidly increasing with
flight speed). Because of this, no air-breathing propulsion system can ever fly faster than its exit jet.

This fact also helps explain the natural operating speed
range of various engines. Recall that the turboprop provides a
small velocity change to a very large mass of air. Thus, its
exit jet has quite a low velocity which limits the system to low-
speed operation. At the other end of the spectrum we have the
turbojet (or pure jet) which provides a large velocity increment
to a relatively small mass of air. Therefore, this device can
operate at much higher flight speeds.

All of the equations that we have developed may also be applied to rockets by simply noting that for this case there is no
inlet. Thus, any term involving the inflow may be dropped from
the equations.

$$F_{net} = \frac{\dot{m}_2 V_2}{g_c} + A_2(p_2 - p_o) \tag{11.48}$$

or

$$F_{net} = \frac{\dot{m} V_j}{g_c} \tag{11.49}$$

Note that the propulsive thrust is independent of the flight speed and thus a rocket can easily fly faster than its exit jet.

B. POWER CONSIDERATIONS

There are three different measures of power which are connected with propulsion systems. They are

1. Input power.
2. Propulsive power.
3. Thrust power.

Consideration of these power quantities enables us to separate the performance of the thermodynamic cycle from that of the propulsion element. The general relationship among these various power quantities is shown in Figure 11.18. The thermodynamic cycle is concerned with input power and propulsive power whereas the propulsive device is the link between the propulsive power and the thrust power.

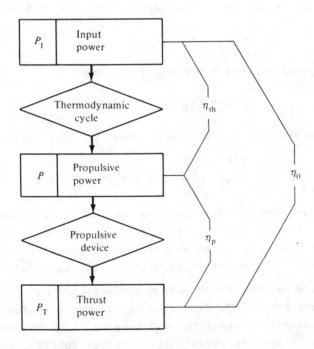

Figure 11.18 Power quantities of a propulsion system

The power input to the working fluid is designated as P_I and is the rate at which heat or chemical energy is supplied to the system. This energy is the input to the thermodynamic cycle.

$$P_I = \dot{m}_f \, (HV) \tag{11.50}$$

The output of the cycle is the input to the propulsion element
and is designated as P and called propulsive power. In the
case of propeller-driven systems the propulsive power is easily
visualized as it is the shaft power supplied to the propeller.
For other systems the propulsive power can be viewed as the
change in kinetic energy of the working medium as it passes
through the system.

$$P = \Delta KE = \frac{\dot{m}_2 V_j^{\,2}}{2g_c} - \frac{\dot{m}_o V_o^{\,2}}{2g_c} \approx \frac{\dot{m}}{2g_c} (V_j^{\,2} - V_o^{\,2}) \tag{11.51}$$

Note the use of the effective exhaust velocity here. This is the
only fair way to compute the propulsive power since there most
likely will be some pressure thrust.

The thrust power output of the propulsive device is the
actual rate of doing useful propulsion work and is designated as
P_T .

$$P_T = F_n V_o \tag{11.52}$$

For an air-breather this becomes (see 11.46)

$$P_T = (\dot{m}_2 V_j - \dot{m}_o V_o) \frac{V_o}{g_c} \tag{11.53}$$

Neglecting the slight difference between $\dot{m}_o$ and $\dot{m}_2$ this is

$$P_T = \frac{\dot{m}_o V_o}{g_c} (V_j - V_o) \tag{11.54}$$

Looking at equation 11.54 we can see that the thrust power of an
air-breather is zero when the flight speed is either zero or equal
to V_j . In the former case we have a high thrust but no motion,
thus no power. In the latter case the thrust is reduced to zero.

Somewhere between these extremes there must be a point of
maximum thrust power. To find this condition, we differentiate
equation 11.54 with respect to V_o keeping V_j constant. Setting this equal to zero reveals that maximum thrust power results
when

$$V_j = 2V_o$$

Remember that this situation applies only to air-breathers.

For the case of rockets the thrust power equals

$$P_T = \frac{\dot{m}V_j V_o}{g_c} \tag{11.55}$$

Here, no maximum is reached as the power continually increases with flight speed.

It is generally easier to compute the propulsive power by noting that the difference between the propulsive power and the thrust power is the lost power P_L , or

$$P = P_T + P_L \tag{11.56}$$

The major loss is the <u>absolute</u> kinetic energy of the exit jet and this is an unavoidable loss, even for a perfect propulsion system. In addition to this, other energy may be unavailable for thrust purposes. For instance, the exhaust jet may not all be directed axially or it may have a swirl component. In any event, the <u>minimum</u> power loss can be computed as follows:

$$V_j - V_o = \text{absolute velocity of exit jet}$$

$$P_{Lmin} = \frac{\dot{m}}{2g_c} (V_j - V_o)^2 \tag{11.57}$$

C. EFFICIENCY CONSIDERATIONS

The identification of the different power quantities P_I , P , and P_T permits various efficiency factors to be defined. These are also indicated in Figure 11.18.

Thermal Efficiency $\qquad \eta_{th} = \dfrac{P}{P_I} \tag{11.58}$

Propulsive Efficiency $\qquad \eta_p = \dfrac{P_T}{P} = \dfrac{P_T}{P_T + P_L} \tag{11.59}$

Overall Efficiency $\qquad \eta_o = \dfrac{P_T}{P_I} = \eta_{th}\eta_p \tag{11.60}$

Thermal efficiency indicates how well the thermodynamic cycle converts the chemical energy of the fuel into work which is available for propulsion. The propulsive efficiency indicates how well this work is actually utilized by the thrust device to propel the vehicle. An alternate form of propulsive efficiency is

shown in terms of the lost power. The overall efficiency is a
performance index for the entire propulsion system. Be careful
to use consistent units when computing any of the above efficiency factors.

Let us examine an ideal propulsion system, that is, one in
which there are no unavoidable losses. From equations 11.59,
11.57, and 11.52 the propulsive efficiency becomes

$$\eta_p = \frac{F_n V_o}{F_n V_o + \frac{\dot{m}_2}{2g_c}(V_j - V_o)^2} \tag{11.61}$$

For air-breathers, substitution of the net thrust from (11.46)
yields

$$\eta_p = \frac{(\dot{m}_2 V_j - \dot{m}_o V_o)\frac{V_o}{g_c}}{(\dot{m}_2 V_j - \dot{m}_o V_o)\frac{V_o}{g_c} + \frac{\dot{m}_2}{2g_c}(V_j - V_o)^2} \tag{11.62}$$

This can be greatly simplified by neglecting the difference between $\dot{m}_o$ and $\dot{m}_2$. It is also helpful to introduce the speed
ratio

$$\nu \equiv \frac{V_o}{V_j} \tag{11.63}$$

Show that under these conditions equation 11.62 can be written
as

$$\boxed{\eta_p = \frac{2\nu}{1 + \nu}} \tag{11.64}$$

This shows that the propulsive efficiency for air-breathers continually increases with flight speed, reaching a maximum when
$\nu = 1$ (or when $V_o = V_j$). This is quite reasonable since under
this condition the absolute velocity of the exit jet is zero and
there is no exit loss. (See equation 11.57.)

At this point you can begin to see some of the problems involved in optimizing air-breathing jet propulsion systems. We
previously showed that maximum thrust power is attained when
$V_j = 2V_o$. Now we see that maximum propulsive efficiency is attained when $V_j = V_o$, but unfortunately for this latter case
the thrust is zero.

Equation 11.64 further confirms the natural operating speed range of the various turbojet engines. Recall that a pure jet provides a large velocity change to a relatively small mass of air. Thus, to have a high propulsive efficiency ($\nu \to 1$) it must fly at high speeds. The fanjet provides a moderate velocity increment to a larger mass of air. Thus, it will be more efficient at medium flight speeds. The turboprop provides a small velocity increment to a very large mass of air. Thus, it is well suited to low-speed operation.

The propulsive efficiency of a rocket can be found by substituting the net thrust from (11.49) into equation 11.61.

$$\eta_p = \frac{\frac{\dot{m}}{g_c} V_j V_o}{\frac{\dot{m}}{g_c} V_j V_o + \frac{\dot{m}}{2g_c}(V_j - V_o)^2} \qquad (11.65)$$

Introduce the speed ratio ν and <u>show</u> that this becomes

$$\boxed{\eta_p = \frac{2\nu}{1 + \nu^2}} \qquad (11.66)$$

Like the equation for the air-breather, this expression is also maximum when $\nu = 1$; only in this case, the condition is actually attainable.

Specific Impulse:

Since the thrust of an engine is dependent on its size, the use of thrust alone as a performance criterion is meaningless. What is significant is the net thrust per unit mass flow rate, which is called specific thrust or "specific impulse."

$$\boxed{I_{sp} \equiv \frac{F_{net}}{\dot{m}}} \qquad (11.67)$$

Note that the units of specific impulse are properly lbf-sec/lbm but these are usually erroneously referred to as "seconds." This criterion is used mostly for rockets and in this case it is seen from (11.49) and (11.67) that the effective ex-

haust velocity becomes quite significant. What are the units of
I_{sp} in the SI system?

$$I_{sp} = \frac{F_{net}}{\dot{m}} = \frac{V_j}{g_c} \tag{11.68}$$

Specific Fuel Consumption:

Whereas specific impulse is a good overall performance indicator for rockets, specific fuel consumption is a comparable performance index for air-breathing engines. For a propeller-driven
engine it is based on shaft power and is called "brake specific
fuel consumption."

$$bsfc = \frac{lbm\ fuel\ per\ hour}{Shaft\ horsepower} = \frac{lbm}{HP-hr} \tag{11.69}$$

For other air-breathers it is based on thrust and is called
"thrust specific fuel consumption"

$$tsfc = \frac{lbm\ fuel\ per\ hour}{lbf\ thrust} = \frac{lbm}{lbf-hr} \tag{11.70}$$

or

$$tsfc = \frac{\dot{m}_f\ (3600)}{F_n} \tag{11.71}$$

By comparing equation 11.71 with 11.60 and 11.50 we see that the
thrust specific fuel consumption also can be written as

$$tsfc = \frac{V_o\ (3600)}{\eta_o\ (HV)} \tag{11.72}$$

and is a direct indication of the overall efficiency. Thus, it
is not surprising to find tsfc as the primary economic parameter for any air-breathing propulsion system. Equation 11.72 also
shows that as we increase flight speeds we must develop more efficient propulsion schemes or the fuel consumption will become
unbearable.

Example 11.5 We continue with Example 11.3 and compute the
thrust and other performance parameters of the turbofan engine.
The following pertinent information is repeated here for convenience:

$$\dot{m}_a = 50\ lbm/sec\ ,\ \dot{m}_a' = 150\ lbm/sec$$

$$f = 0.0191 \quad , \quad HV = 18,900 \text{ Btu/lbm}$$

$$V_o = 882 \text{ ft/sec} \quad , \quad p_o = 546 \text{ psfa} \quad , \quad T_o = 400^\circ R$$

$$V_4' = 1113 \text{ ft/sec} \quad , \quad p_4' = 1157 \text{ psfa} \quad , \quad T_4' = 516^\circ R$$

$$V_6 = 1686 \text{ ft/sec} \quad , \quad p_6 = 778 \text{ psfa} \quad , \quad T_6 = 1183^\circ R$$

We now compute the exit densities and areas.

$$\rho_4' = \frac{p_4'}{RT_4'} = \frac{1157}{(53.3)(516)} = 0.0421 \text{ lbm/ft}^3$$

$$A_4' = \frac{\dot{m}_a'}{\rho_4' V_4'} = \frac{150}{(0.0421)(1113)} = 3.20 \text{ ft}^2$$

$$\rho_6 = \frac{p_6}{RT_6} = \frac{778}{(53.3)(1183)} = 0.01234 \text{ lbm/ft}^3$$

$$A_6 = \frac{\dot{m}_a}{\rho_6 V_6} = \frac{50}{(0.01234)(1686)} = 2.40 \text{ ft}^2$$

The net propulsive thrust is (by 11.43)

$$F_{net} = \frac{\dot{m}_a V_6}{g_c} + A_6(p_6 - p_o) + \frac{\dot{m}_a' V_4'}{g_c} + A_4'(p_4' - p_o) - (\dot{m}_a + \dot{m}_a')\frac{V_o}{g_c}$$

$$= \frac{50(1686)}{32.2} + 2.40(778-546) + \frac{150(1113)}{32.2} + 3.20(1157-546)$$

$$- (50+150)\frac{882}{32.2}$$

$$F_{net} = \underline{4840} \text{ lbf}$$

The thrust horsepower is (by 11.52)

$$P_T = F_n V_o = \frac{4840(882)}{550} = \underline{7760} \text{ HP}$$

The input horsepower is (by 11.50)

$$P_I = \dot{m}_f(HV) = \dot{m}_a(f)(HV) = \frac{50(0.0191)(18,900)(778)}{550} = \underline{25,530} \text{ HP}$$

The overall efficiency is (by 11.60)

$$\eta_o = \frac{P_T}{P_I} = \frac{7760}{25,530} = \underline{30.4\%}$$

Thrust specific fuel consumption is (by 11.71)

$$tsfc = \frac{\dot{m}_f(3600)}{F_n} = \frac{50(0.0191)3600}{4840} = 0.71 \frac{lbm}{lbf\text{-}hr}$$

This specific fuel consumption is slightly low, even for a fanjet engine. Had we changed to a higher value of specific heat in the hot sections (turbine and turbine nozzle) two effects would be noted:

(a) The fuel-air ratio would increase because the enthalpy entering the turbine would increase.
(b) The thrust would rise due to an increased exhaust velocity and exit pressure.

The increase in thrust would be small compared to the increase in fuel-air ratio, and the net effect would be to raise the tsfc to about 0.8.

Example 11.6 We continue and compute the performance parameters for the ramjet of Example 11.4. The following pertinent information is repeated here for convenience:

$$\dot{m}_a = 35.9 \text{ kg/s} , \quad f = 0.0450 , \quad HV = 4.42 \times 10^7 \text{ J/kg}$$

$$M_o = 1.8 , \quad T_o = 218^\circ K , \quad M_5 = 1.588 , \quad T_5 = 1479^\circ K$$

$$V_o = M_o a_o = (1.8)\left[(1.4)(1)(287)(218)\right]^{\frac{1}{2}} = 533 \text{ m/s}$$

$$V_5 = M_5 a_5 = (1.588)\left[(1.4)(1)(287)(1479)\right]^{\frac{1}{2}} = 1224 \text{ m/s}$$

If we neglect the mass of fuel added, the net propulsive thrust is

$$F_{net} = \frac{m}{g_c} (V_5 - V_o) = \frac{35.9}{1} (1224-533) = \underline{24,800} \text{ N}$$

Thrust specific fuel consumption is

$$tsfc = \frac{\dot{m}_f(3600)}{F_n} = \frac{(0.0450)(35.9)(3600)}{24,800} = \underline{0.235} \frac{kg}{N\text{-}hr}$$

This is equivalent to tsfc = 2.3 lbm/lbf-hr which is quite high in comparison to the fanjet of the previous example. This illustrates the uneconomical operation of ramjets at low flight speeds.

Example 11.7 A liquid rocket has a pressure and temperature of 400 psia and 5000°R, respectively, in the combustion chamber and is operating at an altitude where the ambient pressure is 200 psfa. The gases exit through an isentropic converging-diverging nozzle which produces a Mach number of 4.0. Approximate the exhaust gases by taking $\gamma = 1.4$ and a molecular weight of 20. Determine the specific impulse and the effective exhaust velocity.

For $M_2 = 4.0$, $p/p_t = 0.00659$, $T/T_t = 0.2381$

$$p_2 = \frac{p}{p_t} p_t = (0.00659)(400)(144) = 380 \text{ psfa}$$

$$T_2 = \frac{T}{T_t} T_t = (0.2381)(5000) = 1190^\circ R$$

$$\rho_2 = \frac{p_2}{RT_2} = \frac{380(20)}{1545(1190)} = 0.00413 \text{ lbm/ft}^3$$

$$V_2 = M_2 a_2 = 4.0 \left[(1.4)(32.2) \frac{1545}{20} (1190) \right]^{\frac{1}{2}} = 8143 \text{ ft/sec}$$

$$F_{net} = \frac{\dot{m} V_2}{g_c} + A_2(p_2 - p_o) = \frac{\rho_2 A_2 V_2^2}{g_c} + A_2(p_2 - p_o)$$

$$I_{sp} = \frac{F_{net}}{\dot{m}} = \frac{F_{net}}{\rho_2 A_2 V_2} = \frac{V_2}{g_c} + \frac{(p_2 - p_o)}{\rho_2 V_2}$$

$$I_{sp} = \frac{8143}{32.2} + \frac{(380 - 200)}{(0.00413)(8143)} = \underline{258.2} \text{ lbf-sec/lbm}$$

$$V_j = I_{sp} g_c = 258.2(32.2) = \underline{8314} \text{ ft/sec}$$

11.6 SUPERSONIC DIFFUSERS

The deceleration of an air stream in the inlet of a propulsion system causes special problems at supersonic flight speeds. If a subsonic diffuser is used (diverging section) a normal shock will occur at the inlet with an associated loss in stagnation pressure. This loss is small if flight speeds are low, say $M_o < 1.4$. At speeds between $1.4 < M_o < 2.0$ an oblique shock inlet is required (similar to the one used on the ramjet in Example 11.4). Above $M_o = 2.0$ two oblique shocks, as shown in Figure 7.16, are necessary.

The requirement to be met in each case is to keep the total-pressure recovery factor as high as possible. A value of $\eta_r = 0.95$ is considered satisfactory at low supersonic speeds but this becomes increasingly critical as flight speeds increase. Two oblique shocks plus one normal shock are inadequate at speeds above approximately M_o = 2.5. See pages 421–427 of Zucrow (Volume I of reference 24) for the effects of multiple conical shocks. From our studies of varying-area flow we might assume that a converging-diverging section would make a good supersonic diffuser — and indeed it would. Recall that this configuration was used for the exhaust section of a supersonic wind tunnel in Unit 6. However, there are some practical operating difficulties involved in using a <u>fixed-geometry</u> converging-diverging section for a supersonic air inlet.

Suppose we design the inlet diffuser for an airplane that will fly at about M = 1.86 . From the isentropic tables we see that the area ratio corresponding to this Mach number is 1.507. For simplicity, we shall construct the diffuser with an area ratio (inlet area to throat area) of 1.50. The design operation of this diffuser is shown in Figure 11.19. In the discussion below, we follow the operation of this diffuser as the aircraft takes off and accelerates to its design speed.

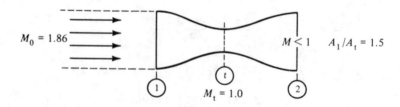

Figure 11.19 Desired operation of converging-diverging diffuser

Note that as the flight speed reaches approximately M_o = 0.43, the diffuser becomes choked with M = 1.0 in the throat. (Check the subsonic portion of the isentropic tables for the above area ratio.) This condition is shown in Figure 11.20(a). Now increase the flight speed to say M_o = 0.6. "Spillage" or external diffusion occurs as indicated in Figure 11.20(b). As M_o is increased to 1.0 there is a further decrease in the "capture area" (area of the flow at the free-stream Mach number that actually enters the diffuser). See Figure 11.20(c).

As we increase M_o to supersonic speeds a "detached" shock wave forms in front of the inlet. Spillage still occurs as shown in Figure 11.20(d). Note that at higher flight speeds less external diffusion is necessary to produce the required M = 0.43

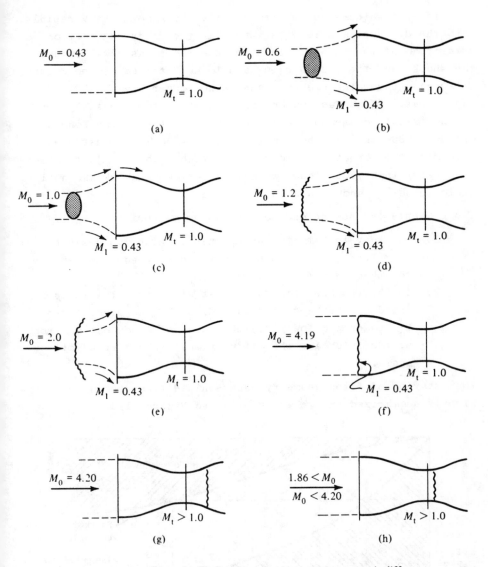

Figure 11.20 Starting a fixed-geometry supersonic diffuser
(area ratio = 1.5)

at the inlet. Thus, the shock moves closer to the inlet as
speeds increase. See Figure 11.20(e). Also note that it is necessary to fly at approximately $M_0 = 4.19$ in order for the shock
to become attached to the inlet. (Check the shock tables to sub-
stantiate this.) This condition, indicated in Figure 11.20(f)
is far above the design flight speed.

If we now increase M_o to 4.2 the shock moves very rapidly into the diffuser and locates itself in the divergent section downstream of the throat. This is referred to as "swallowing the shock" and the diffuser is said to be "started." See Figure 11.20(g). Under these conditions we no longer have Mach 1.0 in the throat. (What Mach number does exist in the throat?) We can now slowly decrease the flight speed to the design condition of M = 1.86 and the shock will move to a position just downstream of the throat and occur at the Mach number of just slightly greater than 1.0. Thus, we have a very weak shock and negligible losses as shown in Figure 11.20(h).

Two comments can now be made on the above performance:

1) In order to "start" the diffuser, which was designed for $M_o = 1.86$, it is necessary to "overspeed" the vehicle to a Mach number of 4.2.

2) If the vehicle slows down just slightly below its design speed (or perhaps minor air disturbances might cause M_o to drop below 1.86) the shock will pop out in front of the inlet and the diffuser must be "started" all over again.

The behavior of fixed-geometry supersonic diffusers can be conveniently summarized in a chart similar to Figure 11.21.

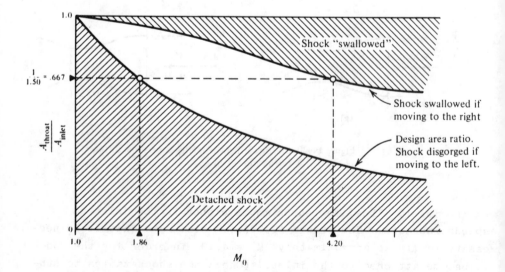

Figure 11.21 Performance of fixed-geometry supersonic diffusers

It should be obvious that the above described operation could not be tolerated and for this reason one does not see fixed-geometry converging-diverging diffusers used for air inlets. At flight speeds above M_o = 2.5 a combination of oblique shocks and a variable-geometry converging-diverging diffuser are required for efficient pressure recovery.

11.7 SUMMARY

An analysis of the ideal Brayton cycle revealed that its thermodynamic efficiency is a function of the pressure ratio as given by

$$\eta_{th} = 1 - \left(\frac{1}{r_p}\right)^{\frac{\gamma-1}{\gamma}} \qquad (11.22)$$

Perhaps the most significant feature of this cycle is that the work input is a large percentage of the work output. Because of this, machine efficiencies are most critical in any power plant operating on the Brayton cycle. Also, to produce a reasonable quantity of net work, large amounts of air must be handled which makes this cycle particularly suitable for turbomachinery.

In discussing the various types of jet propulsion systems it was noted that pure jets move a relatively small amount of air through a large velocity change. On the other hand, propeller systems move a relatively large amount of air through a small velocity increment. Fanjets occupy a middle ground on both criteria.

The net thrust of any propulsive device was found to be

$$F_{net} = \frac{\dot{m}_2 V_2}{g_c} - \frac{\dot{m}_o V_o}{g_c} + A_2(p_2 - p_o) \qquad (11.43)$$

You should learn this equation as it is probably the most important relation in this unit. Also, you should not overlook the various power and efficiency parameters discussed in Section 11.5. Perhaps the most interesting of these is the propulsive efficiency since this is a measure of what the propulsive device is accomplishing, exclusive of the energy producer.

For air-breathers, in terms of the speed ratio $\nu = V_o/V_j$

$$\eta_p = \frac{2\nu}{1 + \nu} \qquad (11.64)$$

Equation 11.64 explains why pure jets operate more efficiently at high speeds, whereas fanjets and propjets fare better at progressively lower speeds. We also see that, for air-breathers, maximum efficiency occurs at minimum thrust.

Rockets are not subject to this dilemma and their propulsive efficiency is

$$\eta_p = \frac{2\nu}{1 + \nu^2} \qquad (11.66)$$

Other important performance indicators are

$$I_{sp} = \frac{F_{net}}{\dot{m}} \qquad (11.67)$$

$$tsfc = \frac{\dot{m}_f(3600)}{F_{net}} \qquad (11.71)$$

Air inlets for supersonic vehicles should have total-pressure recovery factors of 0.95 or above. At lower speeds one uses a subsonic diffuser preceded by ramps or a spike to induce one or more oblique shocks before the normal shock. At high supersonic flight speeds variable-geometry features are additionally required.

11.8 PROBLEMS

In the problems that follow you may assume perfect gas behavior and constant specific heats even though the temperature range may be rather large in some cases. Also neglect any effects of dissociation and assume that all propellants have the properties of air.

1. Conditions entering the compressor of an ideal Brayton cycle are $520°R$ and 5 psia. The compressor pressure ratio is 12 and the maximum allowable cycle temperature is $2400°R$. Assume that air has negligible velocities in the ducting.
   (a) Determine w_t , w_c , w_n , q_a and η_{th} .
   (b) What flow rate is required for a net output of 5000 HP?

2. Rework problem 2 with a compressor efficiency of 89% and a turbine efficiency of 92%.

3. A stationary power plant produces 1×10^7 watts output when operating under the following conditions: Compressor inlet

is $0°C$ and 1 bar absolute, turbine inlet is $1250°K$, cycle
pressure ratio is 10 , fluid is air with negligible velocities. The turbine and compressor efficiencies are both 90%.
Determine the cycle efficiency and the mass flow rate.

4. Assume that all data given in problem 3 remains the same except that the turbine and compressor are 80% efficient.
   (a) Determine the cycle efficiency.
   (b) Compare the net work output and cycle efficiency with
       that of problem 3.
   (c) What value of machine efficiency (assuming $\eta_t = \eta_c$)
       will cause zero net work output from this cycle?

5. Consider an ideal Brayton cycle as shown in Figure 11.2.
   Let $\alpha = T_{t3}/T_{t1}$, the cycle temperature ratio

$$\theta = \left(\frac{p_{t2}}{p_{t1}}\right)^{\frac{\gamma-1}{\gamma}}$$, the cycle pressure ratio parameter

   (a) Show that the net work output can be expressed as

$$w_n = c_p T_{t1}\left[\frac{\theta-1}{\theta}\right](\alpha-\theta)$$

   (b) Show that for a given α the maximum net work occurs
       when $\theta = \sqrt{\alpha}$.
   (c) On the same T-s diagram sketch cycles for a given tem-
       perature ratio but for different pressure ratios. Which
       one is most efficient? Which produces the most net work?

6. An airplane is traveling at 550 mph at an altitude where the
   ambient pressure is 6.5 psia. The exit area of the jet engine is 1.65 ft^2 and the exit jet has a relative velocity
   of 1500 ft/sec. The pressure at the exit plane is found to
   be 10 psia. Air flow is measured at 175 lbm/sec. You may
   neglect the weight of fuel added.
   (a) What is the net propulsive thrust of this engine?
   (b) Determine the effective exhaust velocity.

7. The air flow through a jet engine is 30 kg/s and the fuel
   flow is 1 kg/s. The exhaust gases leave with a relative velocity of 610 m/s. Pressure equilibrium exists over the exit
   plane. Compute the velocity of the airplane if the thrust
   power is 1.12×10^6 watts.

8. A twin-engined jet aircraft requires a total net propulsive
    thrust of 6000 lbf. Each engine consumes air at the rate of
    120 lbm/sec when traveling at 650 ft/sec. Fuel is added in
    each engine at the rate of 3.0 lbm/sec. Assume that pressure
    equilibrium exists across the exit plane and compute the ve-
    locity of the exhaust gases relative to the plane.

9. A boat is propelled by an hydraulic jet. The inlet scoop has
    an area of 0.5 ft^2 and the area of the exit duct is 0.20 ft^2.
    Since the exit velocity will always be subsonic, pressure
    equilibrium exists over the exit plane. No spillage occurs
    at the inlet when the boat is moving through fresh water at
    50 mph.
    (a) Compute the net propulsive force being developed.
    (b) What is the propulsive efficiency?
    (c) How much energy is added to the water as it passes through
        the device? (Assume no losses.)

10. It is proposed to power a mono-rail car by a pulsejet. A net
    propulsive thrust of 5350 N is required when traveling at a
    speed of 210 km/hr. The gases leave the engine with an aver-
    age velocity of 350 m/s. Assume pressure equilibrium exists
    at the outlet plane and neglect the weight of fuel added.
    (a) Compute the mass flow rate required.
    (b) What inlet area is necessary, assuming no spillage occurs?
        (Assume 16°C and one atmosphere.)
    (c) What is the thrust power?
    (d) What is the propulsive efficiency?
    (e) How much energy is added to the air as it passes through
        the engine if the outlet temperature is 980°C?

11. A ramjet flies at M_o = 4.0 at 30,000 ft altitude where
    T_o = 411°R and p_o = 628 psfa . The exhaust nozzle exit
    diameter is 18 inches. The exhaust jet has a velocity of
    5000 ft/sec relative to the missile and is at 1800°R and
    850 psfa. Neglect the fuel added.
    (a) Determine the net propulsive thrust.
    (b) What is the effective exhaust velocity?
    (c) How much thrust power is developed?

12. In Sections 11.4 and 11.5 an example of a fanjet engine analy-
    sis was given. Remove the fan from this engine. Readjust
    the turbine expansion to produce the appropriate compressor
    work. Assume all component efficiencies remain unchanged.

Compute the net propulsive thrust and thrust specific fuel consumption for the purejet engine and compare to that of the fanjet.

13. It has been suggested that an afterburner be added to the fanjet engine used in the example problem in Sections 11.4 and 11.5. Assume that the gas leaves the turbine with a velocity of 400 ft/sec. Enough fuel is added in the afterburner to raise the stagnation temperature to $3500^O R$ with a combustion efficiency of $\eta_{ab} = 0.85$. Determine the cross-sectional area of the afterburner, the conditions at the exit of the afterburner (assume Rayleigh flow), the new conditions at the nozzle exit, the required exit area, and the resultant effect on the performance parameters of the engine. (Neglect mass of fuel.)

14. A ramjet is designed to operate at $M_o = 3.0$ at an altitude of 40,000 ft where the temperature and pressure are $390^O R$ and 400 psfa. The total-pressure recovery factor for the inlet is $\eta_r = p_{t2}/p_{t0} = 0.85$. The velocity is reduced to 300 ft/sec before entering the combustion chamber where the total temperature is raised to $4000^O R$. Combustion efficiency is $\eta_b = 0.96$ and the heating value of the fuel is 18,500 Btu/lbm. The exit nozzle has an efficiency of $\eta_n = 0.95$ and expands the flow through a converging-diverging section to the same area as the combustion chamber (similar to that shown in Figure 11.11). Compute the net propulsive thrust per unit area and the thrust specific fuel consumption. (You may neglect the mass of fuel added.)

15. A rocket sled used for test purposes requires a thrust of 20,000 lbf. The specific impulse is 240 lbf-sec/lbm.
    (a) What is the flow rate?
    (b) Compute the exhaust velocity if the nozzle expands the gases to ambient pressure.

16. The German V-2 had a sea level thrust of 249,000 N, a propellant flow rate of 125 kg/s, an exhaust velocity of 1995 m/s, and the nozzle outlet size was 74 cm in diameter.
    (a) Compute the specific impulse.
    (b) Calculate the pressure at the nozzle outlet.

17. An ideal rocket nozzle was originally designed to expand the exhaust gases to ambient pressure when at sea level and oper-

ating with a combustion chamber pressure of 400 psia and a
temperature of 5000°R. The rocket is now used to propel a
missile which is fired from an airplane at 38,000 ft where
the pressure is 3.27 psia.
(a) Determine the exit area required to produce a thrust of
    1000 lbf.
(b) Compute the exit velocity, the effective exhaust velocity
    and the specific impulse.

18. The combustion chamber of a rocket has stagnation conditions
of 22 bars and 2500°K. Assume that the nozzle is ideal and
expands the flow to the ambient pressure of 0.25 bars.
(a) Determine the nozzle area ratio and exit velocity.
(b) What is the specific impulse?

19. Compare the total-pressure recovery factors for the air inlets described in Problem #14 of Section 7.10.

20. Sketch a supersonic inlet which has one oblique shock followed by a normal shock attached to the entrance of a subsonic
diffuser. Draw streamlines and identify the capture area
(that portion of the free stream that actually enters the diffuser). Now vary the wedge angle and cause the oblique shock
to form at a different angle. Again, determine the capture
area. Show that maximum flow enters the inlet when the
oblique shock just touches the outer lip of the diffuser.

21. Figure 11.21 illustrates the peculiar operating conditions
associated with fixed-geometry supersonic diffusers. Unfortunately this figure was not drawn to scale and, therefore,
cannot be used as a working plot.
(a) Construct an accurate version of Figure 11.21.
(b) If the design flight speed is $M_0 = 1.5$, to what velocity
    must the vehicle be overspeeded in order to start the dif-
    fuser?
(c) Suppose the design speed is $M_0 = 2.0$. How fast must
    the vehicle go to start the diffuser?

22. A converging-diverging supersonic inlet is to be designed
with a variable area. The idea is to swallow the shock when
the vehicle has just reached its design flight speed. Then
the diffuser area ratio will be changed to operate properly
without any shock. Thus, the inlet does not have to be over-
speeded to start. Calculate the maximum and minimum area

ratios that would be required to operate in the above manner
if the flight speed is M_o = 2.80 .

11.9 CHECK TEST

You should be able to complete this test without reference to
material in the unit.

1. We wish to build an electric generator for use at a ski lodge.
    To keep this small and light-weight we have decided to use an
    open Brayton cycle as shown below. Write an expression (in
    terms of properties at 1, 2, 3 and 4) which will represent
    for each pound mass flowing:
    (a) The compressor work input.
    (b) The turbine work output.
    (c) The cycle thermodynamic efficiency.

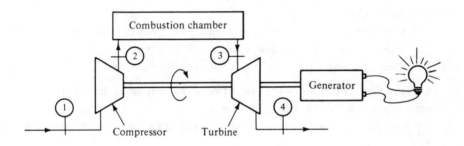

2. If the machine efficiencies are not fairly high, the thermody-
    namic efficiency of a Brayton cycle will be extremely poor.
    What basic characteristic of the Brayton cycle accounts for
    this fact?

3. The conditions entering a turbine are T_t = $1060^{\circ}C$ and
    p_t = 6.5 bars . The turbine efficiency is η_t = 90% and
    the mass flow rate is 45 kg/s. Compute the turbine outlet
    stagnation conditions if the turbine produces 2.08×10^7
    watts of work. Neglect any heat transfer.

4. Draw an h-s diagram for the secondary (fan) air of a turbo-
    fan engine (a real engine — not an ideal one).
    (a) Indicate static and stagnation points if they are signifi-
        cantly different.
    (b) Indicate pertinent velocities, work quantities, etc.

5. Mark the following statements as true or false:
    (a) Thrust power output can be viewed as the change in kinet-
        ic energy of the working medium.
    (b) If the exhaust gases leave a rocket at a speed of
        7000 ft/sec relative to the rocket it would be impossible
        for the rocket to be traveling at 8000 ft/sec relative to
        the ground.
    (c) It is possible to operate a ramjet at 100 percent propul-
        sive efficiency and develop thrust.
    (d) One would expect that a turbofan engine will have a
        higher tsfc than a ramjet engine.

6. A rocket is traveling at 4500 ft/sec at an altitude of
    20,000 ft where the temperature and pressure are 447^OR and
    972 psfa, respectively. The exit diameter of the nozzle is
    24 inches and the exhaust jet has the following characteris-
    tics: T = 1500^OR , p = 1200 psfa , V = 6600 ft/sec (relative
    to the rocket).
    (a) Compute the flow rate and the net propulsive thrust.
    (b) What is the effective exhaust velocity?
    (c) Compute the specific impulse and thrust power.

7. A fixed-geometry converging-diverging supersonic diffuser is
    contemplated for a vehicle having a design Mach number of
    M_o = 1.65 . How fast must the plane fly to "start" this dif-
    fuser?

APPENDICES

1

SUMMARY OF THE ENGLISH ENGINEERING SYSTEM OF UNITS

Force	–	pound force	(lbf)
Mass	–	pound mass	(lbm)
Length	–	foot	(ft)
Time	–	second	(sec)
Temperature	–	Rankine	(R)

NEVER say pound as this is ambiguous! It is either
a pound force (lbf) or a pound mass (lbm).

A one-pound force will give a one-pound mass
an acceleration of 32.174 feet/second^2.

$$F = \frac{ma}{g_c}$$

$$1(lbf) = \frac{1(lbm)\ 32.174(ft/sec^2)}{g_c}$$

Thus $g_c = 32.174$ lbm-ft/lbf-sec^2

Temperature	–	$T(^oR) = T(^oF) + 459.67$
Gas Constant	–	$R = 1545/M.W.$ ft-lbf/lbm-^oR
Pressure	–	1 atmos = 2116.2 lbf/ft^2
Heat to work	–	1 Btu = 778.2 ft-lbf
Power	–	1 Horsepower = 550 ft-lbf/sec
Standard Gravity	–	$g = 32.174$ ft/sec^2

2

SUMMARY OF THE INTERNATIONAL SYSTEM (SI) OF UNITS

Force	–	newton	(N)
Mass	–	kilogram	(kg)
Length	–	meter	(m)
Time	–	second	(s)
Temperature	–	kelvin	(K)

A one-newton force will give a one-kilogram mass an acceleration of 1 meter/second^2.

$$F = \frac{ma}{g_c}$$

$$1(N) = \frac{1(kg)\ 1(m/s^2)}{g_c}$$

Thus $\quad g_c = 1\ kg\text{-}m/N\text{-}s^2$

Temperature	–	$T(^O K) = T(^O C) + 273.15$
Gas Constant	–	$R = 8314/M.W.\ N\text{-}m/kg\text{-}^O K$
Pressure	–	$1\ atmos = 1.013 \times 10^5\ N/m^2$
		$1\ pascal\ (Pa) = 1\ N/m^2$
		$1\ bar\ (bar) = 1 \times 10^5\ N/m^2$
Work	–	$1\ joule\ (J) = 1\ N\text{-}m$
Power	–	$1\ watt\ (W) = 1\ J/s$
Standard Gravity	–	$g = 9.81\ m/s^2$

3

SOME USEFUL CONVERSION FACTORS

(From "The International System of Units," NASA SP-7012, 1973)

To convert from		to	multiply by
foot		meter	3.048×10^{-1}
inch		meter	2.54×10^{-2}
lbf		newton	4.448
lbm		kilogram	4.536×10^{-1}
^{o}R		^{o}K	5.555×10^{-1}
Btu	(q)	joule	1.055×10^{3}
Btu	(q)	kW-hr	2.930×10^{-4}
ft-lbf	(w)	joule	1.356
horsepower		watt	7.457×10^{2}
ft/sec	(V)	m/s	3.048×10^{-1}
mph	(V)	m/s	4.470×10^{-1}
mph	(V)	km/hr	1.609
atmosphere	(p)	N/m^2	1.013×10^{5}
lbf/in^2	(p)	N/m^2	6.895×10^{3}
lbf/ft^2	(p)	N/m^2	4.788×10
lbm/ft^3	(ρ)	kg/m^3	1.602×10
$lbf\text{-}sec/ft^2$	(μ)	$N\text{-}s/m^2$	4.788×10
ft^2/sec	(ν)	m^2/s	9.290×10^{-2}
$Btu/lbm\text{-}^{o}R$	(c_p)	$J/kg\text{-}^{o}K$	4.187×10^{3}
$ft\text{-}lbf/lbm\text{-}^{o}R$	(R)	$N\text{-}m/kg\text{-}^{o}K$	5.381

4

PROPERTIES OF GASES
ENGLISH ENGINEERING SYSTEM

Gas	Symbol	Molecular Weight	$\gamma = \dfrac{c_p}{c_v}$	Gas Constant ft-lbf/lbm-°R R	Specific Heats Btu/lbm-°R c_p	Specific Heats Btu/lbm-°R c_v	Viscosity lbf-sec/ft² μ
Air		28.97	1.40	53.3	0.240	0.171	3.8×10^{-7}
Argon	Ar	39.94	1.67	38.7	0.124	0.074	4.7×10^{-7}
Carbon Dioxide	CO_2	44.01	1.29	35.1	0.203	0.157	3.1×10^{-7}
Carbon Monoxide	CO	28.01	1.40	55.2	0.248	0.177	3.7×10^{-7}
Helium	He	4.00	1.67	386	1.25	0.750	4.2×10^{-7}
Hydrogen	H_2	2.02	1.41	766	3.42	2.43	1.9×10^{-7}
Methane	CH_4	16.04	1.32	96.4	0.532	0.403	2.3×10^{-7}
Nitrogen	N_2	28.02	1.40	55.1	0.248	0.177	3.6×10^{-7}
Oxygen	O_2	32.00	1.40	48.3	0.218	0.156	4.2×10^{-7}
Water Vapor	H_2O	18.02	1.33	85.7	0.445	0.335	2.2×10^{-7}

Tabular values are for normal room temperature and pressure.

5

PROPERTIES OF GASES
INTERNATIONAL SYSTEM (SI)

Gas	Symbol	Molecular Weight	$\gamma = \dfrac{c_p}{c_v}$	Gas Constant $N\text{-}m/kg\text{-}^{\circ}K$	Specific Heats $J/kg\text{-}^{\circ}K$		Viscosity $N\text{-}s/m^2$
				R	c_p	c_v	μ
Air		28.97	1.40	287	1,000	716	1.8×10^{-5}
Argon	Ar	39.94	1.67	208	519	310	2.3×10^{-5}
Carbon Dioxide	CO_2	44.01	1.29	189	850	657	1.5×10^{-5}
Carbon Monoxide	CO	28.01	1.40	297	1,040	741	1.8×10^{-5}
Helium	He	4.00	1.67	2,080	5,230	3,140	2.0×10^{-5}
Hydrogen	H_2	2.02	1.41	4,120	14,300	10,200	9.1×10^{-6}
Methane	CH_4	16.04	1.32	519	2,230	1,690	1.1×10^{-5}
Nitrogen	N_2	28.02	1.40	296	1,040	741	1.7×10^{-5}
Oxygen	O_2	32.00	1.40	260	913	653	2.0×10^{-5}
Water Vapor	H_2O	18.02	1.33	461	1,860	1,400	1.1×10^{-5}

Tabular values are for normal room temperature and pressure.

6

FRICTION-FACTOR CHART

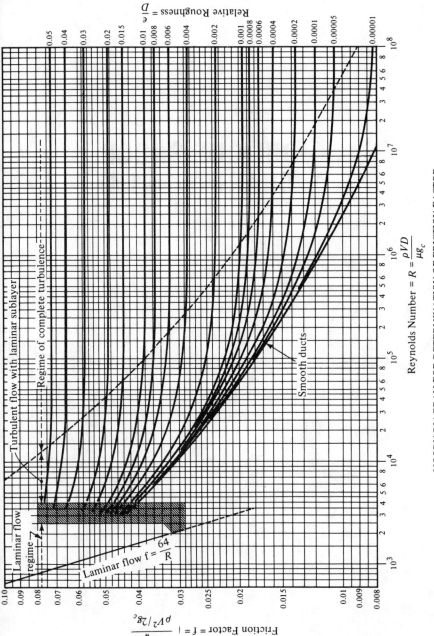

MOODY DIAGRAM FOR DETERMINATION OF FRICTION FACTOR

(Adapted with permission from *Friction Factors for Pipe Flow*
by L.F. Moody, Transactions of ASME. Vol. 66, 1944.)

7

OBLIQUE-SHOCK
CHART ($\gamma = 1.4$)

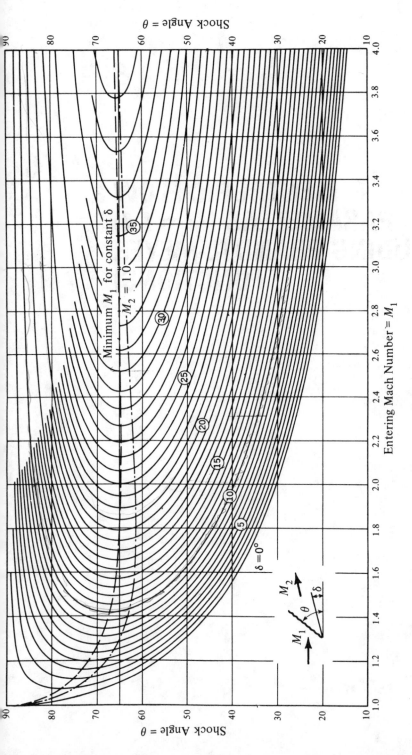

OBLIQUE-SHOCK CHART ($\gamma = 1.4$)

(Adapted with permission from *Gas Dynamics* by A.B. Cambel
and B.H. Jennings, McGraw-Hill, New York, 1958.)

8

ISENTROPIC FLOW PARAMETERS ($\gamma = 1.4$) (INCLUDING PRANDTL-MEYER FUNCTIONS)

M	p/p_t	T/T_t	A/A^*	pA/p_tA^*	ν	μ
0.0	1.00000	1.00000	∞	∞		
0.01	0.99993	0.99998	57.87384	57.86979		
0.02	0.99972	0.99992	28.94213	28.93403		
0.03	0.99937	0.99982	19.30054	19.28839		
0.04	0.99868	0.99968	14.48149	14.46528		
0.05	0.99825	0.99950	11.59144	11.57118		
0.06	0.99748	0.99928	9.66591	9.64159		
0.07	0.99658	0.99902	8.29153	8.26315		
0.08	0.99553	0.99872	7.26161	7.22917		
0.09	0.99435	0.99838	6.46134	6.42484		
0.10	0.99303	0.99800	5.82183	5.78126		
0.11	0.99158	0.99759	5.29923	5.25459		
0.12	0.98998	0.99713	4.86432	4.81560		
0.13	0.98826	0.99663	4.49686	4.44406		
0.14	0.98640	0.99610	4.18240	4.12552		
0.15	0.98441	0.99552	3.91034	3.84937		
0.16	0.98228	0.99491	3.67274	3.60767		
0.17	0.98003	0.99425	3.46351	3.39434		
0.18	0.97765	0.99356	3.27793	3.20465		
0.19	0.97514	0.99283	3.11226	3.03487		
0.20	0.97250	0.99206	2.96352	2.88201		
0.21	0.96973	0.99126	2.82929	2.74366		
0.22	0.96685	0.99041	2.70760	2.61783		
0.23	0.96383	0.98953	2.59681	2.50290		
0.24	0.96070	0.98861	2.49556	2.39750		
0.25	0.95745	0.98765	2.40271	2.30048		
0.26	0.95408	0.98666	2.31729	2.21089		
0.27	0.95060	0.98563	2.23847	2.12789		
0.28	0.94700	0.98456	2.16555	2.05078		
0.29	0.94329	0.98346	2.09793	1.97896		
0.30	0.93947	0.98232	2.03507	1.91188		
0.31	0.93554	0.98114	1.97651	1.84910		
0.32	0.93150	0.97993	1.92185	1.79021		
0.33	0.92736	0.97868	1.87074	1.73486		
0.34	0.92312	0.97740	1.82288	1.68273		
0.35	0.91877	0.97609	1.77797	1.63355		
0.36	0.91433	0.97473	1.73578	1.58707		
0.37	0.90979	0.97335	1.69609	1.54308		
0.38	0.90516	0.97193	1.65870	1.50138		
0.39	0.90043	0.97048	1.62343	1.46179		

M	p/p_t	T/T_t	A/A^*	pA/p_tA^*	ν	μ
0.40	0.89561	0.96899	1.59014	1.42415		
0.41	0.89071	0.96747	1.55867	1.38833		
0.42	0.88572	0.96592	1.52890	1.35419		
0.43	0.88065	0.96434	1.50072	1.32161		
0.44	0.87550	0.96272	1.47401	1.29049		
0.45	0.87027	0.96108	1.44867	1.26073		
0.46	0.86496	0.95940	1.42463	1.23225		
0.47	0.85958	0.95769	1.40180	1.20495		
0.48	0.85413	0.95595	1.38010	1.17878		
0.49	0.84861	0.95418	1.35947	1.15365		
0.50	0.84302	0.95238	1.33984	1.12951		
0.51	0.83737	0.95055	1.32117	1.10630		
0.52	0.83165	0.94869	1.30339	1.08397		
0.53	0.82588	0.94681	1.28645	1.06246		
0.54	0.82005	0.94489	1.27032	1.04173		
0.55	0.81417	0.94295	1.25495	1.02173		
0.56	0.80823	0.94098	1.24029	1.00244		
0.57	0.80224	0.93898	1.22633	0.98381		
0.58	0.79621	0.93696	1.21301	0.96580		
0.59	0.79013	0.93491	1.20031	0.94840		
0.60	0.78400	0.93284	1.18820	0.93155		
0.61	0.77784	0.93073	1.17665	0.91525		
0.62	0.77164	0.92861	1.16565	0.89946		
0.63	0.76540	0.92646	1.15515	0.88416		
0.64	0.75913	0.92428	1.14515	0.86932		
0.65	0.75283	0.92208	1.13562	0.85493		
0.66	0.74650	0.91986	1.12654	0.84096		
0.67	0.74014	0.91762	1.11789	0.82739		
0.68	0.73376	0.91535	1.10965	0.81422		
0.69	0.72735	0.91306	1.10182	0.80141		
0.70	0.72093	0.91075	1.09437	0.78896		
0.71	0.71448	0.90841	1.08729	0.77685		
0.72	0.70803	0.90606	1.08057	0.76507		
0.73	0.70155	0.90369	1.07419	0.75360		
0.74	0.69507	0.90129	1.06814	0.74243		
0.75	0.68857	0.89888	1.06242	0.73155		
0.76	0.68207	0.89644	1.05700	0.72095		
0.77	0.67556	0.89399	1.05188	0.71061		
0.78	0.66905	0.89152	1.04705	0.70053		
0.79	0.66254	0.88903	1.04251	0.69070		

M	p/p_t	T/T_t	A/A^*	pA/p_tA^*	ν	μ
0.80	0.65602	0.88652	1.03823	0.68110		
0.81	0.64951	0.88400	1.03422	0.67173		
0.82	0.64300	0.88146	1.03046	0.66259		
0.83	0.63650	0.87890	1.02696	0.65366		
0.84	0.63000	0.87633	1.02370	0.64493		
0.85	0.62351	0.87374	1.02067	0.63640		
0.86	0.61703	0.87114	1.01787	0.62806		
0.87	0.61057	0.86852	1.01530	0.61991		
0.88	0.60412	0.86589	1.01294	0.61193		
0.89	0.59768	0.86324	1.01080	0.60413		
0.90	0.59126	0.86059	1.00886	0.59650		
0.91	0.58486	0.85791	1.00713	0.58903		
0.92	0.57848	0.85523	1.00560	0.58171		
0.93	0.57211	0.85253	1.00426	0.57455		
0.94	0.56578	0.84982	1.00311	0.56753		
0.95	0.55946	0.84710	1.00215	0.56066		
0.96	0.55317	0.84437	1.00136	0.55392		
0.97	0.54691	0.84162	1.00076	0.54732		
0.98	0.54067	0.83887	1.00034	0.54085		
0.99	0.53446	0.83611	1.00008	0.53451		
1.00	0.52828	0.83333	1.00000	0.52828	0.0	90.0000
1.01	0.52213	0.83055	1.00008	0.52218	0.04472	81.9307
1.02	0.51602	0.82776	1.00033	0.51619	0.12569	78.6351
1.03	0.50994	0.82496	1.00074	0.51031	0.22943	76.1376
1.04	0.50389	0.82215	1.00131	0.50454	0.35098	74.0576
1.05	0.49787	0.81934	1.00203	0.49888	0.48741	72.2472
1.06	0.49189	0.81651	1.00291	0.49332	0.63669	70.6300
1.07	0.48595	0.81368	1.00394	0.48787	0.79729	69.1603
1.08	0.48005	0.81085	1.00512	0.48250	0.96804	67.8084
1.09	0.47418	0.80800	1.00645	0.47724	1.14795	66.5534
1.10	0.46835	0.80515	1.00793	0.47207	1.33620	65.3800
1.11	0.46257	0.80230	1.00955	0.46698	1.53210	64.2767
1.12	0.45682	0.79944	1.01131	0.46199	1.73504	63.2345
1.13	0.45111	0.79657	1.01322	0.45708	1.94448	62.2461
1.14	0.44545	0.79370	1.01527	0.45225	2.15996	61.3056
1.15	0.43983	0.79083	1.01745	0.44751	2.38104	60.4082
1.16	0.43425	0.78795	1.01978	0.44284	2.60735	59.5497
1.17	0.42872	0.78506	1.02224	0.43825	2.83852	58.7267
1.18	0.42322	0.78218	1.02484	0.43374	3.07426	57.9362
1.19	0.41778	0.77929	1.02757	0.42930	3.31425	57.1756

M	p/p_t	T/T_t	A/A^*	pA/p_tA^*	ν	μ
1.20	0.41238	0.77640	1.03044	0.42493	3.55823	56.4427
1.21	0.40702	0.77350	1.03344	0.42063	3.80596	55.7354
1.22	0.40171	0.77061	1.03657	0.41640	4.05720	55.0520
1.23	0.39645	0.76771	1.03983	0.41224	4.31173	54.3909
1.24	0.39123	0.76481	1.04323	0.40814	4.56936	53.7507
1.25	0.38606	0.76190	1.04675	0.40411	4.82989	53.1301
1.26	0.38093	0.75900	1.05041	0.40014	5.09315	52.5280
1.27	0.37586	0.75610	1.05419	0.39622	5.35897	51.9433
1.28	0.37083	0.75319	1.05810	0.39237	5.62720	51.3752
1.29	0.36585	0.75029	1.06214	0.38858	5.89768	50.8226
1.30	0.36091	0.74738	1.06630	0.38484	6.17029	50.2849
1.31	0.35603	0.74448	1.07060	0.38116	6.44488	49.7612
1.32	0.35119	0.74158	1.07502	0.37754	6.72133	49.2509
1.33	0.34640	0.73867	1.07957	0.37396	6.99953	48.7535
1.34	0.34166	0.73577	1.08424	0.37044	7.27937	48.2682
1.35	0.33697	0.73287	1.08904	0.36697	7.56072	47.7946
1.36	0.33233	0.72997	1.09396	0.36355	7.84351	47.3321
1.37	0.32773	0.72707	1.09902	0.36018	8.12762	46.8803
1.38	0.32319	0.72418	1.10419	0.35686	8.41297	46.4387
1.39	0.31869	0.72128	1.10950	0.35359	8.69946	46.0070
1.40	0.31424	0.71839	1.11493	0.35036	8.98702	45.5847
1.41	0.30984	0.71550	1.12048	0.34717	9.27556	45.1715
1.42	0.30549	0.71262	1.12616	0.34403	9.56502	44.7670
1.43	0.30118	0.70973	1.13197	0.34093	9.85531	44.3709
1.44	0.29693	0.70685	1.13790	0.33788	10.14636	43.9830
1.45	0.29272	0.70398	1.14396	0.33486	10.43811	43.6028
1.46	0.28856	0.70110	1.15015	0.33189	10.73050	43.2302
1.47	0.28445	0.69824	1.15646	0.32896	11.02346	42.8649
1.48	0.28039	0.69537	1.16290	0.32606	11.31694	42.5066
1.49	0.27637	0.69251	1.16947	0.32321	11.61087	42.1552
1.50	0.27240	0.68966	1.17617	0.32039	11.90521	41.8103
1.51	0.26848	0.68680	1.18299	0.31761	12.19990	41.4718
1.52	0.26461	0.68396	1.18994	0.31487	12.49489	41.1395
1.53	0.26078	0.68112	1.19702	0.31216	12.79014	40.8132
1.54	0.25700	0.67828	1.20423	0.30949	13.08559	40.4927
1.55	0.25326	0.67545	1.21157	0.30685	13.38121	40.1778
1.56	0.24957	0.67262	1.21904	0.30424	13.67696	39.8683
1.57	0.24593	0.66980	1.22664	0.30167	13.97278	39.5642
1.58	0.24233	0.66699	1.23438	0.29913	14.26865	39.2652
1.59	0.23878	0.66418	1.24224	0.29662	14.56452	38.9713

M	p/p_t	T/T_t	A/A^*	pA/p_tA^*	ν	μ
1.60	0.23527	0.66138	1.25023	0.29414	14.86035	38.6822
1.61	0.23181	0.65858	1.25836	0.29170	15.15612	38.3978
1.62	0.22839	0.65579	1.26663	0.28928	15.45180	38.1181
1.63	0.22501	0.65301	1.27502	0.28690	15.74733	37.8428
1.64	0.22168	0.65023	1.28355	0.28454	16.04271	37.5719
1.65	0.21839	0.64746	1.29222	0.28221	16.33789	37.3052
1.66	0.21515	0.64470	1.30102	0.27991	16.63284	37.0427
1.67	0.21195	0.64194	1.30996	0.27764	16.92755	36.7842
1.68	0.20879	0.63919	1.31904	0.27540	17.22198	36.5296
1.69	0.20567	0.63645	1.32825	0.27318	17.51611	36.2789
1.70	0.20259	0.63371	1.33761	0.27099	17.80991	36.0319
1.71	0.19956	0.63099	1.34710	0.26883	18.10336	35.7885
1.72	0.19656	0.62827	1.35674	0.26669	18.39643	35.5487
1.73	0.19361	0.62556	1.36651	0.26457	18.68911	35.3124
1.74	0.19070	0.62285	1.37643	0.26248	18.98137	35.0795
1.75	0.18782	0.62016	1.38649	0.26042	19.27319	34.8499
1.76	0.18499	0.61747	1.39670	0.25837	19.56456	34.6235
1.77	0.18219	0.61479	1.40705	0.25636	19.85544	34.4003
1.78	0.17944	0.61211	1.41755	0.25436	20.14584	34.1802
1.79	0.17672	0.60945	1.42819	0.25239	20.43571	33.9631
1.80	0.17404	0.60680	1.43898	0.25044	20.72506	33.7490
1.81	0.17140	0.60415	1.44992	0.24851	21.01387	33.5377
1.82	0.16879	0.60151	1.46101	0.24661	21.30211	33.3293
1.83	0.16622	0.59888	1.47225	0.24472	21.58977	33.1237
1.84	0.16369	0.59626	1.48365	0.24286	21.87685	32.9207
1.85	0.16119	0.59365	1.49519	0.24102	22.16332	32.7204
1.86	0.15873	0.59104	1.50689	0.23920	22.44917	32.5227
1.87	0.15631	0.58845	1.51875	0.23739	22.73439	32.3276
1.88	0.15392	0.58586	1.53076	0.23561	23.01896	32.1349
1.89	0.15156	0.58329	1.54293	0.23385	23.30288	31.9447
1.90	0.14924	0.58072	1.55526	0.23211	23.58613	31.7569
1.91	0.14695	0.57816	1.56774	0.23038	23.86871	31.5714
1.92	0.14470	0.57561	1.58039	0.22868	24.15059	31.3882
1.93	0.14247	0.57307	1.59320	0.22699	24.43178	31.2072
1.94	0.14028	0.57054	1.60617	0.22532	24.71226	31.0285
1.95	0.13813	0.56802	1.61931	0.22367	24.99202	30.8519
1.96	0.13600	0.56551	1.63261	0.22203	25.27105	30.6774
1.97	0.13390	0.56301	1.64608	0.22042	25.54935	30.5050
1.98	0.13184	0.56051	1.65972	0.21882	25.82691	30.3347
1.99	0.12981	0.55803	1.67352	0.21724	26.10371	30.1664

M	p/p_t	T/T_t	A/A^*	pA/p_tA^*	ν	μ
2.00	0.12780	0.55556	1.68750	0.21567	26.37976	30.0000
2.01	0.12583	0.55309	1.70165	0.21412	26.65504	29.8356
2.02	0.12389	0.55064	1.71597	0.21259	26.92955	29.6730
2.03	0.12197	0.54819	1.73047	0.21107	27.20328	29.5123
2.04	0.12009	0.54576	1.74514	0.20957	27.47622	29.3535
2.05	0.11823	0.54333	1.75999	0.20808	27.74837	29.1964
2.06	0.11640	0.54091	1.77502	0.20661	28.01973	29.0411
2.07	0.11460	0.53851	1.79022	0.20516	28.29028	28.8875
2.08	0.11282	0.53611	1.80561	0.20371	28.56003	28.7357
2.09	0.11107	0.53373	1.82119	0.20229	28.82896	28.5855
2.10	0.10935	0.53135	1.83694	0.20088	29.09708	28.4369
2.11	0.10766	0.52898	1.85289	0.19948	29.36438	28.2899
2.12	0.10599	0.52663	1.86902	0.19809	29.63085	28.1446
2.13	0.10434	0.52428	1.88533	0.19672	29.89649	28.0008
2.14	0.10273	0.52194	1.90184	0.19537	30.16130	27.8585
2.15	0.10113	0.51962	1.91854	0.19403	30.42527	27.7177
2.16	0.09956	0.51730	1.93544	0.19270	30.68841	27.5785
2.17	0.09802	0.51499	1.95252	0.19138	30.95070	27.4406
2.18	0.09649	0.51269	1.96981	0.19008	31.21215	27.3043
2.19	0.09500	0.51041	1.98729	0.18879	31.47275	27.1693
2.20	0.09352	0.50813	2.00497	0.18751	31.73250	27.0357
2.21	0.09207	0.50586	2.02286	0.18624	31.99139	26.9035
2.22	0.09064	0.50361	2.04094	0.18499	32.24943	26.7726
2.23	0.08923	0.50136	2.05923	0.18375	32.50662	26.6430
2.24	0.08785	0.49912	2.07773	0.18252	32.76294	26.5148
2.25	0.08648	0.49689	2.09644	0.18130	33.01841	26.3878
2.26	0.08514	0.49468	2.11535	0.18010	33.27301	26.2621
2.27	0.08382	0.49247	2.13447	0.17890	33.52676	26.1376
2.28	0.08251	0.49027	2.15381	0.17772	33.77963	26.0144
2.29	0.08123	0.48809	2.17336	0.17655	34.03165	25.8923
2.30	0.07997	0.48591	2.19313	0.17539	34.28279	25.7715
2.31	0.07873	0.48374	2.21312	0.17424	34.53307	25.6518
2.32	0.07751	0.48158	2.23332	0.17310	34.78249	25.5332
2.33	0.07631	0.47944	2.25375	0.17198	35.03103	25.4158
2.34	0.07512	0.47730	2.27440	0.17086	35.27871	25.2995
2.35	0.07396	0.47517	2.29528	0.16975	35.52552	25.1843
2.36	0.07281	0.47305	2.31638	0.16866	35.77146	25.0702
2.37	0.07168	0.47095	2.33771	0.16757	36.01653	24.9572
2.38	0.07057	0.46885	2.35928	0.16649	36.26073	24.8452
2.39	0.06948	0.46676	2.38107	0.16543	36.50406	24.7342

M	p/p_t	T/T_t	A/A^*	pA/p_tA^*	ν	μ
2.40	0.06840	0.46468	2.40310	0.15437	36.74653	24.6243
2.41	0.06734	0.46262	2.42537	0.16332	36.98813	24.5154
2.42	0.06630	0.46056	2.44787	0.16229	37.22886	24.4075
2.43	0.06527	0.45851	2.47061	0.16126	37.46872	24.3005
2.44	0.06426	0.45647	2.49360	0.16024	37.70772	24.1945
2.45	0.06327	0.45444	2.51683	0.15923	37.94585	24.0895
2.46	0.06229	0.45242	2.54031	0.15823	38.18312	23.9854
2.47	0.06133	0.45041	2.56403	0.15724	38.41952	23.8822
2.48	0.06038	0.44841	2.58801	0.15626	38.65507	23.7800
2.49	0.05945	0.44642	2.61224	0.15529	38.88974	23.6786
2.50	0.05853	0.44444	2.63672	0.15432	39.12356	23.5782
2.51	0.05762	0.44247	2.66146	0.15337	39.35652	23.4786
2.52	0.05674	0.44051	2.68645	0.15242	39.58862	23.3799
2.53	0.05586	0.43856	2.71171	0.15148	39.81987	23.2820
2.54	0.05500	0.43662	2.73723	0.15055	40.05026	23.1850
2.55	0.05415	0.43469	2.76301	0.14963	40.27979	23.0888
2.56	0.05332	0.43277	2.78906	0.14871	40.50847	22.9934
2.57	0.05250	0.43085	2.81538	0.14780	40.73630	22.8988
2.58	0.05169	0.42895	2.84197	0.14691	40.96329	22.8051
2.59	0.05090	0.42705	2.86884	0.14602	41.18942	22.7121
2.60	0.05012	0.42517	2.89598	0.14513	41.41471	22.6199
2.61	0.04935	0.42329	2.92339	0.14426	41.63915	22.5284
2.62	0.04859	0.42143	2.95109	0.14339	41.86275	22.4377
2.63	0.04784	0.41957	2.97907	0.14253	42.08551	22.3478
2.64	0.04711	0.41772	3.00733	0.14168	42.30744	22.2586
2.65	0.04639	0.41589	3.03588	0.14083	42.52852	22.1702
2.66	0.04568	0.41406	3.06472	0.13999	42.74877	22.0824
2.67	0.04498	0.41224	3.09385	0.13916	42.96819	21.9954
2.68	0.04429	0.41043	3.12327	0.13834	43.18678	21.9090
2.69	0.04362	0.40863	3.15299	0.13752	43.40454	21.8234
2.70	0.04295	0.40683	3.18301	0.13671	43.62148	21.7385
2.71	0.04229	0.40505	3.21333	0.13591	43.83759	21.6542
2.72	0.04165	0.40328	3.24395	0.13511	44.05288	21.5706
2.73	0.04102	0.40151	3.27488	0.13432	44.26735	21.4876
2.74	0.04039	0.39976	3.30611	0.13354	44.48100	21.4053
2.75	0.03978	0.39801	3.33766	0.13276	44.69384	21.3237
2.76	0.03917	0.39627	3.36952	0.13199	44.90586	21.2427
2.77	0.03858	0.39454	3.40169	0.13123	45.11708	21.1623
2.78	0.03799	0.39282	3.43418	0.13047	45.32749	21.0825
2.79	0.03742	0.39111	3.46699	0.12972	45.53709	21.0034

M	p/p_t	T/T_t	A/A^*	pA/p_tA^*	ν	μ
2.80	0.03685	0.38941	3.50012	0.12897	45.74589	20.9248
2.81	0.03629	0.38771	3.53358	0.12823	45.95389	20.8469
2.82	0.03574	0.38603	3.56737	0.12750	46.16109	20.7695
2.83	0.03520	0.38435	3.60148	0.12678	46.36750	20.6928
2.84	0.03467	0.38268	3.63593	0.12605	46.57312	20.6166
2.85	0.03415	0.38102	3.67072	0.12534	46.77794	20.5410
2.86	0.03363	0.37937	3.70584	0.12463	46.98198	20.4659
2.87	0.03312	0.37773	3.74131	0.12393	47.18523	20.3914
2.88	0.03263	0.37610	3.77711	0.12323	47.38770	20.3175
2.89	0.03213	0.37447	3.81327	0.12254	47.58940	20.2441
2.90	0.03165	0.37286	3.84977	0.12185	47.79031	20.1713
2.91	0.03118	0.37125	3.88662	0.12117	47.99045	20.0990
2.92	0.03071	0.36965	3.92383	0.12049	48.18982	20.0272
2.93	0.03025	0.36806	3.96139	0.11982	48.38842	19.9559
2.94	0.02980	0.36647	3.99932	0.11916	48.58626	19.8852
2.95	0.02935	0.36490	4.03760	0.11850	48.78333	19.8149
2.96	0.02891	0.36333	4.07625	0.11785	48.97965	19.7452
2.97	0.02848	0.36177	4.11527	0.11720	49.17520	19.6760
2.98	0.02805	0.36022	4.15466	0.11655	49.37000	19.6072
2.99	0.02764	0.35868	4.19443	0.11591	49.56405	19.5390
3.00	0.02722	0.35714	4.23457	0.11528	49.75735	19.4712
3.01	0.02682	0.35562	4.27509	0.11465	49.94990	19.4039
3.02	0.02642	0.35410	4.31599	0.11403	50.14171	19.3371
3.03	0.02603	0.35259	4.35728	0.11341	50.33277	19.2708
3.04	0.02564	0.35108	4.39895	0.11279	50.52310	19.2049
3.05	0.02526	0.34959	4.44102	0.11219	50.71270	19.1395
3.06	0.02489	0.34810	4.48347	0.11158	50.90156	19.0745
3.07	0.02452	0.34662	4.52633	0.11098	51.08969	19.0100
3.08	0.02416	0.34515	4.56959	0.11039	51.27710	18.9459
3.09	0.02380	0.34369	4.61325	0.10979	51.46378	18.8823
3.10	0.02345	0.34223	4.65731	0.10921	51.64974	18.8191
3.11	0.02310	0.34078	4.70178	0.10863	51.83499	18.7563
3.12	0.02276	0.33934	4.74667	0.10805	52.01952	18.6939
3.13	0.02243	0.33791	4.79197	0.10748	52.20333	18.6320
3.14	0.02210	0.33648	4.83769	0.10691	52.38644	18.5705
3.15	0.02177	0.33506	4.88383	0.10634	52.56884	18.5094
3.16	0.02146	0.33365	4.93039	0.10578	52.75053	18.4487
3.17	0.02114	0.33225	4.97739	0.10523	52.93153	18.3884
3.18	0.02083	0.33085	5.02481	0.10468	53.11182	18.3285
3.19	0.02053	0.32947	5.07266	0.10413	53.29143	18.2691

M	p/p_t	T/T_t	A/A^*	pA/p_tA^*	ν	μ
3.20	0.02023	0.32808	5.12096	0.10359	53.47033	18.2100
3.21	0.01993	0.32671	5.16969	0.10305	53.64855	18.1512
3.22	0.01964	0.32534	5.21887	0.10251	53.82609	18.0929
3.23	0.01936	0.32398	5.26849	0.10198	54.00294	18.0350
3.24	0.01908	0.32263	5.31857	0.10145	54.17910	17.9774
3.25	0.01880	0.32129	5.36909	0.10093	54.35459	17.9202
3.26	0.01853	0.31995	5.42008	0.10041	54.52941	17.8634
3.27	0.01826	0.31862	5.47152	0.09989	54.70355	17.8069
3.28	0.01799	0.31729	5.52343	0.09938	54.87703	17.7508
3.29	0.01773	0.31597	5.57580	0.09887	55.04983	17.6951
3.30	0.01748	0.31466	5.62865	0.09837	55.22198	17.6397
3.31	0.01722	0.31336	5.68196	0.09787	55.39346	17.5847
3.32	0.01698	0.31206	5.73576	0.09737	55.56428	17.5300
3.33	0.01673	0.31077	5.79003	0.09688	55.73445	17.4756
3.34	0.01649	0.30949	5.84479	0.09639	55.90396	17.4216
3.35	0.01625	0.30821	5.90004	0.09590	56.07283	17.3680
3.36	0.01602	0.30694	5.95577	0.09542	56.24105	17.3147
3.37	0.01579	0.30568	6.01201	0.09494	56.40862	17.2617
3.38	0.01557	0.30443	6.06873	0.09447	56.57556	17.2090
3.39	0.01534	0.30318	6.12596	0.09399	56.74185	17.1567
3.40	0.01512	0.30193	6.18370	0.09353	56.90751	17.1046
3.41	0.01491	0.30070	6.24194	0.09306	57.07254	17.0529
3.42	0.01470	0.29947	6.30070	0.09260	57.23694	17.0016
3.43	0.01449	0.29824	6.35997	0.09214	57.40071	16.9505
3.44	0.01428	0.29702	6.41976	0.09168	57.56385	16.8997
3.45	0.01408	0.29581	6.48007	0.09123	57.72637	16.8493
3.46	0.01388	0.29461	6.54092	0.09078	57.88828	16.7991
3.47	0.01368	0.29341	6.60229	0.09034	58.04957	16.7493
3.48	0.01349	0.29222	6.66419	0.08989	58.21024	16.6997
3.49	0.01330	0.29103	6.72664	0.08945	58.37030	16.6505
3.50	0.01311	0.28986	6.78962	0.08902	58.52976	16.6015
3.51	0.01293	0.28868	6.85315	0.08858	58.68861	16.5529
3.52	0.01274	0.28751	6.91723	0.08815	58.84685	16.5045
3.53	0.01256	0.28635	6.98186	0.08773	59.00450	16.4564
3.54	0.01239	0.28520	7.04705	0.08730	59.16155	16.4086
3.55	0.01221	0.28405	7.11281	0.08688	59.31801	16.3611
3.56	0.01204	0.28291	7.17912	0.08646	59.47387	16.3139
3.57	0.01188	0.28177	7.24601	0.08605	59.62914	16.2669
3.58	0.01171	0.28064	7.31346	0.08563	59.78383	16.2202
3.59	0.01155	0.27952	7.38150	0.08522	59.93793	16.1738

M	p/p_t	T/T_t	A/A^*	pA/p_tA^*	ν	μ
3.60	0.01138	0.27840	7.45011	0.08482	60.09146	16.1276
3.61	0.01123	0.27728	7.51931	0.08441	60.24440	16.0817
3.62	0.01107	0.27618	7.58910	0.08401	60.39677	16.0361
3.63	0.01092	0.27507	7.65948	0.08361	60.54856	15.9907
3.64	0.01076	0.27398	7.73045	0.08322	60.69978	15.9456
3.65	0.01062	0.27289	7.80203	0.08282	60.85044	15.9008
3.66	0.01047	0.27180	7.87421	0.08243	61.00052	15.8562
3.67	0.01032	0.27073	7.94700	0.08205	61.15005	15.8119
3.68	0.01018	0.26965	8.02040	0.08166	61.29902	15.7678
3.69	0.01004	0.26858	8.09442	0.08128	61.44742	15.7239
3.70	0.00990	0.26752	8.16907	0.08090	61.59527	15.6803
3.71	0.00977	0.26647	8.24433	0.08052	61.74257	15.6370
3.72	0.00963	0.26542	8.32023	0.08014	61.88932	15.5939
3.73	0.00950	0.26437	8.39676	0.07977	62.03552	15.5510
3.74	0.00937	0.26333	8.47393	0.07940	62.18118	15.5084
3.75	0.00924	0.26230	8.55174	0.07904	62.32629	15.4660
3.76	0.00912	0.26127	8.63020	0.07867	62.47086	15.4239
3.77	0.00899	0.26024	8.70931	0.07831	62.61490	15.3819
3.78	0.00887	0.25922	8.78907	0.07795	62.75840	15.3402
3.79	0.00875	0.25821	8.86950	0.07759	62.90136	15.2988
3.80	0.00863	0.25720	8.95059	0.07723	63.04380	15.2575
3.81	0.00851	0.25620	9.03234	0.07688	63.18571	15.2165
3.82	0.00840	0.25520	9.11477	0.07653	63.32709	15.1757
3.83	0.00828	0.25421	9.19788	0.07618	63.46795	15.1351
3.84	0.00817	0.25322	9.28167	0.07584	63.60829	15.0948
3.85	0.00806	0.25224	9.36614	0.07549	63.74811	15.0547
3.86	0.00795	0.25126	9.45131	0.07515	63.88741	15.0147
3.87	0.00784	0.25029	9.53717	0.07481	64.02620	14.9750
3.88	0.00774	0.24932	9.62373	0.07447	64.16448	14.9355
3.89	0.00763	0.24836	9.71100	0.07414	64.30225	14.8962
3.90	0.00753	0.24740	9.79897	0.07381	64.43952	14.8572
3.91	0.00743	0.24645	9.88766	0.07348	64.57628	14.8183
3.92	0.00733	0.24550	9.97707	0.07315	64.71254	14.7796
3.93	0.00723	0.24456	10.06720	0.07282	64.84829	14.7412
3.94	0.00714	0.24362	10.15806	0.07250	64.98356	14.7029
3.95	0.00704	0.24269	10.24965	0.07217	65.11832	14.6649
3.96	0.00695	0.24176	10.34197	0.07185	65.25260	14.6270
3.97	0.00686	0.24084	10.43504	0.07154	65.38638	14.5893
3.98	0.00676	0.23992	10.52886	0.07122	65.51968	14.5519
3.99	0.00667	0.23900	10.62343	0.07091	65.65249	14.5146

M	p/p_t	T/T_t	A/A^*	pA/p_tA^*	ν	μ
4.00	0.00059	0.23810	10.71875	0.07059	65.78482	14.4775
4.10	0.00577	0.22925	11.71465	0.06758	67.08200	14.1170
4.20	0.00506	0.22085	12.79164	0.06475	68.33324	13.7741
4.30	0.00445	0.21286	13.95490	0.06209	69.54063	13.4477
4.40	0.00392	0.20525	15.20987	0.05959	70.70616	13.1366
4.50	0.00346	0.19802	16.56219	0.05723	71.83174	12.8396
4.60	0.00305	0.19113	18.01779	0.05500	72.91915	12.5559
4.70	0.00270	0.18457	19.58283	0.05290	73.97012	12.2845
4.80	0.00239	0.17832	21.26371	0.05091	74.98627	12.0247
4.90	0.00213	0.17235	23.06712	0.04903	75.96915	11.7757
5.00	0.00189	0.16667	25.00000	0.04725	76.92021	11.5370
5.10	0.00168	0.16124	27.06957	0.04556	77.84087	11.3077
5.20	0.00150	0.15605	29.28333	0.04396	78.73243	11.0875
5.30	0.00134	0.15110	31.64905	0.04244	79.59616	10.8757
5.40	0.00120	0.14637	34.17481	0.04100	80.43323	10.6719
5.50	0.00107	0.14184	36.86896	0.03963	81.24479	10.4757
5.60	0.000964	0.13751	39.74018	0.03832	82.03190	10.2866
5.70	0.000866	0.13337	42.79743	0.03708	82.79558	10.1042
5.80	0.000779	0.12940	46.05000	0.03589	83.53681	9.9282
5.90	0.000702	0.12560	49.50747	0.03476	84.25649	9.7583
6.00	0.000633	0.12195	53.17978	0.03368	84.95550	9.5941
6.10	0.000572	0.11846	57.07718	0.03265	85.63467	9.4353
6.20	0.000517	0.11510	61.21023	0.03167	86.29479	9.2818
6.30	0.000468	0.11188	65.58987	0.03073	86.93661	9.1332
6.40	0.000425	0.10879	70.22736	0.02982	87.56084	8.9893
6.50	0.000385	0.10582	75.13431	0.02896	88.16816	8.8499
6.60	0.000350	0.10297	80.32271	0.02814	88.75922	8.7147
6.70	0.000319	0.10022	85.80487	0.02734	89.33463	8.5837
6.80	0.000290	0.09758	91.59351	0.02658	89.89499	8.4565
6.90	0.000265	0.09504	97.70169	0.02586	90.44084	8.3331
7.00	0.000242	0.09259	104.14286	0.02516	90.97273	8.2132
7.50	0.000155	0.08163	141.84148	0.02205	93.43967	7.6623
8.00	0.000102	0.07246	190.10937	0.01947	95.62467	7.1808
8.50	0.0000690	0.06472	251.086167	0.01732	97.57220	6.7563
9.00	0.0000474	0.05814	327.189300	0.01550	99.31810	6.3794
9.50	0.0000331	0.05249	421.131373	0.01396	100.89148	6.0423
10.00	0.0000236	0.04762	535.937500	0.01263	102.31625	5.7392
∞	0.0	0.0	∞	0.0	130.4541	0.0

9

NORMAL-SHOCK
PARAMETERS ($\gamma = 1.4$)

M_1	M_2	p_2/p_1	T_2/T_1	$\Delta V/a_1$	p_{t2}/p_{t1}	p_{t2}/p_1
1.00	1.00000	1.00000	1.00000	0.0	1.00000	1.89293
1.01	0.99013	1.02345	1.00664	0.01658	1.00000	1.91521
1.02	0.98052	1.04713	1.01325	0.03301	0.99999	1.93790
1.03	0.97115	1.07105	1.01981	0.04927	0.99997	1.96097
1.04	0.96203	1.09520	1.02634	0.06538	0.99992	1.98442
1.05	0.95313	1.11958	1.03284	0.08135	0.99985	2.00825
1.06	0.94445	1.14420	1.03931	0.09717	0.99975	2.03245
1.07	0.93598	1.16905	1.04575	0.11285	0.99961	2.05702
1.08	0.92771	1.19413	1.05217	0.12840	0.99943	2.08194
1.09	0.91965	1.21945	1.05856	0.14381	0.99920	2.10722
1.10	0.91177	1.24500	1.06494	0.15909	0.99893	2.13285
1.11	0.90408	1.27078	1.07129	0.17425	0.99860	2.15882
1.12	0.89656	1.29680	1.07763	0.18929	0.99821	2.18513
1.13	0.88922	1.32305	1.08396	0.20420	0.99777	2.21178
1.14	0.88204	1.34953	1.09027	0.21901	0.99726	2.23877
1.15	0.87502	1.37625	1.09658	0.23370	0.99669	2.26608
1.16	0.86816	1.40320	1.10287	0.24828	0.99605	2.29372
1.17	0.86145	1.43038	1.10916	0.26275	0.99535	2.32169
1.18	0.85488	1.45780	1.11544	0.27712	0.99457	2.34998
1.19	0.84846	1.48545	1.12172	0.29139	0.99372	2.37858
1.20	0.84217	1.51333	1.12799	0.30556	0.99280	2.40750
1.21	0.83601	1.54145	1.13427	0.31963	0.99180	2.43674
1.22	0.82999	1.56980	1.14054	0.33361	0.99073	2.46628
1.23	0.82408	1.59838	1.14682	0.34749	0.98958	2.49613
1.24	0.81830	1.62720	1.15309	0.36129	0.98836	2.52629
1.25	0.81264	1.65625	1.15937	0.37500	0.98706	2.55676
1.26	0.80709	1.68553	1.16566	0.38862	0.98568	2.58753
1.27	0.80164	1.71505	1.17195	0.40217	0.98422	2.61860
1.28	0.79631	1.74480	1.17825	0.41562	0.98268	2.64996
1.29	0.79108	1.77478	1.18456	0.42901	0.98107	2.68163
1.30	0.78596	1.80500	1.19087	0.44231	0.97937	2.71359
1.31	0.78093	1.83545	1.19720	0.45553	0.97760	2.74585
1.32	0.77600	1.86613	1.20353	0.46869	0.97575	2.77840
1.33	0.77116	1.89705	1.20988	0.48177	0.97382	2.81125
1.34	0.76641	1.92820	1.21624	0.49478	0.97182	2.84438
1.35	0.76175	1.95958	1.22261	0.50772	0.96974	2.87781
1.36	0.75718	1.99120	1.22900	0.52059	0.96758	2.91152
1.37	0.75269	2.02305	1.23540	0.53339	0.96534	2.94552
1.38	0.74829	2.05513	1.24181	0.54614	0.96304	2.97981
1.39	0.74396	2.08745	1.24825	0.55881	0.96065	3.01438

M_1	M_2	p_2/p_1	T_2/T_1	$\Delta V/a_1$	p_{t2}/p_{t1}	p_{t2}/p_1
1.40	0.73971	2.12000	1.25469	0.57143	0.95819	3.04924
1.41	0.73554	2.15278	1.26116	0.58398	0.95566	3.08438
1.42	0.73144	2.18580	1.26764	0.59648	0.95306	3.11980
1.43	0.72741	2.21905	1.27414	0.60892	0.95039	3.15551
1.44	0.72345	2.25253	1.28066	0.62130	0.94765	3.19149
1.45	0.71956	2.28625	1.28720	0.63362	0.94484	3.22776
1.46	0.71574	2.32020	1.29377	0.64589	0.94196	3.26431
1.47	0.71198	2.35438	1.30035	0.65811	0.93901	3.30113
1.48	0.70829	2.38880	1.30695	0.67027	0.93600	3.33823
1.49	0.70466	2.42345	1.31357	0.68238	0.93293	3.37562
1.50	0.70109	2.45833	1.32022	0.69444	0.92979	3.41327
1.51	0.69758	2.49345	1.32688	0.70646	0.92659	3.45121
1.52	0.69413	2.52880	1.33357	0.71842	0.92332	3.48942
1.53	0.69073	2.56438	1.34029	0.73034	0.92000	3.52791
1.54	0.68739	2.60020	1.34703	0.74221	0.91662	3.56667
1.55	0.68410	2.63625	1.35379	0.75403	0.91319	3.60570
1.56	0.68087	2.67253	1.36057	0.76581	0.90970	3.64501
1.57	0.67768	2.70905	1.36738	0.77755	0.90615	3.68459
1.58	0.67455	2.74580	1.37422	0.78924	0.90255	3.72445
1.59	0.67147	2.78278	1.38108	0.80089	0.89890	3.76457
1.60	0.66844	2.82000	1.38797	0.81250	0.89520	3.80497
1.61	0.66545	2.85745	1.39488	0.82407	0.89145	3.84564
1.62	0.66251	2.89513	1.40182	0.83560	0.88765	3.88658
1.63	0.65962	2.93305	1.40879	0.84709	0.88381	3.92780
1.64	0.65677	2.97120	1.41578	0.85854	0.87992	3.96928
1.65	0.65396	3.00958	1.42280	0.86995	0.87599	4.01103
1.66	0.65119	3.04820	1.42985	0.88133	0.87201	4.05305
1.67	0.64847	3.08705	1.43693	0.89266	0.86800	4.09535
1.68	0.64579	3.12613	1.44403	0.90397	0.86394	4.13791
1.69	0.64315	3.16545	1.45117	0.91524	0.85985	4.18074
1.70	0.64054	3.20500	1.45833	0.92647	0.85572	4.22383
1.71	0.63798	3.24478	1.46552	0.93767	0.85156	4.26720
1.72	0.63545	3.28480	1.47274	0.94884	0.84736	4.31083
1.73	0.63296	3.32505	1.47999	0.95997	0.84312	4.35473
1.74	0.63051	3.36553	1.48727	0.97107	0.83886	4.39890
1.75	0.62809	3.40625	1.49458	0.98214	0.83457	4.44334
1.76	0.62570	3.44720	1.50192	0.99318	0.83024	4.48804
1.77	0.62335	3.48838	1.50929	1.00419	0.82589	4.53301
1.78	0.62104	3.52980	1.51669	1.01517	0.82151	4.57825
1.79	0.61875	3.57145	1.52412	1.02612	0.81711	4.62375

M_1	M_2	p_2/p_1	T_2/T_1	$\Delta V/a_1$	p_{t2}/p_{t1}	p_{t2}/p_1
1.80	0.61650	3.61333	1.53158	1.03704	0.81268	4.66952
1.81	0.61428	3.65545	1.53907	1.04793	0.80823	4.71555
1.82	0.61209	3.69780	1.54659	1.05879	0.80376	4.76185
1.83	0.60993	3.74038	1.55415	1.06963	0.79927	4.80841
1.84	0.60780	3.78320	1.56173	1.08043	0.79476	4.85524
1.85	0.60570	3.82625	1.56935	1.09122	0.79023	4.90234
1.86	0.60363	3.86953	1.57700	1.10197	0.78569	4.94970
1.87	0.60158	3.91305	1.58468	1.11270	0.78112	4.99732
1.88	0.59957	3.95680	1.59239	1.12340	0.77655	5.04521
1.89	0.59758	4.00078	1.60014	1.13408	0.77196	5.09336
1.90	0.59562	4.04500	1.60792	1.14474	0.76736	5.14178
1.91	0.59368	4.08945	1.61573	1.15537	0.76274	5.19046
1.92	0.59177	4.13413	1.62357	1.16597	0.75812	5.23940
1.93	0.58988	4.17905	1.63144	1.17655	0.75349	5.28861
1.94	0.58802	4.22420	1.63935	1.18711	0.74884	5.33808
1.95	0.58618	4.26958	1.64729	1.19765	0.74420	5.38782
1.96	0.58437	4.31520	1.65527	1.20816	0.73954	5.43782
1.97	0.58258	4.36105	1.66328	1.21865	0.73488	5.48808
1.98	0.58082	4.40713	1.67132	1.22912	0.73021	5.53860
1.99	0.57907	4.45345	1.67939	1.23957	0.72555	5.58939
2.00	0.57735	4.50000	1.68750	1.25000	0.72087	5.64044
2.01	0.57565	4.54678	1.69564	1.26041	0.71620	5.69175
2.02	0.57397	4.59380	1.70382	1.27079	0.71153	5.74333
2.03	0.57231	4.64105	1.71203	1.28116	0.70685	5.79517
2.04	0.57068	4.63853	1.72027	1.29150	0.70218	5.84727
2.05	0.56906	4.73625	1.72855	1.30183	0.69751	5.89963
2.06	0.56747	4.78420	1.73686	1.31214	0.69284	5.95226
2.07	0.56589	4.83238	1.74521	1.32242	0.68817	6.00514
2.08	0.56433	4.88080	1.75359	1.33269	0.68351	6.05829
2.09	0.56280	4.92945	1.76200	1.34294	0.67885	6.11170
2.10	0.56128	4.97833	1.77045	1.35317	0.67420	6.16537
2.11	0.55978	5.02745	1.77893	1.36339	0.66956	6.21931
2.12	0.55829	5.07680	1.78745	1.37358	0.66492	6.27351
2.13	0.55683	5.12638	1.79601	1.38376	0.66029	6.32796
2.14	0.55538	5.17620	1.80459	1.39393	0.65567	6.38268
2.15	0.55395	5.22625	1.81322	1.40407	0.65105	6.43766
2.16	0.55254	5.27653	1.82188	1.41420	0.64645	6.49290
2.17	0.55115	5.32705	1.83057	1.42431	0.64185	6.54841
2.18	0.54977	5.37780	1.83930	1.43440	0.63727	6.60417
2.19	0.54840	5.42878	1.84806	1.44448	0.63270	6.66019

M_1	M_2	p_2/p_1	T_2/T_1	$\Delta V/a_1$	p_{t2}/p_{t1}	p_{t2}/p_1
2.20	0.54706	5.48000	1.85686	1.45455	0.62814	6.71648
2.21	0.54572	5.53145	1.86569	1.46459	0.62359	6.77303
2.22	0.54441	5.58313	1.87456	1.47462	0.61905	6.82983
2.23	0.54311	5.63505	1.88347	1.48464	0.61453	6.88690
2.24	0.54182	5.68720	1.89241	1.49464	0.61002	6.94423
2.25	0.54055	5.73958	1.90138	1.50463	0.60553	7.00182
2.26	0.53930	5.79220	1.91040	1.51460	0.60105	7.05967
2.27	0.53805	5.84505	1.91944	1.52456	0.59659	7.11778
2.28	0.53683	5.89813	1.92853	1.53450	0.59214	7.17616
2.29	0.53561	5.95145	1.93765	1.54443	0.58771	7.23479
2.30	0.53441	6.00500	1.94680	1.55435	0.58329	7.29368
2.31	0.53322	6.05878	1.95599	1.56425	0.57890	7.35283
2.32	0.53205	6.11280	1.96522	1.57414	0.57452	7.41225
2.33	0.53089	6.16705	1.97448	1.58401	0.57015	7.47192
2.34	0.52974	6.22153	1.98378	1.59387	0.56581	7.53185
2.35	0.52861	6.27625	1.99311	1.60372	0.56148	7.59205
2.36	0.52749	6.33120	2.00249	1.61356	0.55718	7.65250
2.37	0.52638	6.38638	2.01189	1.62338	0.55289	7.71321
2.38	0.52528	6.44180	2.02134	1.63319	0.54862	7.77419
2.39	0.52419	6.49745	2.03082	1.64299	0.54437	7.83542
2.40	0.52312	6.55333	2.04033	1.65278	0.54014	7.89691
2.41	0.52206	6.60945	2.04988	1.66255	0.53594	7.95867
2.42	0.52100	6.66580	2.05947	1.67231	0.53175	8.02068
2.43	0.51996	6.72238	2.06910	1.68206	0.52758	8.08295
2.44	0.51894	6.77920	2.07876	1.69180	0.52344	8.14549
2.45	0.51792	6.83625	2.08846	1.70153	0.51931	8.20828
2.46	0.51691	6.89353	2.09819	1.71125	0.51521	8.27133
2.47	0.51592	6.95105	2.10797	1.72095	0.51113	8.33464
2.48	0.51493	7.00880	2.11777	1.73065	0.50707	8.39821
2.49	0.51395	7.06678	2.12762	1.74033	0.50303	8.46205
2.50	0.51299	7.12500	2.13750	1.75000	0.49901	8.52614
2.51	0.51203	7.18345	2.14742	1.75966	0.49502	8.59049
2.52	0.51109	7.24213	2.15737	1.76931	0.49105	8.65510
2.53	0.51015	7.30105	2.16737	1.77895	0.48711	8.71996
2.54	0.50923	7.36020	2.17739	1.78858	0.48318	8.78509
2.55	0.50831	7.41958	2.18746	1.79820	0.47928	8.85048
2.56	0.50741	7.47920	2.19756	1.80781	0.47540	8.91613
2.57	0.50651	7.53905	2.20770	1.81741	0.47155	8.98203
2.58	0.50562	7.59913	2.21788	1.82700	0.46772	9.04820
2.59	0.50474	7.65945	2.22809	1.83658	0.46391	9.11462

M_1	M_2	p_2/p_1	T_2/T_1	$\Delta V/a_1$	p_{t2}/p_{t1}	p_{t2}/p_1
2.60	0.50387	7.72000	2.23834	1.84615	0.46012	9.18131
2.61	0.50301	7.78078	2.24863	1.85572	0.45636	9.24825
2.62	0.50216	7.84180	2.25896	1.86527	0.45263	9.31545
2.63	0.50131	7.90305	2.26932	1.87481	0.44891	9.38291
2.64	0.50048	7.96453	2.27972	1.88434	0.44522	9.45064
2.65	0.49965	8.02625	2.29015	1.89387	0.44156	9.51862
2.66	0.49883	8.08820	2.30063	1.90338	0.43792	9.58685
2.67	0.49802	8.15038	2.31114	1.91289	0.43430	9.65535
2.68	0.49722	8.21280	2.32168	1.92239	0.43070	9.72411
2.69	0.49642	8.27545	2.33227	1.93188	0.42714	9.79312
2.70	0.49563	8.33833	2.34289	1.94136	0.42359	9.86240
2.71	0.49485	8.40145	2.35355	1.95083	0.42007	9.93193
2.72	0.49408	8.46480	2.36425	1.96029	0.41657	10.00173
2.73	0.49332	8.52838	2.37498	1.96975	0.41310	10.07178
2.74	0.49256	8.59220	2.38576	1.97920	0.40965	10.14209
2.75	0.49181	8.65625	2.39657	1.98864	0.40623	10.21266
2.76	0.49107	8.72053	2.40741	1.99807	0.40283	10.28349
2.77	0.49033	8.78505	2.41830	2.00749	0.39945	10.35457
2.78	0.48960	8.84980	2.42922	2.01691	0.39610	10.42592
2.79	0.48888	8.91478	2.44018	2.02631	0.39277	10.49752
2.80	0.48817	8.98000	2.45117	2.03571	0.38946	10.56939
2.81	0.48746	9.04545	2.46221	2.04511	0.38618	10.64151
2.82	0.48676	9.11113	2.47328	2.05449	0.38293	10.71389
2.83	0.48606	9.17705	2.48439	2.06387	0.37969	10.78653
2.84	0.48538	9.24320	2.49554	2.07324	0.37649	10.85943
2.85	0.48469	9.30958	2.50672	2.08260	0.37330	10.93258
2.86	0.48402	9.37620	2.51794	2.09196	0.37014	11.00600
2.87	0.48335	9.44305	2.52920	2.10131	0.36700	11.07967
2.88	0.48269	9.51013	2.54050	2.11065	0.36389	11.15361
2.89	0.48203	9.57745	2.55183	2.11998	0.36080	11.22780
2.90	0.48138	9.64500	2.56321	2.12931	0.35773	11.30225
2.91	0.48073	9.71278	2.57462	2.13863	0.35469	11.37695
2.92	0.48010	9.78080	2.58607	2.14795	0.35167	11.45192
2.93	0.47946	9.84905	2.59755	2.15725	0.34867	11.52715
2.94	0.47884	9.91753	2.60908	2.16655	0.34570	11.60263
2.95	0.47821	9.98625	2.62064	2.17585	0.34275	11.67837
2.96	0.47760	10.05520	2.63224	2.18514	0.33982	11.75438
2.97	0.47699	10.12438	2.64387	2.19442	0.33692	11.83064
2.98	0.47638	10.19380	2.65555	2.20369	0.33404	11.90715
2.99	0.47578	10.26345	2.66726	2.21296	0.33118	11.98393

M_1	M_2	p_2/p_1	T_2/T_1	$\Delta V/a_1$	p_{t2}/p_{t1}	p_{t2}/p_1
3.00	0.47519	10.33333	2.67901	2.22222	0.32834	12.06096
3.01	0.47460	10.40345	2.69080	2.23148	0.32553	12.13826
3.02	0.47402	10.47380	2.70263	2.24073	0.32274	12.21581
3.03	0.47344	10.54438	2.71449	2.24997	0.31997	12.29362
3.04	0.47287	10.61520	2.72639	2.25921	0.31723	12.37169
3.05	0.47230	10.68625	2.73833	2.26844	0.31450	12.45002
3.06	0.47174	10.75753	2.75031	2.27767	0.31180	12.52860
3.07	0.47118	10.82905	2.76233	2.28689	0.30912	12.60745
3.08	0.47063	10.90080	2.77438	2.29610	0.30646	12.68655
3.09	0.47008	10.97278	2.78647	2.30531	0.30383	12.76591
3.10	0.46953	11.04500	2.79860	2.31452	0.30121	12.84553
3.11	0.46899	11.11745	2.81077	2.32371	0.29862	12.92540
3.12	0.46846	11.19013	2.82298	2.33291	0.29605	13.00554
3.13	0.46793	11.26305	2.83522	2.34209	0.29350	13.08593
3.14	0.46741	11.33620	2.84750	2.35127	0.29097	13.16659
3.15	0.46689	11.40958	2.85982	2.36045	0.28846	13.24750
3.16	0.46637	11.48320	2.87218	2.36962	0.28597	13.32866
3.17	0.46586	11.55705	2.88458	2.37879	0.28350	13.41009
3.18	0.46535	11.63113	2.89701	2.38795	0.28106	13.49178
3.19	0.46485	11.70545	2.90948	2.39710	0.27863	13.57372
3.20	0.46435	11.78000	2.92199	2.40625	0.27623	13.65592
3.21	0.46385	11.85478	2.93454	2.41539	0.27384	13.73838
3.22	0.46336	11.92980	2.94713	2.42453	0.27148	13.82110
3.23	0.46288	12.00505	2.95975	2.43367	0.26914	13.90407
3.24	0.46240	12.08053	2.97241	2.44280	0.26681	13.98731
3.25	0.46192	12.15625	2.98511	2.45192	0.26451	14.07080
3.26	0.46144	12.23220	2.99785	2.46104	0.26222	14.15455
3.27	0.46097	12.30838	3.01063	2.47016	0.25996	14.23856
3.28	0.46051	12.38480	3.02345	2.47927	0.25771	14.32283
3.29	0.46004	12.46145	3.03630	2.48837	0.25548	14.40735
3.30	0.45959	12.53833	3.04919	2.49747	0.25328	14.49214
3.31	0.45913	12.61545	3.06212	2.50657	0.25109	14.57718
3.32	0.45868	12.69280	3.07509	2.51566	0.24892	14.66248
3.33	0.45823	12.77038	3.08809	2.52475	0.24677	14.74804
3.34	0.45779	12.84820	3.10114	2.53383	0.24463	14.83385
3.35	0.45735	12.92625	3.11422	2.54291	0.24252	14.91992
3.36	0.45691	13.00453	3.12734	2.55198	0.24043	15.00626
3.37	0.45648	13.08305	3.14050	2.56105	0.23835	15.09285
3.38	0.45605	13.16180	3.15370	2.57012	0.23629	15.17969
3.39	0.45562	13.24078	3.16693	2.57918	0.23425	15.26680

M_1	M_2	p_2/p_1	T_2/T_1	$\Delta V/a_1$	p_{t2}/p_{t1}	p_{t2}/p_1
3.40	0.45520	13.32000	3.18021	2.58824	0.23223	15.35417
3.41	0.45478	13.39945	3.19352	2.59729	0.23022	15.44179
3.42	0.45436	13.47913	3.20687	2.60634	0.22823	15.52967
3.43	0.45395	13.55905	3.22026	2.61538	0.22626	15.61781
3.44	0.45354	13.63920	3.23369	2.62442	0.22431	15.70620
3.45	0.45314	13.71958	3.24715	2.63345	0.22237	15.79486
3.46	0.45273	13.80020	3.26065	2.64249	0.22045	15.88377
3.47	0.45233	13.88105	3.27420	2.65151	0.21855	15.97294
3.48	0.45194	13.96213	3.28778	2.66054	0.21667	16.06237
3.49	0.45154	14.04345	3.30139	2.66956	0.21480	16.15206
3.50	0.45115	14.12500	3.31505	2.67857	0.21295	16.24200
3.51	0.45077	14.20678	3.32875	2.68758	0.21111	16.33220
3.52	0.45038	14.28880	3.34248	2.69659	0.20929	16.42266
3.53	0.45000	14.37105	3.35625	2.70559	0.20749	16.51338
3.54	0.44962	14.45353	3.37006	2.71460	0.20570	16.60436
3.55	0.44925	14.53625	3.38391	2.72359	0.20393	16.69559
3.56	0.44887	14.61920	3.39780	2.73258	0.20218	16.78709
3.57	0.44850	14.70238	3.41172	2.74157	0.20044	16.87884
3.58	0.44814	14.78580	3.42569	2.75056	0.19871	16.97085
3.59	0.44777	14.86945	3.43969	2.75954	0.19701	17.06311
3.60	0.44741	14.95333	3.45373	2.76852	0.19531	17.15564
3.61	0.44705	15.03745	3.46781	2.77749	0.19363	17.24842
3.62	0.44670	15.12180	3.48192	2.78646	0.19197	17.34146
3.63	0.44635	15.20638	3.49608	2.79543	0.19032	17.43476
3.64	0.44600	15.29120	3.51027	2.80440	0.18869	17.52831
3.65	0.44565	15.37625	3.52451	2.81336	0.18707	17.62213
3.66	0.44530	15.46153	3.53878	2.82231	0.18547	17.71620
3.67	0.44496	15.54705	3.55309	2.83127	0.18388	17.81053
3.68	0.44462	15.63280	3.56743	2.84022	0.18230	17.90512
3.69	0.44428	15.71878	3.58182	2.84916	0.18074	17.99996
3.70	0.44395	15.80500	3.59624	2.85811	0.17919	18.09507
3.71	0.44362	15.89145	3.61071	2.86705	0.17766	18.19043
3.72	0.44329	15.97813	3.62521	2.87599	0.17614	18.28605
3.73	0.44296	16.06505	3.63975	2.88492	0.17464	18.38192
3.74	0.44263	16.15220	3.65433	2.89385	0.17314	18.47806
3.75	0.44231	16.23958	3.66894	2.90278	0.17166	18.57445
3.76	0.44199	16.32720	3.68360	2.91170	0.17020	18.67110
3.77	0.44167	16.41505	3.69829	2.92062	0.16875	18.76801
3.78	0.44136	16.50313	3.71302	2.92954	0.16731	18.86518
3.79	0.44104	16.59145	3.72779	2.93846	0.16588	18.96260

M_1	M_2	p_2/p_1	T_2/T_1	$\Delta V/a_1$	p_{t2}/p_{t1}	p_{t2}/p_1
3.80	0.44073	16.68000	3.74260	2.94737	0.16447	19.06029
3.81	0.44042	16.76878	3.75745	2.95628	0.16307	19.15823
3.82	0.44012	16.85780	3.77234	2.96518	0.16168	19.25642
3.83	0.43981	16.94705	3.78726	2.97409	0.16031	19.35488
3.84	0.43951	17.03653	3.80223	2.98299	0.15895	19.45359
3.85	0.43921	17.12625	3.81723	2.99188	0.15760	19.55257
3.86	0.43891	17.21620	3.83227	3.00078	0.15626	19.65180
3.87	0.43862	17.30638	3.84735	3.00967	0.15493	19.75128
3.88	0.43832	17.39680	3.86246	3.01856	0.15362	19.85103
3.89	0.43803	17.48745	3.87762	3.02744	0.15232	19.95103
3.90	0.43774	17.57833	3.89281	3.03632	0.15103	20.05129
3.91	0.43746	17.66945	3.90805	3.04520	0.14975	20.15181
3.92	0.43717	17.76060	3.92332	3.05408	0.14848	20.25259
3.93	0.43689	17.85238	3.93863	3.06296	0.14723	20.35362
3.94	0.43661	17.94420	3.95398	3.07183	0.14598	20.45491
3.95	0.43633	18.03625	3.96936	3.08070	0.14475	20.55646
3.96	0.43605	18.12853	3.98479	3.08956	0.14353	20.65827
3.97	0.43577	18.22105	4.00025	3.09843	0.14232	20.76034
3.98	0.43550	18.31380	4.01575	3.10729	0.14112	20.86266
3.99	0.43523	18.40678	4.03130	3.11614	0.13993	20.96524
4.00	0.43496	18.50000	4.04687	3.12500	0.13876	21.06808
4.10	0.43236	19.44500	4.20479	3.21341	0.12756	22.11065
4.20	0.42994	20.41333	4.36657	3.30159	0.11733	23.17899
4.30	0.42767	21.40500	4.53221	3.38953	0.10800	24.27311
4.40	0.42554	22.42000	4.70171	3.47727	0.09948	25.39300
4.50	0.42355	23.45833	4.87509	3.56481	0.09170	26.53867
4.60	0.42168	24.52000	5.05233	3.65217	0.08459	27.71010
4.70	0.41992	25.60500	5.23343	3.73936	0.07809	28.90729
4.80	0.41826	26.71333	5.41842	3.82639	0.07214	30.13026
4.90	0.41670	27.84500	5.60727	3.91327	0.06670	31.37898
5.00	0.41523	29.00000	5.80000	4.00000	0.06172	32.65347
5.10	0.41384	30.17833	5.99660	4.08660	0.05715	33.95373
5.20	0.41252	31.38000	6.19709	4.17308	0.05297	35.27974
5.30	0.41127	32.60500	6.40144	4.25943	0.04913	36.63152
5.40	0.41009	33.85333	6.60968	4.34566	0.04560	38.00906
5.50	0.40897	35.12500	6.82180	4.43182	0.04236	39.41235
5.60	0.40791	36.42000	7.03779	4.51786	0.03933	40.84141
5.70	0.40690	37.73833	7.25767	4.60380	0.03664	42.29622
5.80	0.40594	39.08000	7.48143	4.68966	0.03412	43.77679
5.90	0.40503	40.44500	7.70907	4.77542	0.03179	45.28312

M_1	M_2	p_2/p_1	T_2/T_1	$\Delta V/a_1$	p_{t2}/p_{t1}	p_{t2}/p_1
6.00	0.40416	41.83333	7.94059	4.86111	0.02965	46.81521
6.10	0.40333	43.24500	8.17599	4.94672	0.02767	48.37305
6.20	0.40254	44.68000	8.41528	5.03226	0.02584	49.95665
6.30	0.40179	46.13833	8.65845	5.11772	0.02416	51.56600
6.40	0.40107	47.62000	8.90550	5.20312	0.02259	53.20111
6.50	0.40038	49.12500	9.15643	5.28846	0.02115	54.86198
6.60	0.39972	50.65333	9.41126	5.37374	0.01981	56.54860
6.70	0.39909	52.20500	9.66996	5.45896	0.01857	58.26097
6.80	0.39849	53.78000	9.93255	5.54412	0.01741	59.99910
6.90	0.39791	55.37833	10.19903	5.62923	0.01634	61.76299
7.00	0.39736	57.00000	10.46939	5.71429	0.01535	63.55263
7.50	0.39491	65.45833	11.87948	6.13889	0.01133	72.88713
8.00	0.39289	74.50000	13.38672	6.56250	0.00849	82.86547
8.50	0.39121	84.12500	14.99113	6.98529	0.00645	93.48763
9.00	0.38980	94.33333	16.69273	7.40741	0.00496	104.75360
9.50	0.38860	105.12500	18.49152	7.82895	0.00387	116.66339
10.00	0.38758	116.50000	20.38750	8.25000	0.00304	129.21697
∞	0.37796	∞	∞	∞	0.0	∞

10

FANNO FLOW
PARAMETERS ($\gamma = 1.4$)

M	T/T^*	p/p^*	p_t/p_t^*	V/V^*	fL_{max}/D	S_{max}/R
0.0	1.20000	∞	∞	0.0	∞	∞
0.01	1.19998	109.54342	57.87384	0.01095	7134.40454	4.05827
0.02	1.19990	54.77006	28.94213	0.02191	1778.44988	3.36530
0.03	1.19978	36.51155	19.30054	0.03286	787.08139	2.96013
0.04	1.19962	27.38175	14.48149	0.04381	440.35221	2.67287
0.05	1.19940	21.90343	11.59144	0.05476	280.02031	2.45027
0.06	1.19914	18.25085	9.66591	0.06570	193.03108	2.26861
0.07	1.19883	15.64155	8.29153	0.07664	140.65501	2.11523
0.08	1.19847	13.68431	7.26161	0.08758	106.71822	1.98260
0.09	1.19806	12.16177	6.46134	0.09851	83.49612	1.86584
0.10	1.19760	10.94351	5.82183	0.10944	66.92156	1.76161
0.11	1.19710	9.94656	5.29923	0.12035	54.68790	1.66756
0.12	1.19655	9.11559	4.86432	0.13126	45.40796	1.58193
0.13	1.19596	8.41230	4.49686	0.14217	38.20700	1.50338
0.14	1.19531	7.80932	4.18240	0.15306	32.51131	1.43089
0.15	1.19462	7.28659	3.91034	0.16395	27.93197	1.36363
0.16	1.19389	6.82907	3.67274	0.17482	24.19783	1.30094
0.17	1.19310	6.42525	3.46351	0.18569	21.11518	1.24228
0.18	1.19227	6.06618	3.27793	0.19654	18.54265	1.18721
0.19	1.19140	5.74480	3.11226	0.20739	16.37516	1.13535
0.20	1.19048	5.45545	2.96352	0.21822	14.53327	1.08638
0.21	1.18951	5.19355	2.82929	0.22904	12.95602	1.04003
0.22	1.18850	4.95537	2.70760	0.23984	11.59605	0.99606
0.23	1.18744	4.73781	2.59681	0.25063	10.41609	0.95428
0.24	1.18633	4.53829	2.49556	0.26141	9.38648	0.91451
0.25	1.18519	4.35465	2.40271	0.27217	8.48341	0.87660
0.26	1.18399	4.18505	2.31729	0.28291	7.68757	0.84040
0.27	1.18276	4.02795	2.23847	0.29364	6.98317	0.80579
0.28	1.18147	3.88199	2.16555	0.30435	6.35721	0.77268
0.29	1.18015	3.74602	2.09793	0.31504	5.79891	0.74095
0.30	1.17878	3.61906	2.03507	0.32572	5.29925	0.71053
0.31	1.17737	3.50022	1.97651	0.33637	4.85066	0.68133
0.32	1.17592	3.38874	1.92185	0.34701	4.44674	0.65329
0.33	1.17442	3.28396	1.87074	0.35762	4.08205	0.62634
0.34	1.17288	3.18529	1.82288	0.36822	3.75195	0.60042
0.35	1.17130	3.09219	1.77797	0.37879	3.45245	0.57547
0.36	1.16968	3.00422	1.73578	0.38935	3.18012	0.55146
0.37	1.16802	2.92094	1.69609	0.39988	2.93198	0.52832
0.38	1.16632	2.84200	1.65870	0.41039	2.70545	0.50603
0.39	1.16457	2.76706	1.62343	0.42087	2.49828	0.48454

M	T/T^*	p/p^*	p_t/p_t^*	V/V^*	fL_{max}/D	S_{max}/R
0.40	1.16279	2.69582	1.59014	0.43133	2.30849	0.46382
0.41	1.16097	2.62801	1.55867	0.44177	2.13436	0.44384
0.42	1.15911	2.56338	1.52890	0.45218	1.97437	0.42455
0.43	1.15721	2.50171	1.50072	0.46257	1.82715	0.40594
0.44	1.15527	2.44280	1.47401	0.47293	1.69152	0.38798
0.45	1.15329	2.38648	1.44367	0.48326	1.56643	0.37065
0.46	1.15128	2.33256	1.42463	0.49357	1.45091	0.35391
0.47	1.14923	2.28089	1.40180	0.50385	1.34413	0.33775
0.48	1.14714	2.23135	1.38010	0.51410	1.24534	0.32215
0.49	1.14502	2.18378	1.35947	0.52433	1.15385	0.30709
0.50	1.14286	2.13809	1.33984	0.53452	1.06906	0.29255
0.51	1.14066	2.09415	1.32117	0.54469	0.99041	0.27852
0.52	1.13843	2.05187	1.30339	0.55483	0.91742	0.26497
0.53	1.13617	2.01116	1.28645	0.56493	0.84962	0.25189
0.54	1.13387	1.97192	1.27032	0.57501	0.78663	0.23927
0.55	1.13154	1.93407	1.25495	0.58506	0.72805	0.22709
0.56	1.12918	1.89755	1.24029	0.59507	0.67357	0.21535
0.57	1.12678	1.86228	1.22633	0.60505	0.62287	0.20402
0.58	1.12435	1.82820	1.21301	0.61501	0.57568	0.19310
0.59	1.12189	1.79525	1.20031	0.62492	0.53174	0.18258
0.60	1.11940	1.76336	1.18820	0.63481	0.49082	0.17244
0.61	1.11688	1.73250	1.17665	0.64466	0.45271	0.16267
0.62	1.11433	1.70261	1.16565	0.65448	0.41720	0.15328
0.63	1.11175	1.67364	1.15515	0.66427	0.38412	0.14423
0.64	1.10914	1.64556	1.14515	0.67402	0.35330	0.13553
0.65	1.10650	1.61831	1.13562	0.68374	0.32459	0.12718
0.66	1.10383	1.59187	1.12654	0.69342	0.29785	0.11915
0.67	1.10114	1.56620	1.11789	0.70307	0.27295	0.11144
0.68	1.09842	1.54126	1.10965	0.71268	0.24978	0.10405
0.69	1.09567	1.51702	1.10182	0.72225	0.22820	0.09696
0.70	1.09290	1.49345	1.09437	0.73179	0.20814	0.09018
0.71	1.09010	1.47053	1.08729	0.74129	0.18948	0.08369
0.72	1.08727	1.44823	1.08057	0.75076	0.17215	0.07749
0.73	1.08442	1.42652	1.07419	0.76019	0.15605	0.07157
0.74	1.08155	1.40537	1.06814	0.76958	0.14112	0.06592
0.75	1.07865	1.38478	1.06242	0.77894	0.12728	0.06055
0.76	1.07573	1.36470	1.05700	0.78825	0.11447	0.05543
0.77	1.07279	1.34514	1.05188	0.79753	0.10262	0.05058
0.78	1.06982	1.32605	1.04705	0.80677	0.09167	0.04598
0.79	1.06684	1.30744	1.04251	0.81597	0.08158	0.04163

M	T/T^*	p/p^*	p_t/p_t^*	V/V^*	fL_{max}/D	S_{max}/R
0.80	1.06383	1.28928	1.03823	0.82514	0.07229	0.03752
0.81	1.06080	1.27155	1.03422	0.83426	0.06376	0.03365
0.82	1.05775	1.25423	1.03046	0.84335	0.05593	0.03001
0.83	1.05469	1.23732	1.02696	0.85239	0.04878	0.02660
0.84	1.05160	1.22080	1.02370	0.86140	0.04226	0.02342
0.85	1.04849	1.20466	1.02067	0.87037	0.03633	0.02046
0.86	1.04537	1.18888	1.01787	0.87929	0.03097	0.01771
0.87	1.04223	1.17344	1.01530	0.88818	0.02613	0.01518
0.88	1.03907	1.15835	1.01294	0.89703	0.02179	0.01286
0.89	1.03589	1.14358	1.01080	0.90583	0.01793	0.01074
0.90	1.03270	1.12913	1.00886	0.91460	0.01451	0.00882
0.91	1.02950	1.11499	1.00713	0.92332	0.01151	0.00711
0.92	1.02627	1.10114	1.00560	0.93201	0.00891	0.00558
0.93	1.02304	1.08759	1.00426	0.94065	0.00669	0.00425
0.94	1.01978	1.07430	1.00311	0.94925	0.00482	0.00310
0.95	1.01652	1.06129	1.00215	0.95781	0.00328	0.00214
0.96	1.01324	1.04854	1.00136	0.96633	0.00206	0.00136
0.97	1.00995	1.03604	1.00076	0.97481	0.00113	0.00076
0.98	1.00664	1.02379	1.00034	0.98325	0.00049	0.00034
0.99	1.00333	1.01178	1.00008	0.99165	0.00012	0.00008
1.00	1.00000	1.00000	1.00000	1.00000	0.00000	0.00000
1.01	0.99666	0.98844	1.00008	1.00831	0.00012	0.00008
1.02	0.99331	0.97711	1.00033	1.01658	0.00046	0.00033
1.03	0.98995	0.96598	1.00074	1.02481	0.00101	0.00074
1.04	0.98658	0.95507	1.00131	1.03300	0.00177	0.00130
1.05	0.98320	0.94435	1.00203	1.04114	0.00271	0.00203
1.06	0.97982	0.93383	1.00291	1.04925	0.00384	0.00290
1.07	0.97642	0.92349	1.00394	1.05731	0.00513	0.00393
1.08	0.97302	0.91335	1.00512	1.06533	0.00658	0.00511
1.09	0.96960	0.90338	1.00645	1.07331	0.00819	0.00643
1.10	0.96618	0.89359	1.00793	1.08124	0.00994	0.00789
1.11	0.96276	0.88397	1.00955	1.08913	0.01182	0.00950
1.12	0.95932	0.87451	1.01131	1.09699	0.01382	0.01125
1.13	0.95589	0.86522	1.01322	1.10479	0.01595	0.01313
1.14	0.95244	0.85608	1.01527	1.11256	0.01819	0.01515
1.15	0.94899	0.84710	1.01745	1.12029	0.02053	0.01730
1.16	0.94554	0.83826	1.01978	1.12797	0.02293	0.01959
1.17	0.94208	0.82958	1.02224	1.13561	0.02552	0.02200
1.18	0.93861	0.82103	1.02484	1.14321	0.02814	0.02454
1.19	0.93515	0.81263	1.02757	1.15077	0.03085	0.02720

M	T/T^*	p/p^*	p_t/p_t^*	V/V^*	fL_{max}/D	S_{max}/R
1.20	0.93168	0.80436	1.03044	1.15828	0.03364	0.02999
1.21	0.92820	0.79623	1.03344	1.16575	0.03650	0.03289
1.22	0.92473	0.78822	1.03657	1.17319	0.03943	0.03592
1.23	0.92125	0.78034	1.03983	1.18057	0.04242	0.03906
1.24	0.91777	0.77258	1.04323	1.18792	0.04547	0.04232
1.25	0.91429	0.76495	1.04675	1.19523	0.04858	0.04569
1.26	0.91080	0.75743	1.05041	1.20249	0.05174	0.04918
1.27	0.90732	0.75003	1.05419	1.20972	0.05495	0.05277
1.28	0.90383	0.74274	1.05810	1.21690	0.05820	0.05647
1.29	0.90035	0.73556	1.06214	1.22404	0.06150	0.06028
1.30	0.89686	0.72848	1.06630	1.23114	0.06483	0.06420
1.31	0.89338	0.72152	1.07060	1.23819	0.06820	0.06822
1.32	0.88989	0.71465	1.07502	1.24521	0.07161	0.07234
1.33	0.88641	0.70789	1.07957	1.25218	0.07504	0.07656
1.34	0.88292	0.70122	1.08424	1.25912	0.07850	0.08088
1.35	0.87944	0.69466	1.08904	1.26601	0.08199	0.08529
1.36	0.87596	0.68818	1.09396	1.27286	0.08550	0.08981
1.37	0.87249	0.68180	1.09902	1.27968	0.08904	0.09441
1.38	0.86901	0.67551	1.10419	1.28645	0.09259	0.09911
1.39	0.86554	0.66931	1.10950	1.29318	0.09615	0.10391
1.40	0.86207	0.66320	1.11493	1.29987	0.09974	0.10879
1.41	0.85860	0.65717	1.12048	1.30652	0.10334	0.11376
1.42	0.85514	0.65122	1.12616	1.31313	0.10694	0.11882
1.43	0.85168	0.64536	1.13197	1.31970	0.11056	0.12396
1.44	0.84822	0.63958	1.13790	1.32623	0.11419	0.12919
1.45	0.84477	0.63387	1.14396	1.33272	0.11782	0.13450
1.46	0.84133	0.62825	1.15015	1.33917	0.12146	0.13989
1.47	0.83788	0.62269	1.15646	1.34558	0.12511	0.14537
1.48	0.83445	0.61722	1.16290	1.35195	0.12875	0.15092
1.49	0.83101	0.61181	1.16947	1.35828	0.13240	0.15655
1.50	0.82759	0.60648	1.17617	1.36458	0.13605	0.16226
1.51	0.82416	0.60122	1.18299	1.37083	0.13970	0.16805
1.52	0.82075	0.59602	1.18994	1.37705	0.14335	0.17391
1.53	0.81734	0.59089	1.19702	1.38322	0.14699	0.17984
1.54	0.81393	0.58583	1.20423	1.38936	0.15063	0.18584
1.55	0.81054	0.58084	1.21157	1.39546	0.15427	0.19192
1.56	0.80715	0.57591	1.21904	1.40152	0.15790	0.19807
1.57	0.80376	0.57104	1.22664	1.40755	0.16152	0.20428
1.58	0.80038	0.56623	1.23438	1.41353	0.16514	0.21057
1.59	0.79701	0.56148	1.24224	1.41948	0.16875	0.21692

M	T/T^*	p/p^*	p_t/p_t^*	V/V^*	fL_{max}/D	S_{max}/R
1.60	0.79365	0.55679	1.25023	1.42539	0.17236	0.22333
1.61	0.79030	0.55216	1.25836	1.43127	0.17595	0.22981
1.62	0.78695	0.54759	1.26663	1.43710	0.17954	0.23636
1.63	0.78361	0.54308	1.27502	1.44290	0.18311	0.24296
1.64	0.78027	0.53862	1.28355	1.44866	0.18667	0.24963
1.65	0.77695	0.53421	1.29222	1.45439	0.19023	0.25636
1.66	0.77363	0.52986	1.30102	1.46008	0.19377	0.26315
1.67	0.77033	0.52556	1.30996	1.46573	0.19729	0.27000
1.68	0.76703	0.52131	1.31904	1.47135	0.20081	0.27690
1.69	0.76374	0.51711	1.32825	1.47693	0.20431	0.28386
1.70	0.76046	0.51297	1.33761	1.48247	0.20780	0.29088
1.71	0.75718	0.50887	1.34710	1.48798	0.21128	0.29795
1.72	0.75392	0.50482	1.35674	1.49345	0.21474	0.30508
1.73	0.75067	0.50082	1.36651	1.49889	0.21819	0.31226
1.74	0.74742	0.49686	1.37643	1.50429	0.22162	0.31949
1.75	0.74419	0.49295	1.38649	1.50966	0.22504	0.32678
1.76	0.74096	0.48909	1.39670	1.51499	0.22844	0.33411
1.77	0.73774	0.48527	1.40705	1.52029	0.23182	0.34149
1.78	0.73454	0.48149	1.41755	1.52555	0.23519	0.34893
1.79	0.73134	0.47776	1.42819	1.53078	0.23855	0.35641
1.80	0.72816	0.47407	1.43898	1.53598	0.24189	0.36394
1.81	0.72498	0.47042	1.44992	1.54114	0.24521	0.37151
1.82	0.72181	0.46681	1.46101	1.54626	0.24851	0.37913
1.83	0.71866	0.46324	1.47225	1.55136	0.25180	0.38680
1.84	0.71551	0.45972	1.48365	1.55642	0.25507	0.39450
1.85	0.71238	0.45623	1.49519	1.56145	0.25832	0.40226
1.86	0.70925	0.45278	1.50689	1.56644	0.26156	0.41005
1.87	0.70614	0.44937	1.51875	1.57140	0.26478	0.41789
1.88	0.70304	0.44600	1.53076	1.57633	0.26798	0.42576
1.89	0.69995	0.44266	1.54293	1.58123	0.27116	0.43368
1.90	0.69686	0.43936	1.55526	1.58609	0.27433	0.44164
1.91	0.69379	0.43610	1.56774	1.59092	0.27748	0.44964
1.92	0.69073	0.43287	1.58039	1.59572	0.28061	0.45767
1.93	0.68769	0.42967	1.59320	1.60049	0.28372	0.46574
1.94	0.68465	0.42651	1.60617	1.60523	0.28681	0.47385
1.95	0.68162	0.42339	1.61931	1.60993	0.28989	0.48200
1.96	0.67861	0.42029	1.63261	1.61460	0.29295	0.49018
1.97	0.67561	0.41724	1.64608	1.61925	0.29599	0.49840
1.98	0.67262	0.41421	1.65972	1.62386	0.29901	0.50665
1.99	0.66964	0.41121	1.67352	1.62844	0.30201	0.51493

M	T/T^*	p/p^*	p_t/p_t^*	V/V^*	fL_{max}/D	S_{max}/R
2.00	0.66667	0.40825	1.68750	1.63299	0.30500	0.52325
2.01	0.66371	0.40532	1.70165	1.63751	0.30796	0.53160
2.02	0.66076	0.40241	1.71597	1.64201	0.31091	0.53998
2.03	0.65783	0.39954	1.73047	1.64647	0.31384	0.54839
2.04	0.65491	0.39670	1.74514	1.65090	0.31676	0.55683
2.05	0.65200	0.39388	1.75999	1.65530	0.31965	0.56531
2.06	0.64910	0.39110	1.77502	1.65967	0.32253	0.57381
2.07	0.64621	0.38834	1.79022	1.66402	0.32538	0.58234
2.08	0.64334	0.38562	1.80561	1.66833	0.32822	0.59090
2.09	0.64047	0.38292	1.82119	1.67262	0.33105	0.59949
2.10	0.63762	0.38024	1.83694	1.67687	0.33385	0.60810
2.11	0.63478	0.37760	1.85289	1.68110	0.33664	0.61674
2.12	0.63195	0.37498	1.86902	1.68530	0.33940	0.62541
2.13	0.62914	0.37239	1.88533	1.68947	0.34215	0.63411
2.14	0.62633	0.36982	1.90184	1.69362	0.34489	0.64282
2.15	0.62354	0.36728	1.91854	1.69774	0.34760	0.65157
2.16	0.62076	0.36476	1.93544	1.70183	0.35030	0.66033
2.17	0.61799	0.36227	1.95252	1.70589	0.35298	0.66912
2.18	0.61523	0.35980	1.96981	1.70992	0.35564	0.67794
2.19	0.61249	0.35736	1.98729	1.71393	0.35828	0.68677
2.20	0.60976	0.35494	2.00497	1.71791	0.36091	0.69563
2.21	0.60704	0.35255	2.02286	1.72187	0.36352	0.70451
2.22	0.60433	0.35017	2.04094	1.72579	0.36611	0.71341
2.23	0.60163	0.34782	2.05923	1.72970	0.36869	0.72233
2.24	0.59895	0.34550	2.07773	1.73357	0.37124	0.73128
2.25	0.59627	0.34319	2.09644	1.73742	0.37378	0.74024
2.26	0.59361	0.34091	2.11535	1.74125	0.37631	0.74922
2.27	0.59096	0.33865	2.13447	1.74504	0.37881	0.75822
2.28	0.58833	0.33641	2.15381	1.74982	0.38130	0.76724
2.29	0.58570	0.33420	2.17336	1.75257	0.38377	0.77628
2.30	0.58309	0.33200	2.19313	1.75629	0.38623	0.78533
2.31	0.58049	0.32983	2.21312	1.75999	0.38867	0.79440
2.32	0.57790	0.32767	2.23332	1.76366	0.39109	0.80349
2.33	0.57532	0.32554	2.25375	1.76731	0.39350	0.81260
2.34	0.57276	0.32342	2.27440	1.77093	0.39589	0.82172
2.35	0.57021	0.32133	2.29528	1.77453	0.39826	0.83085
2.36	0.56767	0.31925	2.31638	1.77811	0.40062	0.84001
2.37	0.56514	0.31720	2.33771	1.78166	0.40296	0.84917
2.38	0.56262	0.31516	2.35928	1.78519	0.40529	0.85835
2.39	0.56011	0.31314	2.38107	1.78869	0.40760	0.86755

M	T/T^*	p/p^*	p_t/p_t^*	V/V^*	fL_{max}/D	S_{max}/R
2.40	0.55762	0.31114	2.40310	1.79218	0.40989	0.87676
2.41	0.55514	0.30916	2.42537	1.79563	0.41217	0.88598
2.42	0.55267	0.30720	2.44787	1.79907	0.41443	0.89522
2.43	0.55021	0.30525	2.47061	1.80248	0.41668	0.90447
2.44	0.54777	0.30332	2.49360	1.80587	0.41891	0.91373
2.45	0.54533	0.30141	2.51683	1.80924	0.42112	0.92300
2.46	0.54291	0.29952	2.54031	1.81258	0.42332	0.93229
2.47	0.54050	0.29765	2.56403	1.81591	0.42551	0.94158
2.48	0.53810	0.29579	2.58801	1.81921	0.42768	0.95089
2.49	0.53571	0.29394	2.61224	1.82249	0.42984	0.96021
2.50	0.53333	0.29212	2.63672	1.82574	0.43198	0.96954
2.51	0.53097	0.29031	2.66146	1.82898	0.43410	0.97887
2.52	0.52862	0.28852	2.68645	1.83219	0.43621	0.98822
2.53	0.52627	0.28674	2.71171	1.83538	0.43831	0.99758
2.54	0.52394	0.28498	2.73723	1.83855	0.44039	1.00695
2.55	0.52163	0.28323	2.76301	1.84170	0.44246	1.01632
2.56	0.51932	0.28150	2.78906	1.84483	0.44451	1.02571
2.57	0.51702	0.27978	2.81538	1.84794	0.44655	1.03510
2.58	0.51474	0.27808	2.84197	1.85103	0.44858	1.04450
2.59	0.51247	0.27640	2.86884	1.85410	0.45059	1.05391
2.60	0.51020	0.27473	2.89598	1.85714	0.45259	1.06332
2.61	0.50795	0.27307	2.92339	1.86017	0.45457	1.07274
2.62	0.50571	0.27143	2.95109	1.86318	0.45654	1.08217
2.63	0.50349	0.26980	2.97907	1.86616	0.45850	1.09161
2.64	0.50127	0.26818	3.00733	1.86913	0.46044	1.10105
2.65	0.49906	0.26658	3.03588	1.87208	0.46237	1.11050
2.66	0.49687	0.26500	3.06472	1.87501	0.46429	1.11996
2.67	0.49469	0.26342	3.09385	1.87792	0.46619	1.12942
2.68	0.49251	0.26186	3.12327	1.88081	0.46808	1.13888
2.69	0.49035	0.26032	3.15299	1.88368	0.46996	1.14835
2.70	0.48820	0.25878	3.18301	1.88653	0.47182	1.15783
2.71	0.48606	0.25726	3.21333	1.88936	0.47367	1.16731
2.72	0.48393	0.25575	3.24395	1.89218	0.47551	1.17679
2.73	0.48182	0.25426	3.27488	1.89497	0.47733	1.18628
2.74	0.47971	0.25278	3.30611	1.89775	0.47915	1.19577
2.75	0.47761	0.25131	3.33766	1.90051	0.48095	1.20527
2.76	0.47553	0.24985	3.36952	1.90325	0.48273	1.21477
2.77	0.47345	0.24840	3.40169	1.90598	0.48451	1.22427
2.78	0.47139	0.24697	3.43418	1.90868	0.48627	1.23378
2.79	0.46933	0.24555	3.46699	1.91137	0.48803	1.24329

M	T/T^*	p/p^*	p_t/p_t^*	V/V^*	fL_{max}/D	S_{max}/R
2.80	0.46729	0.24414	3.50012	1.91404	0.48976	1.2528C
2.81	0.46526	0.24274	3.53358	1.91669	0.49149	1.26231
2.82	0.46323	0.24135	3.56737	1.91933	0.49321	1.27183
2.83	0.46122	0.23998	3.60148	1.92195	0.49491	1.28135
2.84	0.45922	0.23861	3.63593	1.92455	0.49660	1.29087
2.85	0.45723	0.23726	3.67072	1.92714	0.49828	1.30039
2.86	0.45525	0.23592	3.70584	1.92970	0.49995	1.30991
2.87	0.45328	0.23459	3.74131	1.93225	0.50161	1.31943
2.88	0.45132	0.23326	3.77711	1.93479	0.50326	1.32896
2.89	0.44937	0.23195	3.81327	1.93731	0.50489	1.33849
2.90	0.44743	0.23066	3.84977	1.93981	0.50652	1.34801
2.91	0.44550	0.22937	3.88662	1.94230	0.50813	1.35754
2.92	0.44358	0.22809	3.92383	1.94477	0.50973	1.36707
2.93	0.44167	0.22682	3.96139	1.94722	0.51132	1.37660
2.94	0.43977	0.22556	3.99932	1.94966	0.51290	1.38612
2.95	0.43788	0.22431	4.03760	1.95208	0.51447	1.39565
2.96	0.43600	0.22307	4.07625	1.95449	0.51603	1.40518
2.97	0.43413	0.22185	4.11527	1.95688	0.51758	1.41471
2.98	0.43226	0.22063	4.15466	1.95925	0.51912	1.42423
2.99	0.43041	0.21942	4.19443	1.96162	0.52064	1.43376
3.00	0.42857	0.21822	4.23457	1.96396	0.52216	1.44328
3.01	0.42674	0.21703	4.27509	1.96629	0.52367	1.45286
3.02	0.42492	0.21585	4.31599	1.96861	0.52516	1.46233
3.03	0.42310	0.21467	4.35728	1.97091	0.52665	1.47185
3.04	0.42130	0.21351	4.39895	1.97319	0.52813	1.4813
3.05	0.41951	0.21236	4.44102	1.97547	0.52959	1.49088
3.06	0.41772	0.21121	4.48347	1.97772	0.53105	1.50040
3.07	0.41595	0.21008	4.52633	1.97997	0.53249	1.5099
3.08	0.41418	0.20895	4.56959	1.98219	0.53393	1.5194
3.09	0.41242	0.20783	4.61325	1.98441	0.53536	1.5289
3.10	0.41068	0.20672	4.65731	1.98661	0.53678	1.53846
3.11	0.40894	0.20562	4.70178	1.98879	0.53818	1.54796
3.12	0.40721	0.20453	4.74667	1.99097	0.53958	1.55744
3.13	0.40549	0.20344	4.79197	1.99313	0.54097	1.56694
3.14	0.40378	0.20237	4.83769	1.99527	0.54235	1.5764
3.15	0.40208	0.20130	4.88383	1.99740	0.54372	1.5859
3.16	0.40038	0.20024	4.93039	1.99952	0.54509	1.5954
3.17	0.39870	0.19919	4.97739	2.00162	0.54644	1.6049
3.18	0.39702	0.19814	5.02481	2.00372	0.54778	1.6143
3.19	0.39536	0.19711	5.07266	2.00579	0.54912	1.6238

M	T/T^*	p/p^*	p_t/p_t^*	V/V^*	fL_{max}/D	S_{max}/R
3.20	0.39370	0.19608	5.12096	2.00786	0.55044	1.63334
3.21	0.39205	0.19506	5.16969	2.00991	0.55176	1.64281
3.22	0.39041	0.19405	5.21887	2.01195	0.55307	1.65228
3.23	0.38878	0.19304	5.26849	2.01398	0.55437	1.66174
3.24	0.38716	0.19204	5.31857	2.01599	0.55566	1.67120
3.25	0.38554	0.19105	5.36909	2.01799	0.55694	1.68066
3.26	0.38394	0.19007	5.42008	2.01998	0.55822	1.69011
3.27	0.38234	0.18909	5.47152	2.02196	0.55948	1.69956
3.28	0.38075	0.18812	5.52343	2.02392	0.56074	1.70900
3.29	0.37917	0.18716	5.57580	2.02587	0.56199	1.71844
3.30	0.37760	0.18621	5.62865	2.02781	0.56323	1.72787
3.31	0.37603	0.18526	5.68196	2.02974	0.56446	1.73730
3.32	0.37448	0.18432	5.73576	2.03165	0.56569	1.74672
3.33	0.37293	0.18339	5.79003	2.03356	0.56691	1.75614
3.34	0.37139	0.18246	5.84479	2.03545	0.56812	1.76555
3.35	0.36986	0.18154	5.90004	2.03733	0.56932	1.77496
3.36	0.36833	0.18063	5.95577	2.03920	0.57051	1.78436
3.37	0.36682	0.17972	6.01201	2.04106	0.57170	1.79376
3.38	0.36531	0.17882	6.06873	2.04290	0.57287	1.80315
3.39	0.36381	0.17793	6.12596	2.04474	0.57404	1.81254
3.40	0.36232	0.17704	6.18370	2.04656	0.57521	1.82192
3.41	0.36083	0.17616	6.24194	2.04837	0.57636	1.83129
3.42	0.35936	0.17528	6.30070	2.05017	0.57751	1.84066
3.43	0.35789	0.17441	6.35997	2.05196	0.57865	1.85002
3.44	0.35643	0.17355	6.41976	2.05374	0.57978	1.85938
3.45	0.35498	0.17270	6.48007	2.05551	0.58091	1.86873
3.46	0.35353	0.17185	6.54092	2.05727	0.58203	1.87808
3.47	0.35209	0.17100	6.60229	2.05901	0.58314	1.88742
3.48	0.35066	0.17016	6.66419	2.06075	0.58424	1.89675
3.49	0.34924	0.16933	6.72664	2.06247	0.58534	1.90608
3.50	0.34783	0.16851	6.78962	2.06419	0.58643	1.91540
3.51	0.34642	0.16768	6.85315	2.06589	0.58751	1.92471
3.52	0.34502	0.16687	6.91723	2.06759	0.58859	1.93402
3.53	0.34362	0.16606	6.98186	2.06927	0.58966	1.94332
3.54	0.34224	0.16526	7.04705	2.07094	0.59072	1.95261
3.55	0.34086	0.16446	7.11281	2.07261	0.59178	1.96190
3.56	0.33949	0.16367	7.17912	2.07426	0.59282	1.97118
3.57	0.33813	0.16288	7.24601	2.07590	0.59387	1.98045
3.58	0.33677	0.16210	7.31346	2.07754	0.59490	1.98972
3.59	0.33542	0.16132	7.38150	2.07916	0.59593	1.99898

M	T/T^*	p/p^*	p_t/p_t^*	V/V^*	fL_{max}/D	S_{max}/R
3.60	0.33408	0.16055	7.45011	2.08077	0.59695	2.00823
3.61	0.33274	0.15979	7.51931	2.08238	0.59797	2.01747
3.62	0.33141	0.15903	7.58910	2.08397	0.59898	2.02671
3.63	0.33009	0.15827	7.65948	2.08556	0.59998	2.03594
3.64	0.32877	0.15752	7.73045	2.08713	0.60098	2.04517
3.65	0.32747	0.15678	7.80203	2.08870	0.60197	2.05438
3.66	0.32616	0.15604	7.87421	2.09026	0.60296	2.06359
3.67	0.32487	0.15531	7.94700	2.09180	0.60394	2.07279
3.68	0.32358	0.15458	8.02040	2.09334	0.60491	2.08199
3.69	0.32230	0.15385	8.09442	2.09487	0.60588	2.09118
3.70	0.32103	0.15313	8.16907	2.09639	0.60684	2.10035
3.71	0.31976	0.15242	8.24433	2.09790	0.60779	2.10953
3.72	0.31850	0.15171	8.32023	2.09941	0.60874	2.11869
3.73	0.31724	0.15100	8.39676	2.10090	0.60968	2.12785
3.74	0.31600	0.15030	8.47393	2.10238	0.61062	2.13699
3.75	0.31475	0.14961	8.55174	2.10386	0.61155	2.14613
3.76	0.31352	0.14892	8.63020	2.10533	0.61247	2.15527
3.77	0.31229	0.14823	8.70931	2.10679	0.61339	2.16439
3.78	0.31107	0.14755	8.78907	2.10824	0.61431	2.17351
3.79	0.30985	0.14687	8.86950	2.10968	0.61522	2.18262
3.80	0.30864	0.14620	8.95059	2.11111	0.61612	2.19172
3.81	0.30744	0.14553	9.03234	2.11254	0.61702	2.20081
3.82	0.30624	0.14487	9.11477	2.11395	0.61791	2.20990
3.83	0.30505	0.14421	9.19788	2.11536	0.61879	2.21897
3.84	0.30387	0.14355	9.28167	2.11676	0.61968	2.22804
3.85	0.30269	0.14290	9.36614	2.11815	0.62055	2.23710
3.86	0.30151	0.14225	9.45131	2.11954	0.62142	2.24615
3.87	0.30035	0.14161	9.53717	2.12091	0.62229	2.25520
3.88	0.29919	0.14097	9.62373	2.12228	0.62315	2.26423
3.89	0.29803	0.14034	9.71100	2.12364	0.62400	2.27326
3.90	0.29688	0.13971	9.79897	2.12499	0.62485	2.28228
3.91	0.29574	0.13908	9.88766	2.12634	0.62569	2.29129
3.92	0.29460	0.13846	9.97707	2.12767	0.62653	2.30029
3.93	0.29347	0.13784	10.06720	2.12900	0.62737	2.30928
3.94	0.29235	0.13723	10.15806	2.13032	0.62819	2.31827
3.95	0.29123	0.13662	10.24965	2.13163	0.62902	2.32724
3.96	0.29011	0.13602	10.34197	2.13294	0.62984	2.33621
3.97	0.28900	0.13541	10.43504	2.13424	0.63065	2.34517
3.98	0.28790	0.13482	10.52886	2.13553	0.63146	2.35412
3.99	0.28681	0.13422	10.62343	2.13681	0.63227	2.36306

M	T/T^*	p/p^*	p_t/p_t^*	V/V^*	fL_{max}/D	S_{max}/R
4.00	0.28571	0.13363	10.71875	2.13809	0.63306	2.37199
4.10	0.27510	0.12793	11.71465	2.15046	0.64080	2.46084
4.20	0.26502	0.12257	12.79164	2.16215	0.64810	2.54879
4.30	0.25543	0.11753	13.95490	2.17321	0.65499	2.63583
4.40	0.24631	0.11279	15.20987	2.18368	0.66149	2.72194
4.50	0.23762	0.10833	16.56219	2.19360	0.66763	2.80712
4.60	0.22936	0.10411	18.01779	2.20300	0.67345	2.89136
4.70	0.22148	0.10013	19.58283	2.21192	0.67895	2.97465
4.80	0.21398	0.09637	21.26371	2.22038	0.68417	3.05700
4.90	0.20683	0.09281	23.06712	2.22842	0.68911	3.13841
5.00	0.20000	0.08944	25.00000	2.23607	0.69330	3.21888
5.10	0.19349	0.08625	27.06957	2.24334	0.69826	3.29841
5.20	0.18727	0.08322	29.28333	2.25026	0.70249	3.37702
5.30	0.18132	0.08034	31.64905	2.25685	0.70652	3.45471
5.40	0.17564	0.07761	34.17481	2.26313	0.71035	3.53149
5.50	0.17021	0.07501	36.86896	2.26913	0.71400	3.60737
5.60	0.16502	0.07254	39.74018	2.27484	0.71748	3.68236
5.70	0.16004	0.07018	42.79743	2.28030	0.72080	3.75648
5.80	0.15528	0.06794	46.05000	2.28552	0.72397	3.82973
5.90	0.15072	0.06580	49.50747	2.29051	0.72699	3.90212
6.00	0.14634	0.06376	53.17978	2.29528	0.72988	3.97368
6.10	0.14215	0.06181	57.07718	2.29984	0.73264	4.04440
6.20	0.13812	0.05994	61.21023	2.30421	0.73528	4.11431
6.30	0.13426	0.05816	65.58987	2.30840	0.73780	4.18342
6.40	0.13055	0.05646	70.22736	2.31241	0.74022	4.25174
6.50	0.12698	0.05482	75.13431	2.31626	0.74254	4.31928
6.60	0.12356	0.05326	80.32271	2.31996	0.74477	4.38605
6.70	0.12026	0.05176	85.80487	2.32351	0.74690	4.45208
6.80	0.11710	0.05032	91.59351	2.32691	0.74895	4.51736
6.90	0.11405	0.04894	97.70169	2.33019	0.75091	4.58192
7.00	0.11111	0.04762	104.14286	2.33333	0.75280	4.64576
7.50	0.09796	0.04173	141.84148	2.34738	0.76121	4.95471
8.00	0.08696	0.03666	190.10937	2.35907	0.76819	5.24760
8.50	0.07767	0.03279	251.08617	2.36889	0.77404	5.52580
9.00	0.06977	0.02935	327.18930	2.37722	0.77899	5.79054
9.50	0.06299	0.02642	421.13137	2.38433	0.78320	6.04294
10.00	0.05714	0.02390	535.93750	2.39046	0.78683	6.28402
∞	0.0	0.0	∞	2.4495	0.82153	∞

11

RAYLEIGH FLOW
PARAMETERS ($\gamma = 1.4$)

M	T_t/T_t^*	T/T^*	p/p^*	p_t/p_t^*	V/V^*	S_{max}/R
0.0	0.0	0.0	2.40000	1.26790	0.0	∞
0.01	0.00048	0.00058	2.39966	1.26779	0.00024	26.98422
0.02	0.00192	0.00230	2.39866	1.26752	0.00096	22.13471
0.03	0.00431	0.00517	2.39698	1.26708	0.00216	19.30065
0.04	0.00765	0.00917	2.39464	1.26646	0.00383	17.29274
0.05	0.01192	0.01430	2.39163	1.26567	0.00598	15.73828
0.06	0.01712	0.02053	2.38796	1.26470	0.00860	14.47123
0.07	0.02322	0.02784	2.38365	1.26356	0.01168	13.40303
0.08	0.03022	0.03621	2.37869	1.26226	0.01522	12.48081
0.09	0.03807	0.04562	2.37309	1.26078	0.01922	11.67046
0.10	0.04678	0.05602	2.36686	1.25915	0.02367	10.94870
0.11	0.05630	0.06739	2.36002	1.25735	0.02856	10.29890
0.12	0.06661	0.07970	2.35257	1.25539	0.03388	9.70879
0.13	0.07768	0.09290	2.34453	1.25329	0.03962	9.16904
0.14	0.08947	0.10695	2.33590	1.25103	0.04578	8.67240
0.15	0.10196	0.12181	2.32671	1.24863	0.05235	8.21311
0.16	0.11511	0.13743	2.31696	1.24608	0.05931	7.78653
0.17	0.12888	0.15377	2.30667	1.24340	0.06666	7.38886
0.18	0.14324	0.17078	2.29586	1.24059	0.07439	7.01694
0.19	0.15814	0.18841	2.28454	1.23765	0.08247	6.66813
0.20	0.17355	0.20661	2.27273	1.23460	0.09091	6.34018
0.21	0.18943	0.22533	2.26044	1.23142	0.09969	6.03118
0.22	0.20574	0.24452	2.24770	1.22814	0.10879	5.73946
0.23	0.22244	0.26413	2.23451	1.22475	0.11821	5.46359
0.24	0.23948	0.28411	2.22091	1.22126	0.12792	5.20232
0.25	0.25684	0.30440	2.20690	1.21767	0.13793	4.95454
0.26	0.27446	0.32496	2.19250	1.21400	0.14821	4.71926
0.27	0.29231	0.34573	2.17774	1.21025	0.15876	4.49561
0.28	0.31035	0.36667	2.16263	1.20642	0.16955	4.28281
0.29	0.32855	0.38774	2.14719	1.20251	0.18058	4.08016
0.30	0.34686	0.40887	2.13144	1.19855	0.19183	3.88703
0.31	0.36525	0.43004	2.11539	1.19452	0.20329	3.70283
0.32	0.38369	0.45119	2.09908	1.19045	0.21495	3.52706
0.33	0.40214	0.47228	2.08250	1.18632	0.22678	3.35922
0.34	0.42056	0.49327	2.06569	1.18215	0.23879	3.19888
0.35	0.43894	0.51413	2.04866	1.17795	0.25096	3.04565
0.36	0.45723	0.53482	2.03142	1.17371	0.26327	2.89915
0.37	0.47541	0.55529	2.01400	1.16945	0.27572	2.75904
0.38	0.49346	0.57553	1.99641	1.16517	0.28828	2.62500
0.39	0.51134	0.59549	1.97866	1.16088	0.30095	2.49673

M	T_t/T_t^*	T/T^*	p/p^*	p_t/p_t^*	V/V^*	S_{max}/R
0.40	0.52903	0.61515	1.96078	1.15658	0.31373	2.37397
0.41	0.54651	0.63448	1.94278	1.15227	0.32658	2.25645
0.42	0.56376	0.65346	1.92468	1.14796	0.33951	2.14394
0.43	0.58076	0.67205	1.90649	1.14366	0.35251	2.03622
0.44	0.59748	0.69025	1.88822	1.13936	0.36556	1.93306
0.45	0.61393	0.70804	1.86989	1.13508	0.37865	1.83429
0.46	0.63007	0.72538	1.85151	1.13082	0.39178	1.73970
0.47	0.64589	0.74228	1.83310	1.12659	0.40493	1.64912
0.48	0.66139	0.75871	1.81466	1.12238	0.41810	1.56239
0.49	0.67655	0.77466	1.79622	1.11820	0.43127	1.47935
0.50	0.69136	0.79012	1.77778	1.11405	0.44444	1.39985
0.51	0.70581	0.80509	1.75935	1.10995	0.45761	1.32374
0.52	0.71990	0.81955	1.74095	1.10588	0.47075	1.25091
0.53	0.73361	0.83351	1.72258	1.10186	0.48387	1.18121
0.54	0.74695	0.84695	1.70425	1.09789	0.49696	1.11453
0.55	0.75991	0.85987	1.68599	1.09397	0.51001	1.05076
0.56	0.77249	0.87227	1.66778	1.09011	0.52302	0.98977
0.57	0.78468	0.88416	1.64964	1.08630	0.53597	0.93148
0.58	0.79648	0.89552	1.63159	1.08256	0.54887	0.87577
0.59	0.80789	0.90637	1.61362	1.07887	0.56170	0.82255
0.60	0.81892	0.91670	1.59574	1.07525	0.57447	0.77174
0.61	0.82957	0.92653	1.57797	1.07170	0.58716	0.72323
0.62	0.83983	0.93584	1.56031	1.06822	0.59978	0.67696
0.63	0.84970	0.94466	1.54275	1.06481	0.61232	0.63284
0.64	0.85920	0.95298	1.52532	1.06147	0.62477	0.59078
0.65	0.86833	0.96081	1.50801	1.05821	0.63713	0.55073
0.66	0.87708	0.96816	1.49083	1.05503	0.64941	0.51260
0.67	0.88547	0.97503	1.47379	1.05193	0.66158	0.47634
0.68	0.89350	0.98144	1.45688	1.04890	0.67366	0.44187
0.69	0.90118	0.98739	1.44011	1.04596	0.68564	0.40913
0.70	0.90850	0.99290	1.42349	1.04310	0.69751	0.37807
0.71	0.91548	0.99796	1.40701	1.04033	0.70928	0.34861
0.72	0.92212	1.00260	1.39069	1.03764	0.72093	0.32072
0.73	0.92843	1.00682	1.37452	1.03504	0.73248	0.29433
0.74	0.93442	1.01062	1.35851	1.03253	0.74392	0.26940
0.75	0.94009	1.01403	1.34266	1.03010	0.75524	0.24587
0.76	0.94546	1.01706	1.32696	1.02777	0.76645	0.22370
0.77	0.95052	1.01970	1.31143	1.02552	0.77755	0.20283
0.78	0.95528	1.02198	1.29606	1.02337	0.78853	0.18324
0.79	0.95975	1.02390	1.28086	1.02131	0.79939	0.16486

M	T_t/T_t^*	T/T^*	p/p^*	p_t/p_t^*	V/V^*	S_{max}/R
0.80	0.96395	1.02548	1.26582	1.01934	0.81013	0.14767
0.81	0.96787	1.02672	1.25095	1.01747	0.82075	0.13162
0.82	0.97152	1.02763	1.23625	1.01569	0.83125	0.11668
0.83	0.97492	1.02823	1.22171	1.01400	0.84164	0.10280
0.84	0.97807	1.02853	1.20734	1.01241	0.85190	0.08995
0.85	0.98097	1.02854	1.19314	1.01091	0.86204	0.07810
0.86	0.98363	1.02826	1.17911	1.00951	0.87207	0.06722
0.87	0.98607	1.02771	1.16524	1.00820	0.88197	0.05727
0.88	0.98828	1.02689	1.15154	1.00699	0.89175	0.04822
0.89	0.99028	1.02583	1.13801	1.00587	0.90142	0.04004
0.90	0.99207	1.02452	1.12465	1.00486	0.91097	0.03270
0.91	0.99366	1.02297	1.11145	1.00393	0.92039	0.02618
0.92	0.99506	1.02120	1.09842	1.00311	0.92970	0.02044
0.93	0.99627	1.01922	1.08555	1.00238	0.93889	0.01547
0.94	0.99729	1.01702	1.07285	1.00175	0.94797	0.01124
0.95	0.99814	1.01463	1.06030	1.00122	0.95693	0.00771
0.96	0.99883	1.01205	1.04793	1.00078	0.96577	0.00488
0.97	0.99935	1.00929	1.03571	1.00044	0.97450	0.00271
0.98	0.99971	1.00636	1.02365	1.00019	0.98311	0.00119
0.99	0.99993	1.00326	1.01174	1.00005	0.99161	0.00029
1.00	1.00000	1.00000	1.00000	1.00000	1.00000	0.00000
1.01	0.99993	0.99659	0.98841	1.00005	1.00828	0.00029
1.02	0.99973	0.99304	0.97698	1.00019	1.01645	0.00114
1.03	0.99940	0.98936	0.96569	1.00044	1.02450	0.00254
1.04	0.99895	0.98554	0.95456	1.00078	1.03246	0.00447
1.05	0.99838	0.98161	0.94358	1.00122	1.04030	0.00690
1.06	0.99769	0.97755	0.93275	1.00175	1.04804	0.00983
1.07	0.99690	0.97339	0.92206	1.00238	1.05567	0.01324
1.08	0.99601	0.96913	0.91152	1.00311	1.06320	0.01711
1.09	0.99501	0.96477	0.90112	1.00394	1.07063	0.02143
1.10	0.99392	0.96031	0.89087	1.00486	1.07795	0.02618
1.11	0.99275	0.95577	0.88075	1.00588	1.08518	0.03135
1.12	0.99148	0.95115	0.87078	1.00699	1.09230	0.03692
1.13	0.99013	0.94645	0.86094	1.00821	1.09933	0.04288
1.14	0.98871	0.94169	0.85123	1.00952	1.10626	0.04922
1.15	0.98721	0.93685	0.84166	1.01093	1.11310	0.05593
1.16	0.98564	0.93196	0.83222	1.01243	1.11984	0.06298
1.17	0.98400	0.92701	0.82292	1.01403	1.12649	0.07038
1.18	0.98230	0.92200	0.81374	1.01573	1.13305	0.07812
1.19	0.98054	0.91695	0.80468	1.01752	1.13951	0.08617

M	T_t/T_t^*	T/T^*	p/p^*	p_t/p_t^*	V/V^*	S_{max}/R
1.20	0.97872	0.91185	0.79576	1.01942	1.14589	0.09453
1.21	0.97684	0.90671	0.78695	1.02140	1.15218	0.10318
1.22	0.97492	0.90153	0.77827	1.02349	1.15838	0.11213
1.23	0.97294	0.89632	0.76971	1.02567	1.16449	0.12135
1.24	0.97092	0.89108	0.76127	1.02795	1.17052	0.13085
1.25	0.96886	0.88581	0.75294	1.03033	1.17647	0.14060
1.26	0.96675	0.88052	0.74473	1.03280	1.18233	0.15061
1.27	0.96461	0.87521	0.73663	1.03537	1.18812	0.16086
1.28	0.96243	0.86988	0.72865	1.03803	1.19382	0.17135
1.29	0.96022	0.86453	0.72078	1.04080	1.19945	0.18206
1.30	0.95798	0.85917	0.71301	1.04366	1.20499	0.19299
1.31	0.95571	0.85380	0.70536	1.04662	1.21046	0.20413
1.32	0.95341	0.84843	0.69780	1.04968	1.21585	0.21548
1.33	0.95108	0.84305	0.69036	1.05283	1.22117	0.22702
1.34	0.94873	0.83766	0.68301	1.05608	1.22642	0.23876
1.35	0.94637	0.83227	0.67577	1.05943	1.23159	0.25068
1.36	0.94398	0.82689	0.66863	1.06288	1.23669	0.26277
1.37	0.94157	0.82151	0.66158	1.06642	1.24173	0.27504
1.38	0.93914	0.81613	0.65464	1.07007	1.24669	0.28747
1.39	0.93671	0.81076	0.64778	1.07381	1.25158	0.30006
1.40	0.93425	0.80539	0.64103	1.07765	1.25641	0.31281
1.41	0.93179	0.80004	0.63436	1.08159	1.26117	0.32570
1.42	0.92931	0.79469	0.62779	1.08563	1.26587	0.33874
1.43	0.92683	0.78936	0.62130	1.08977	1.27050	0.35191
1.44	0.92434	0.78405	0.61491	1.09401	1.27507	0.36522
1.45	0.92184	0.77874	0.60860	1.09835	1.27957	0.37865
1.46	0.91933	0.77346	0.60237	1.10278	1.28402	0.39221
1.47	0.91682	0.76819	0.59623	1.10732	1.28840	0.40589
1.48	0.91431	0.76294	0.59018	1.11196	1.29273	0.41968
1.49	0.91179	0.75771	0.58421	1.11670	1.29700	0.43358
1.50	0.90928	0.75250	0.57831	1.12155	1.30120	0.44758
1.51	0.90676	0.74732	0.57250	1.12649	1.30536	0.46169
1.52	0.90424	0.74215	0.56676	1.13153	1.30945	0.47589
1.53	0.90172	0.73701	0.56111	1.13668	1.31350	0.49019
1.54	0.89920	0.73189	0.55552	1.14193	1.31748	0.50458
1.55	0.89669	0.72680	0.55002	1.14729	1.32142	0.51905
1.56	0.89418	0.72173	0.54458	1.15274	1.32530	0.53361
1.57	0.89168	0.71669	0.53922	1.15830	1.32913	0.54824
1.58	0.88917	0.71168	0.53393	1.16397	1.33291	0.56295
1.59	0.88668	0.70669	0.52871	1.16974	1.33663	0.57774

M	T_t/T_t^*	T/T^*	p/p^*	p_t/p_t^*	V/V^*	S_{max}/R
1.60	0.88419	0.70174	0.52356	1.17561	1.34031	0.59259
1.61	0.88170	0.69680	0.51848	1.18159	1.34394	0.60752
1.62	0.87922	0.69190	0.51346	1.18768	1.34753	0.62250
1.63	0.87675	0.68703	0.50851	1.19387	1.35106	0.63755
1.64	0.87429	0.68219	0.50363	1.20017	1.35455	0.65265
1.65	0.87184	0.67738	0.49880	1.20657	1.35800	0.66781
1.66	0.86939	0.67259	0.49405	1.21309	1.36140	0.68303
1.67	0.86696	0.66784	0.48935	1.21971	1.36475	0.69829
1.68	0.86453	0.66312	0.48472	1.22644	1.36806	0.71360
1.69	0.86212	0.65843	0.48014	1.23328	1.37133	0.72896
1.70	0.85971	0.65377	0.47562	1.24024	1.37455	0.74436
1.71	0.85731	0.64914	0.47117	1.24730	1.37774	0.75981
1.72	0.85493	0.64455	0.46677	1.25447	1.38088	0.77529
1.73	0.85256	0.63999	0.46242	1.26175	1.38398	0.79081
1.74	0.85019	0.63545	0.45813	1.26915	1.38705	0.80636
1.75	0.84784	0.63095	0.45390	1.27666	1.39007	0.82195
1.76	0.84551	0.62649	0.44972	1.28428	1.39306	0.83757
1.77	0.84318	0.62205	0.44559	1.29202	1.39600	0.85322
1.78	0.84087	0.61765	0.44152	1.29987	1.39891	0.86889
1.79	0.83857	0.61328	0.43750	1.30784	1.40179	0.88459
1.80	0.83628	0.60894	0.43353	1.31592	1.40462	0.90031
1.81	0.83400	0.60464	0.42960	1.32413	1.40743	0.91606
1.82	0.83174	0.60036	0.42573	1.33244	1.41019	0.93183
1.83	0.82949	0.59612	0.42191	1.34088	1.41292	0.94761
1.84	0.82726	0.59191	0.41813	1.34943	1.41562	0.96342
1.85	0.82504	0.58774	0.41440	1.35811	1.41829	0.97924
1.86	0.82283	0.58359	0.41072	1.36690	1.42092	0.99507
1.87	0.82064	0.57948	0.40708	1.37582	1.42351	1.01092
1.88	0.81845	0.57540	0.40349	1.38486	1.42608	1.02678
1.89	0.81629	0.57136	0.39994	1.39402	1.42862	1.04265
1.90	0.81414	0.56734	0.39643	1.40330	1.43112	1.05853
1.91	0.81200	0.56336	0.39297	1.41271	1.43359	1.07441
1.92	0.80987	0.55941	0.38955	1.42224	1.43604	1.09031
1.93	0.80776	0.55549	0.38617	1.43190	1.43845	1.10621
1.94	0.80567	0.55160	0.38283	1.44168	1.44083	1.12211
1.95	0.80358	0.54774	0.37954	1.45159	1.44319	1.13802
1.96	0.80152	0.54392	0.37628	1.46164	1.44551	1.15393
1.97	0.79946	0.54012	0.37306	1.47180	1.44781	1.16984
1.98	0.79742	0.53636	0.36988	1.48210	1.45008	1.18575
1.99	0.79540	0.53263	0.36674	1.49253	1.45233	1.20167

M	T_t/T_t^*	T/T^*	p/p^*	p_t/p_t^*	V/V^*	S_{max}/R
2.00	0.79339	0.52893	0.36364	1.50310	1.45455	1.21758
2.01	0.79139	0.52525	0.36057	1.51379	1.45674	1.23348
2.02	0.78941	0.52161	0.35754	1.52462	1.45890	1.24939
2.03	0.78744	0.51800	0.35454	1.53558	1.46104	1.26529
2.04	0.78549	0.51442	0.35158	1.54668	1.46315	1.28118
2.05	0.78355	0.51087	0.34866	1.55791	1.46524	1.29707
2.06	0.78162	0.50735	0.34577	1.56928	1.46731	1.31296
2.07	0.77971	0.50386	0.34291	1.58079	1.46935	1.32883
2.08	0.77782	0.50040	0.34009	1.59244	1.47136	1.34470
2.09	0.77593	0.49696	0.33730	1.60423	1.47336	1.36056
2.10	0.77406	0.49356	0.33454	1.61616	1.47533	1.37641
2.11	0.77221	0.49018	0.33182	1.62823	1.47727	1.39225
2.12	0.77037	0.48684	0.32912	1.64045	1.47920	1.40807
2.13	0.76854	0.48352	0.32646	1.65281	1.48110	1.42389
2.14	0.76673	0.48023	0.32382	1.66531	1.48298	1.43970
2.15	0.76493	0.47696	0.32122	1.67796	1.48484	1.45549
2.16	0.76314	0.47373	0.31865	1.69076	1.48668	1.47127
2.17	0.76137	0.47052	0.31610	1.70371	1.48850	1.48703
2.18	0.75961	0.46734	0.31359	1.71680	1.49029	1.50278
2.19	0.75787	0.46418	0.31110	1.73005	1.49207	1.51852
2.20	0.75613	0.46106	0.30864	1.74345	1.49383	1.53424
2.21	0.75442	0.45796	0.30621	1.75700	1.49556	1.54994
2.22	0.75271	0.45488	0.30381	1.77070	1.49728	1.56563
2.23	0.75102	0.45184	0.30143	1.78456	1.49898	1.58130
2.24	0.74934	0.44882	0.29908	1.79858	1.50066	1.59696
2.25	0.74768	0.44582	0.29675	1.81275	1.50232	1.61259
2.26	0.74602	0.44285	0.29446	1.82708	1.50396	1.62821
2.27	0.74438	0.43990	0.29218	1.84157	1.50558	1.64381
2.28	0.74276	0.43698	0.28993	1.85623	1.50719	1.65939
2.29	0.74114	0.43409	0.28771	1.87104	1.50878	1.67496
2.30	0.73954	0.43122	0.28551	1.88602	1.51035	1.69050
2.31	0.73795	0.42838	0.28333	1.90116	1.51190	1.70602
2.32	0.73638	0.42555	0.28118	1.91647	1.51344	1.72152
2.33	0.73482	0.42276	0.27905	1.93195	1.51496	1.73700
2.34	0.73326	0.41998	0.27695	1.94759	1.51646	1.75246
2.35	0.73173	0.41723	0.27487	1.96340	1.51795	1.76790
2.36	0.73020	0.41451	0.27281	1.97939	1.51942	1.78332
2.37	0.72868	0.41181	0.27077	1.99554	1.52088	1.79872
2.38	0.72718	0.40913	0.26875	2.01187	1.52232	1.81409
2.39	0.72569	0.40647	0.26676	2.02837	1.52374	1.82944

M	T_t/T_t^*	T/T^*	p/p^*	p_t/p_t^*	V/V^*	S_{max}/R
2.40	0.72421	0.40384	0.26478	2.04505	1.52515	1.84477
2.41	0.72275	0.40122	0.26283	2.06191	1.52655	1.86008
2.42	0.72129	0.39864	0.26090	2.07895	1.52793	1.87536
2.43	0.71985	0.39607	0.25899	2.09616	1.52929	1.89062
2.44	0.71842	0.39352	0.25710	2.11356	1.53065	1.90585
2.45	0.71699	0.39100	0.25522	2.13114	1.53198	1.92106
2.46	0.71558	0.38850	0.25337	2.14891	1.53331	1.93625
2.47	0.71419	0.38602	0.25154	2.16685	1.53461	1.95141
2.48	0.71280	0.38356	0.24973	2.18499	1.53591	1.96655
2.49	0.71142	0.38112	0.24793	2.20332	1.53719	1.98167
2.50	0.71006	0.37870	0.24615	2.22183	1.53846	1.99676
2.51	0.70871	0.37630	0.24440	2.24054	1.53972	2.01182
2.52	0.70736	0.37392	0.24266	2.25944	1.54096	2.02686
2.53	0.70603	0.37157	0.24093	2.27853	1.54219	2.04187
2.54	0.70471	0.36923	0.23923	2.29782	1.54341	2.05686
2.55	0.70340	0.36691	0.23754	2.31730	1.54461	2.07183
2.56	0.70210	0.36461	0.23587	2.33699	1.54581	2.08676
2.57	0.70081	0.36233	0.23422	2.35687	1.54699	2.10167
2.58	0.69952	0.36007	0.23258	2.37696	1.54816	2.11656
2.59	0.69826	0.35783	0.23096	2.39725	1.54931	2.13142
2.60	0.69700	0.35561	0.22936	2.41774	1.55046	2.14625
2.61	0.69575	0.35341	0.22777	2.43844	1.55159	2.16106
2.62	0.69451	0.35122	0.22620	2.45935	1.55272	2.17584
2.63	0.69328	0.34906	0.22464	2.48047	1.55383	2.19059
2.64	0.69206	0.34691	0.22310	2.50179	1.55493	2.20532
2.65	0.69084	0.34478	0.22158	2.52334	1.55602	2.22002
2.66	0.68964	0.34266	0.22007	2.54509	1.55710	2.23470
2.67	0.68845	0.34057	0.21857	2.56706	1.55816	2.24934
2.68	0.68727	0.33849	0.21709	2.58925	1.55922	2.26396
2.69	0.68610	0.33643	0.21562	2.61166	1.56027	2.27856
2.70	0.68494	0.33439	0.21417	2.63429	1.56131	2.29312
2.71	0.68378	0.33236	0.21273	2.65714	1.56233	2.30766
2.72	0.68264	0.33035	0.21131	2.68021	1.56335	2.32217
2.73	0.68150	0.32836	0.20990	2.70351	1.56436	2.33666
2.74	0.68037	0.32638	0.20850	2.72704	1.56536	2.35111
2.75	0.67926	0.32442	0.20712	2.75080	1.56634	2.36554
2.76	0.67815	0.32248	0.20575	2.77478	1.56732	2.37995
2.77	0.67705	0.32055	0.20439	2.79900	1.56829	2.39432
2.78	0.67595	0.31864	0.20305	2.82346	1.56925	2.40867
2.79	0.67487	0.31674	0.20172	2.84815	1.57020	2.42299

M	T_t/T_t^*	T/T^*	p/p^*	p_t/p_t^*	V/V^*	S_{max}/R
2.80	0.67380	0.31486	0.20040	2.87308	1.57114	2.43728
2.81	0.67273	0.31299	0.19910	2.89825	1.57207	2.45154
2.82	0.67167	0.31114	0.19780	2.92366	1.57300	2.46578
2.83	0.67062	0.30931	0.19652	2.94931	1.57391	2.47999
2.84	0.66958	0.30749	0.19525	2.97521	1.57482	2.49417
2.85	0.66855	0.30568	0.19399	3.00136	1.57572	2.50833
2.86	0.66752	0.30389	0.19275	3.02775	1.57661	2.52245
2.87	0.66651	0.30211	0.19151	3.05440	1.57749	2.53655
2.88	0.66550	0.30035	0.19029	3.08129	1.57836	2.55062
2.89	0.66450	0.29860	0.18908	3.10844	1.57923	2.56467
2.90	0.66350	0.29687	0.18788	3.13585	1.58008	2.57868
2.91	0.66252	0.29515	0.18669	3.16352	1.58093	2.59267
2.92	0.66154	0.29344	0.18551	3.19145	1.58178	2.60663
2.93	0.66057	0.29175	0.18435	3.21963	1.58261	2.62057
2.94	0.65960	0.29007	0.18319	3.24809	1.58343	2.63447
2.95	0.65865	0.28841	0.18205	3.27680	1.58425	2.64835
2.96	0.65770	0.28675	0.18091	3.30579	1.58506	2.66220
2.97	0.65676	0.28512	0.17979	3.33505	1.58587	2.67602
2.98	0.65583	0.28349	0.17867	3.36457	1.58666	2.68981
2.99	0.65490	0.28188	0.17757	3.39437	1.58745	2.70358
3.00	0.65398	0.28028	0.17647	3.42445	1.58824	2.71732
3.01	0.65307	0.27869	0.17539	3.45481	1.58901	2.73103
3.02	0.65216	0.27711	0.17431	3.48544	1.58978	2.74472
3.03	0.65126	0.27555	0.17324	3.51636	1.59054	2.75837
3.04	0.65037	0.27400	0.17219	3.54756	1.59129	2.77200
3.05	0.64949	0.27246	0.17114	3.57905	1.59204	2.78560
3.06	0.64861	0.27094	0.17010	3.61082	1.59278	2.79918
3.07	0.64774	0.26942	0.16908	3.64289	1.59352	2.81272
3.08	0.64687	0.26792	0.16806	3.67524	1.59425	2.82624
3.09	0.64601	0.26643	0.16705	3.70790	1.59497	2.83974
3.10	0.64516	0.26495	0.16604	3.74084	1.59568	2.85320
3.11	0.64432	0.26349	0.16505	3.77409	1.59639	2.86664
3.12	0.64348	0.26203	0.16407	3.80764	1.59709	2.88005
3.13	0.64265	0.26059	0.16309	3.84149	1.59779	2.89343
3.14	0.64182	0.25915	0.16212	3.87565	1.59848	2.90679
3.15	0.64100	0.25773	0.16117	3.91011	1.59917	2.92011
3.16	0.64018	0.25632	0.16022	3.94488	1.59985	2.93342
3.17	0.63938	0.25492	0.15927	3.97997	1.60052	2.94669
3.18	0.63857	0.25353	0.15834	4.01537	1.60119	2.95994
3.19	0.63778	0.25215	0.15741	4.05108	1.60185	2.97316

M	T_t/T_t^*	T/T^*	p/p^*	p_t/p_t^*	V/V^*	S_{max}/R
3.20	0.63699	0.25078	0.15649	4.08712	1.60250	2.98635
3.21	0.63621	0.24943	0.15558	4.12347	1.60315	2.99952
3.22	0.63543	0.24808	0.15468	4.16015	1.60380	3.01266
3.23	0.63465	0.24674	0.15379	4.19715	1.60444	3.02577
3.24	0.63389	0.24541	0.15290	4.23449	1.60507	3.03885
3.25	0.63313	0.24410	0.15202	4.27215	1.60570	3.05191
3.26	0.63237	0.24279	0.15115	4.31014	1.60632	3.06495
3.27	0.63162	0.24149	0.15028	4.34847	1.60694	3.07795
3.28	0.63088	0.24021	0.14942	4.38714	1.60755	3.09093
3.29	0.63014	0.23893	0.14857	4.42614	1.60816	3.10388
3.30	0.62940	0.23766	0.14773	4.46549	1.60877	3.11681
3.31	0.62868	0.23640	0.14689	4.50518	1.60936	3.12971
3.32	0.62795	0.23515	0.14606	4.54522	1.60996	3.14258
3.33	0.62724	0.23391	0.14524	4.58561	1.61054	3.15543
3.34	0.62652	0.23268	0.14442	4.62635	1.61113	3.16825
3.35	0.62582	0.23146	0.14361	4.66744	1.61170	3.18105
3.36	0.62512	0.23025	0.14281	4.70889	1.61228	3.19382
3.37	0.62442	0.22905	0.14201	4.75070	1.61285	3.20656
3.38	0.62373	0.22785	0.14122	4.79287	1.61341	3.21928
3.39	0.62304	0.22667	0.14044	4.83540	1.61397	3.23197
3.40	0.62236	0.22549	0.13966	4.87830	1.61453	3.24463
3.41	0.62168	0.22432	0.13889	4.92157	1.61508	3.25727
3.42	0.62101	0.22317	0.13813	4.96521	1.61562	3.26988
3.43	0.62034	0.22201	0.13737	5.00923	1.61616	3.28247
3.44	0.61968	0.22087	0.13662	5.05362	1.61670	3.29503
3.45	0.61902	0.21974	0.13587	5.09839	1.61723	3.30757
3.46	0.61837	0.21861	0.13513	5.14355	1.61776	3.32008
3.47	0.61772	0.21750	0.13440	5.18909	1.61829	3.33257
3.48	0.61708	0.21639	0.13367	5.23501	1.61881	3.34503
3.49	0.61644	0.21529	0.13295	5.28133	1.61932	3.35746
3.50	0.61580	0.21419	0.13223	5.32804	1.61983	3.36987
3.51	0.61517	0.21311	0.13152	5.37514	1.62034	3.38225
3.52	0.61455	0.21203	0.13081	5.42264	1.62085	3.39461
3.53	0.61393	0.21096	0.13011	5.47054	1.62135	3.40695
3.54	0.61331	0.20990	0.12942	5.51885	1.62184	3.41926
3.55	0.61270	0.20885	0.12873	5.56756	1.62233	3.43154
3.56	0.61209	0.20780	0.12805	5.61668	1.62282	3.44380
3.57	0.61149	0.20676	0.12737	5.66621	1.62331	3.45603
3.58	0.61089	0.20573	0.12670	5.71615	1.62379	3.46824
3.59	0.61029	0.20470	0.12603	5.76652	1.62427	3.48043

M	T_t/T_t^*	T/T^*	p/p^*	p_t/p_t^*	V/V^*	S_{max}/R
3.60	0.60970	0.20369	0.12537	5.81730	1.62474	3.49259
3.61	0.60911	0.20268	0.12471	5.86850	1.62521	3.50472
3.62	0.60853	0.20167	0.12406	5.92013	1.62567	3.51683
3.63	0.60795	0.20068	0.12341	5.97219	1.62614	3.52892
3.64	0.60738	0.19969	0.12277	6.02468	1.62660	3.54098
3.65	0.60681	0.19871	0.12213	6.07761	1.62705	3.55302
3.66	0.60624	0.19773	0.12150	6.13097	1.62750	3.56503
3.67	0.60568	0.19677	0.12087	6.18477	1.62795	3.57702
3.68	0.60512	0.19581	0.12024	6.23902	1.62840	3.58899
3.69	0.60456	0.19485	0.11963	6.29371	1.62884	3.60093
3.70	0.60401	0.19390	0.11901	6.34884	1.62928	3.61285
3.71	0.60346	0.19296	0.11840	6.40443	1.62971	3.62474
3.72	0.60292	0.19203	0.11780	6.46048	1.63014	3.63661
3.73	0.60238	0.19110	0.11720	6.51698	1.63057	3.64845
3.74	0.60184	0.19018	0.11660	6.57394	1.63100	3.66028
3.75	0.60131	0.18926	0.11601	6.63137	1.63142	3.67207
3.76	0.60078	0.18836	0.11543	6.68926	1.63184	3.68385
3.77	0.60025	0.18745	0.11484	6.74763	1.63225	3.69560
3.78	0.59973	0.18656	0.11427	6.80646	1.63267	3.70733
3.79	0.59921	0.18567	0.11369	6.86578	1.63308	3.71903
3.80	0.59870	0.18478	0.11312	6.92557	1.63348	3.73071
3.81	0.59819	0.18391	0.11256	6.98584	1.63389	3.74237
3.82	0.59768	0.18303	0.11200	7.04660	1.63429	3.75401
3.83	0.59717	0.18217	0.11144	7.10784	1.63469	3.76562
3.84	0.59667	0.18131	0.11089	7.16958	1.63508	3.77721
3.85	0.59617	0.18045	0.11034	7.23181	1.63547	3.78877
3.86	0.59568	0.17961	0.10979	7.29454	1.63586	3.80031
3.87	0.59519	0.17876	0.10925	7.35777	1.63625	3.81183
3.88	0.59470	0.17793	0.10871	7.42151	1.63663	3.82333
3.89	0.59421	0.17709	0.10818	7.48575	1.63701	3.83481
3.90	0.59373	0.17627	0.10765	7.55050	1.63739	3.84626
3.91	0.59325	0.17545	0.10713	7.61577	1.63777	3.85769
3.92	0.59278	0.17463	0.10661	7.68156	1.63814	3.86909
3.93	0.59231	0.17383	0.10609	7.74786	1.63851	3.88048
3.94	0.59184	0.17302	0.10557	7.81469	1.63888	3.89184
3.95	0.59137	0.17222	0.10506	7.88205	1.63924	3.90318
3.96	0.59091	0.17143	0.10456	7.94993	1.63960	3.91450
3.97	0.59045	0.17064	0.10405	8.01835	1.63996	3.92579
3.98	0.58999	0.16986	0.10355	8.08731	1.64032	3.93706
3.99	0.58954	0.16908	0.10306	8.15681	1.64067	3.94831

M	T_t/T_t^*	T/T^*	p/p^*	p_t/p_t^*	V/V^*	S_{max}/R
4.00	0.58909	0.16831	0.10256	8.22685	1.64103	3.95954
4.10	0.58473	0.16086	0.09782	8.95794	1.64441	4.07064
4.20	0.58065	0.15388	0.09340	9.74729	1.64757	4.17961
4.30	0.57682	0.14734	0.08927	10.59854	1.65052	4.28652
4.40	0.57322	0.14119	0.08540	11.51554	1.65329	4.39143
4.50	0.56982	0.13540	0.08177	12.50226	1.65588	4.49440
4.60	0.56663	0.12996	0.07837	13.56288	1.65831	4.59550
4.70	0.56362	0.12483	0.07517	14.70174	1.66059	4.69477
4.80	0.56078	0.12000	0.07217	15.92337	1.66274	4.79229
4.90	0.55809	0.11543	0.06934	17.23245	1.66476	4.88809
5.00	0.55556	0.11111	0.06667	18.63390	1.66667	4.98224
5.10	0.55315	0.10703	0.06415	20.13279	1.66847	5.07477
5.20	0.55088	0.10316	0.06177	21.73439	1.67017	5.16575
5.30	0.54872	0.09950	0.05951	23.44420	1.67178	5.25522
5.40	0.54667	0.09602	0.05738	25.26788	1.67330	5.34322
5.50	0.54473	0.09272	0.05536	27.21132	1.67474	5.42979
5.60	0.54288	0.08958	0.05345	29.28063	1.67611	5.51498
5.70	0.54112	0.08660	0.05163	31.48210	1.67741	5.59883
5.80	0.53944	0.08376	0.04990	33.82228	1.67864	5.68138
5.90	0.53785	0.08106	0.04826	36.30790	1.67982	5.76265
6.00	0.53633	0.07849	0.04669	38.94594	1.68093	5.84270
6.10	0.53488	0.07603	0.04520	41.74362	1.68200	5.92155
6.20	0.53349	0.07369	0.04378	44.70837	1.68301	5.99924
6.30	0.53217	0.07145	0.04243	47.84787	1.68398	6.07579
6.40	0.53091	0.06931	0.04114	51.17004	1.68490	6.15124
6.50	0.52970	0.06726	0.03990	54.68303	1.68579	6.22562
6.60	0.52854	0.06531	0.03872	58.39527	1.68663	6.29896
6.70	0.52743	0.06343	0.03759	62.31541	1.68744	6.37128
6.80	0.52637	0.06164	0.03651	66.45238	1.68821	6.44261
6.90	0.52535	0.05991	0.03547	70.81536	1.68895	6.51298
7.00	0.52438	0.05826	0.03448	75.41379	1.68966	6.58240
7.50	0.52004	0.05094	0.03009	102.28748	1.69279	6.91625
8.00	0.51647	0.04491	0.02649	136.62352	1.69536	7.22982
8.50	0.51349	0.03988	0.02349	179.92363	1.69750	7.52538
9.00	0.51098	0.03565	0.02098	233.88395	1.69930	7.80482
9.50	0.50885	0.03205	0.01885	300.40722	1.70082	8.06978
10.00	0.50702	0.02897	0.01702	381.61488	1.70213	8.32165
∞	0.48980	0.0	0.0	∞	1.7143	∞

12

SELECTED
REFERENCES

Reference numbers referred to in the text correspond to those listed below.

Calculus

1. Leithold, L., "The Calculus with Analytic Geometry," Second Edition, Harper and Row, New York, 1972.

2. Loomis, L., "Calculus," Addison-Wesley, Reading, Mass, 1974.

Thermodynamics

3. Holman, J. P., "Thermodynamics," McGraw-Hill, New York, 1969.

4. Mooney, D. A., "Mechanical Engineering Thermodynamics," Prentice-Hall, Englewood Cliffs, N. J., 1953.

5. Reynolds, W. C. and Perkins, H. C., "Engineering Thermodynamics," McGraw-Hill, New York, 1970.

6. Saad, M. A., "Thermodynamics for Engineers," Prentice-Hall, Englewood Cliffs, N. J., 1966.

7. Sonntag, R. E. and Van Wylen, G. J., "Introduction to Thermodynamics: Classical and Statistical," John Wiley and Sons, New York, 1971.

8. Zemansky, M. W., Abbott, M. M. and Van Ness, H. C., "Basic Engineering Thermodynamics," Second Edition, McGraw-Hill, New York, 1975.

Fluid Mechanics

9. Pao, R. H. F., "Fluid Mechanics," John Wiley and Sons, New York, 1961.

10. Shames, I. H., "Mechanics of Fluids," McGraw-Hill, New York, 1962.

11. Streeter, V. L., "Fluid Mechanics," Fifth Edition, McGraw-Hill, New York, 1971.

12. Vennard, J. K., "Elementary Fluid Mechanics," Fourth Edition, John Wiley and Sons, New York, 1961.

Gas Dynamics

13. Cambel, A. B. and Jennings, B. H., "Gas Dynamics," McGraw-Hill, New York, 1958.

14. Chapman, A. J. and Walker, W. F., "Introductory Gas Dynamics,"
    Holt, Rinehart and Winston, New York, 1971.

15. Hall, N. A., "Thermodynamics of Fluid Flow," Prentice-Hall,
    Englewood Cliffs, N. J., 1951.

16. John, J. E. A., "Gas Dynamics," Allyn and Bacon, Boston, 1969.

17. Liepmann, H. W. and Roshko, A., "Elements of Gasdynamics,"
    John Wiley and Sons, New York, 1957.

18. Rotty, R. M., "Introduction to Gas Dynamics," John Wiley and
    Sons, New York, 1962.

19. Shapiro, A. H., "The Dynamics and Thermodynamics of Compressi-
    ble Fluid Flow," Volumes I and II, Ronald Press, New York,
    1953.

20. Thompson, P. A., "Compressible-Fluid Dynamics," McGraw-Hill,
    New York, 1972.

Tables

21. Keenan, J. H. and Kaye, J., "Gas Tables," John Wiley and Sons,
    New York, 1948.

Propulsion

22. Hesse, W. J. and Mumford, N. V. S., Jr., "Jet Propulsion for
    Aerospace Applications," Second Edition, Pitman, New York,
    1964.

23. Hill, P. G. and Peterson, C. R., "Mechanics and Thermodynamics
    of Propulsion," Addison-Wesley, Reading, Mass., 1965.

24. Zucrow, M. J., "Aircraft and Missile Propulsion," Volumes I
    and II, John Wiley and Sons, New York, 1958.

13

ANSWERS
TO PROBLEMS

UNIT 2

(2)(a) $U_m/2$, (b) $U_m/3$, (c) $2U_m/3$; (3) 13/2 ; (4) $\rho AE_m U_m/3$;
(5)(a) 38.9 ft/sec , (b) $1400/D^2$; (6) 44.4 ft/sec ; (7) 19,010
HP ; (8) 111.2 HP ; (9)(a) 1906 m/s , (b) 5.07 kg/s ; (10) -0.0147
Btu/lbm ; (11)(a) 78.1 m/s , (b) 4.18 ; (12)(a) 2875 ft/sec ,
(b) 1.15 ; (13)(a) 661 m/s , (b) 0.0625 bars abs ; (14)(a) 382
Btu/sec , (b) 0.03% . Check Test: (3) $7\rho AB_m U_m/30$; (5) $\dot{m}_2\beta_2 +$
$\dot{m}_3\beta_3 - \dot{m}_1\beta_1$.

UNIT 3

(4) 246.2 ft/sec ; (5)(a) -450 J/kg , (b) $0.11^{o}K$; (6)(a) 2261
ft/sec , (b) $732^{o}F$, (c) 103.1 psia ; (7) Shaft work input ;
(9)(a) 7.51 ft-lbf/lbm , (b) 2.87 psig ; (10) 54.4 m ; (11)(a)
46.6 ft-lbf/lbm , (b) Flow from two to one ; (12) 14.82 cm ;
(14)(a) 7200 A lbf , (b) 1.50 lbf/ft^2 ; (15)(a) 1.50 bars ,
(b) 7810 N , (c) -56,800 J/kg ; (16)(a) 80 ft/sec , 6.37 psig ,
(b) 3600 lbf ; (17)(a) 32.1 ft/sec , (b) 174.9 lbm/sec , (c) 151
lbf ; (18) 5000 N ; (19) 4.36 ft^2 ; (20) 180° . Check Test:
(4) b ; (5)(a) $q = w_s = 0$, yes , (b) no losses ; (6) s .

UNIT 4

(1) 1128 ft/sec , 4290 ft/sec , 4880 ft/sec , 4680 ft/sec ;
(3)(a) 295 ft/sec , (b) 298 ft/sec , (c) 1291 ft/sec , 1492
ft/sec , (d) At low Mach numbers ; (4) 0.564 ; (5)(a) 0.700 ,
(b) 2.8 kg/m^3 ; (6) 2.1 , 402 psia ; (7) 1266 m/s ; (8) $524^{o}R$,
1779 psfa ; (9) 1.28×10^5 N/m^2 , $330^{o}K$, 491 m/s ; (10) $M = \infty$;
(11) Flows towards 50 psia , 0.0204 Btu/lbm-^{o}R ; (12)(a) $457^{o}K$,
448 m/s , (b) 9.65 bars abs , (c) 0.370 ; (13)(a) $451^{o}R$, 20.95
psia , (b) 0.0254 Btu/lbm-^{o}R , (c) 1571 lbf ; (14)(a) 156.8 m/s ,
(b) 32.5 J/kg-^{o}K , (c) 0.763 ; (15)(a) 85.8 lbm/sec , (b) 1.91 ,
$578^{o}R$, 2140 ft/sec , 0.0758 lbm/ft^3 , 0.528 ft^2 , (c) -6960 lbf .
Check Test: (2)(a) Into , (b) $M_2 < M_1$; (3)(a) T , (b) F , (c) F ,
(d) T , (e) T .

UNIT 5

(1)(a) 0.18 , 94.9 psia , (b) 2.94 , $320^{o}R$; (2) 2.20, 1.64 ;
(3)(a) 0.50 , 35.6 psia , $788^{o}R$, (b) Nozzle , (c) 0.67 , 26.3
psia , $723^{o}R$; (4) $239^{o}K$; (5)(a) 0.607 , 685 ft/sec , 23.1 psia ;
(b) 0.342 , 395 ft/sec , 30.4 psia , (c) 0.855 ; (7)(a) 0.00797

Btu/lbm-$^\circ$R , (b) 0.1502 ; (8)(a) 52.33 J/kg-$^\circ$K , (b) 16.43 cm ;
(9)(a) 26.5 lbm/sec , (b) No change , (c) 53.0 lbm/sec ; (10)(a)
320 m/s , (b) 0.808 kg/sec , (c) 0.844 kg/sec ; (11) 671°R ,
0.768 , 975 ft/sec ; (12)(a) 77.9 psia , (b) 3.77 psia , (c)
0.0406 lbm/ft^3 , 2050 ft/sec ; (13)(a) 38.6 cm^2 , (b) 9.14
kg/sec ; (14) 430 ft/sec ; (15)(a) 140.4 lbm/sec , (b) 0.491 ft^2 ,
(c) 0.787 ft^2 ; (16)(b) 3.53 cm^2 , (c) 4.09 cm^2 ; (17)(a) 1.71 ,
(b) 91.9% , (c) 0.01152 Btu/lbm-$^\circ$R ; (18)(a) 163.9°K , 1.10 bars
abs , 8.61 bars abs , (b) 2.096 , (c) 0.1276 m^2 , (d) 300 kg/sec ;
(19)(a) 23.7 psia , (b) 97.4% , (c) 4.14 . Check Test: (3)
$T_2^* > T_1^*$; (6)(a) 132.1 psia , (b) 0.514 lbm/ft^3 , 1001 ft/sec ,
(c) 0.43 .

UNIT 6

(1)(b) 0.01421 Btu/lbm-$^\circ$R , (c) 0.0646 Btu/lbm-$^\circ$R , 0.1237
Btu/lbm-$^\circ$R ; (2) 84.0 psia ; (3)(a) $(\gamma-1)/2\gamma$, (b) $\rho_2/\rho_1 =$
$(\gamma+1)/(\gamma-1)$; (4) 2.47 , 3.35 ; (5)(a) 2.88 , (b) 1.529 ; (6) 0.69
2.45 ; (7)(a) 0.965 , 0.417 , 0.0585 , (b) 144.8 psia , 62.6 psia
8.78 psia, (c) 15.54 psia , 36.0 psia , 256 psia ; (8)(a) 19.30
cm^2 , (b) 10.52x10^5 N/m^2 , (c) 18.65x10^5 N/m^2 ; (9) 1.30 ft^2 ;
(10)(a) 0.119 , 0.623 , (b) 0.0287 Btu/lbm-$^\circ$R ; (11) 0.498 ;
(12)(a) 4.6 in^2 , (b) 5.35 in^2 , (c) 79 psia , (d) 6.58 in^2 ,
(e) 1.79 ; (13)(a) 3.56 , (b) 0.475 ; (14) 0.67 or 1.405 ; (15)(a)
0.973 , 0.375 , 0.0471 , (b) 0.43 , (c) 2.64 , 2.50 ; (16)(a)
0.271 , (b) 0.0455 Btu/lbm-$^\circ$R , (c) 2.48 , (d) 0.281 ; (17)(a)
54.6 in^2 , (b) 18.39 lbm/sec , (c) 109.4 in^2 , (d) 7.34 psia ,
(e) 9.24 psia , (f) 742 HP . Check Test: (2)(a) Increases ,
(b) Decreases , (c) Decreases , (d) Increases ; (3) 0.973 , 0.376
0.0473 ; (5)(a) 1.625 , (b) From two to one ; (6)(a) 0.380 ,
450 ft/sec , (b) 0.0282 Btu/lbm-$^\circ$R .

UNIT 7

(1)(a) 725°R , 42.0 psia , 922 ft/sec , (b) 0.00787 Btu/lbm-$^\circ$R ;
(2) 1.024x10^6 $^\circ$K , 1.756x10^6 $^\circ$K , 20,500 bars , 135,000 bars ;
(3) 531°R , 19.75 psia , 348 ft/sec ; (4)(a) 957 ft/sec , (b)
658°R , 34.5 psia ; (5)(a) 310°K , 1.219x10^4 N/m^2 , 50.3 m/s ,
(b) 328°K , 1.48x10^4 N/m^2 , 340 m/s ; (6)(a) 1453 ft/sec , 2517
ft/sec , 959 ft/sec , 2517 ft/sec , (b) 619°R , 18.05 psia , (c)
9.1° ; (7)(a) 1.68 , 25.6° , (b) 560°K , 6.10 bars , (c) Weak ;
(8)(a) 52° , 77° , (b) 1013°R , 32.7 psia , 1198°R , 51.3 psia ;

9)(a) 2.06 , (b) All M > 2.06 cause attached shock ; (10)(a) 1.8 ,
b) For M > 1.57 ; (11)(a) 949 m/s , (b) 706°K ; (12)(a) 1928
t/sec , (b) 1045 ft/sec ; (14)(a) 821°R , 2340 psfa , 0.0220
tu/lbm-°R , (b) 826°R , 2468 psfa , 0.0200 Btu/lbm-°R ; (15)(a)
.272 , 166°K , 5.6° , (b) 5.6° , (c) 2.01 , 184.5°K , 1.43 bars ;
16)(a) 1.453 , 696°R , 24.8 psia , (b) Oblique shock with
= 10° , (c) 1.031 , 816°R , 42.7 psia ; (17)(a) 0.783 , 58° ,
b) 6.72 , 0.837 ; (18) 1.032 , 15.92 , 2.61 , 40° .   Check Test:
1)(a) $p_1 = p_1'$, (b) $T_{t1}' < T_{t2}'$, (c) None , (d) $u_2' > u_1'$,
$_2' = u_2$; (2)(a) Greater than , (b)(i) Decreases , (ii) De-
reases ; (6) 1667 ft/sec ; (7)(a) 53.1° , 20° , (b) 625°R , 14.1
sia , 1.23 .

NIT 8

1) 2.60 , 398°R , 936°R , 5.78 psia , 115 psia ; (2)(a) 1.65 ,
.04 , (b) 34.2° , 52.3° ; (3)(a) 174.5°K , 8.76×10^3 N/m^2 ;
4) 1.39 ; (5) 12.1° ; (6)(a) 2.361 , 1.986 , 11.03 , (b) 1.813 ,
.51 , 9.33 ; (d) No ; (7)(a) 6.00 psia , 16.59 psia , (b) 12,020
bf , 2120 lbf ; (8)(c) 6.851 psia , 19.09 psia , 3.346 psia ,
0.483 psia , L = 56.6 lbf/ft of span , D = 13.86 lbf/ft of span ;
10)(a) 2.44 , 392°R , (b) $\Delta\nu$ = 14.2° ; (11)(b) 241°K , 1.0 bar ,
09 m/s ; (12)(c) 1.86 , 20° , 2.67 , 40.5° from centerline ;
13)(a) 15° , (b) 1.691 , 4.143 p_a , (c) Expansion , (d) 2.607 ,
$_a$, 0.865 T_1 , 39.1° from original flow ; (14)(a) 1.0 bar ,
.766 , 6.55$_0$, 1.4 bar , 1.536 , 0° , 1.0 bar , 1.761 , 6.6° ;
15)(b) ∞ , (c) 130.45° , 104.1° , 53.5° , 28.1° , (d) 3604
t/sec ;

16)(a) $\dfrac{L_2}{L_1} = \dfrac{1}{M_2}\left[\dfrac{\gamma+1}{2}\right]^{\frac{\gamma+1}{2(1-\gamma)}}\left[1 + \dfrac{\gamma-1}{2} M_2^2\right]^{\frac{\gamma+1}{2(\gamma-1)}}$, (b) 1.343 .

heck Test: (4) 5.74° ; (5) 845 lbf/ft^2 .

NIT 9

1) 2.22×10^5 N/m^2 , 0.386 ; (2) 76.1 psia , 138.6 lbm/ft^2-sec ;
3)(a) 21.7 D , (b) 55.6% , 87.1% , 20.3% ; (c) 0.0630 Btu/lbm-°R ,
d) -0.59% , -5.9% , -5.4% , 0.00279 Btu/lbm-°R ; (4)(a) 22.1 ft ,
b) 528°R , 24.6 psia , 1072 ft/sec ; (5)(a) 0.0313 , (b) 2730
/m^2 ; (6)(a) 551°R , 0.60 , (b) From two to one, (c) 0.423 ;
7)(a) 157.8°K , 2.98×10^4 N/m^2 , 442°K , 10.95×10^5 N/m^2 , (b)
.0157 ; (8)(a) 556°R , 30.4 psia , 284 ft/sec , (b) 15.06

psia ; (9)(a) 453°R , 8.79 psia , (b) 77.3 ft ; (11)(a) 0.690 ,
0.877 , 1128 ft/sec , 876°R , 38 psia , (b) 0.0205 , 0.0012 ft ;
(12)(a) 324°K , 1.792 bars , 347°K , 2.27 bars ; 121.8^{O}K , 0.214
bars , 347°K , 8.33 bars ; (b) 1959 HP , 4260 HP ; (13)(a) 0.216
(b) 495°R , 10.65 psia , (c) 17.82 ft ; (14) 229°K , 5.33x10^4
N/m^2 ; (15)(b) 0.513 , 0.699 , (c) 0.758 ; (16)(a)(i) 144.4 psia
(ii) 51.7 psia , (iii) 40.8 psia ; (b) 15.2 psia ; (17)(b) 0.013:
(c) 289.4 J/kg-^{O}K ; (18)(b) M = 0.50 , (c) 26.87 bars , (d) 0.40;
0.825 ; (19)(a) 25.96 psia , (b) 39.48 psia . Check Test: (3)
43.5 psia ; (4) 94.3 psia to 31.4 psia .

UNIT 10

(1)(a) 1217°R , 1839°R , (b) 112.6 Btu/lbm added , (2) 1.792x10^5
J/kg removed ; (3) 0.848 , 2.826 , 0.223 ; (4)(a) 3.365 , 2.43x10
N/m^2 , 126.3^{O}K , (b) -890 J/kg-^{O}K ; (5)(a) 767°R , 114.7 psia ,
1112°R , 421 psia , (b) 68.1 Btu/lbm added ; (8)(a) 6.39x10^5 J/k
(b) 892°K , 0.567 atmos ; (9)(b) 2.00 , 600°R , 59.78 psia ; (c)
630°R , 21.0 psia , 756.2^{O}R , 39.8 psia , (d) 38.7 Btu/lbm ;
(10)(a) 2182°R , 172.5 psia ; (11)(a) 1.57x10^4 J/kg added , (b)
6.97x10^4 J/kg removed , (c) No ; (13) 36.5 Btu/lbm removed ;
(14)(b) 0.686 , (c) 1.628x10^5 J/kg ; (15)(a) 47.4 psia , (b) 66.
Btu/lbm added , (c) Less than one , 279 Btu/lbm ; (17)(a) True ,
False ; (18)(a) $A_3 > A_4$, (b) $V_3 < V_4$, $A_3 > A_4$; (21)(a) $A_3 > A_4$
Check Test: (4)(a) 746°R , (b) 53.1 Btu/lbm added.

UNIT 11

(1)(a) 292.8 Btu/lbm , 129 Btu/lbm , 163.8 Btu/lbm , 322 Btu/lbm
50.8% , (b) 21.6 lbm/sec ; (2)(a) 269.4 Btu/lbm , 145 Btu/lbm ,
124.4 Btu/lbm , 306 Btu/lbm , 40.6% , (b) 28.4 lbm/sec ; (3) 37.
38.5 kg/sec ; (4)(a) 24.9% , (c) 64.9% ; (6)(a) 4600 lbf , (b)
1653 ft/sec ; (7) 564 m/s ; (8) 1419 ft/sec ; (9)(a) 7820 lbf ,
(b) 57.1% , (c) 438 ft-lbf/lbm ; (10)(a) 18.34 kg/s , (b) 0.257
m^2 , (c) 3.12x10^5 watts , (d) 28.6% , (e) 10.24x10^5 J/kg ; (11)(a
2880 lbf , (b) 5160 ft/sec , (c) 20,800 HP ; (12) 3290 lbf , 1.0£
lbm fuel/lbf-hr ; (13) 6.34 ft^3 , M = 0.382 , 1309 psfa , 3400°R
742 psfa , 2920°R , 3.96 ft^2 ; 6550 lbf , 1.41 lbm fuel/lbf-hr ;
(14) 4240 lbf/ft^2 , 2.20 lbm fuel/lbf-hr ; (15)(a) 83.3 lbm/sec ,
(b) 7730 ft/sec ; (16)(a) 1992 N-s/kg , (b) p_o - 872 N/m^2 ;
(17)(a) 0.0402 ft^2 , (b) 6060 ft/sec , 6490 ft/sec , 201
lbf-sec/lbm ; (18)(a) 7.46 , 1904 m/s , (b) 1904 N-s/kg ; (19)(a)

0.725 , (b) 0.747 ; (21)(b) M_o = 1.83 , (c) Cannot be started ;
(22) 3.5 to 1.36 .   Check Test: (3) 871^oK , 1.184 bars ; (5)(a)
F , (b) F , (c) F , (d) F ; (6)(a) 311 lbm/sec , 64,500 lbf ,
(b) 6670 ft/sec , (c) 207 lbf-sec/lbm, 5.28×10^5 HP ; (7) M_o = 2.36 .

INDEX